Decision, probability,
and utility

Decision, probability, and utility

Selected readings

Edited by
Peter Gärdenfors and Nils-Eric Sahlin

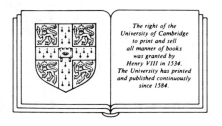

The right of the
University of Cambridge
to print and sell
all manner of books
was granted by
Henry VIII in 1534.
The University has printed
and published continuously
since 1584.

Cambridge University Press

Cambridge
New York   New Rochelle   Melbourne   Sydney

Published by the Press Syndicate of the University of Cambridge
The Pitt Building, Trumpington Street, Cambridge CB2 1RP
32 East 57th Street, New York, NY 10022, USA
10 Stamford Road, Oakleigh, Melbourne 3166, Australia

First published 1988

Printed in the United States of America

Library of Congress Cataloging-in-Publication Data
Decision, probability, and utility.
Bibliography: p.
Includes index.
1. Bayesian statistical decision theory.
I. Gärdenfors, Peter. II. Sahlin, Nils-Eric.
QA279.4.D43   1987   519.5'42   86–33402

British Library Cataloguing in Publication Data
Decision, probability, and utility:
selected readings.
1. Statistical decision
I. Gärdenfors, Peter II. Sahlin, Nils-Eric
519.5'42   QA279.4

ISBN 0 521 33399 1 hard covers
ISBN 0 521 33658 9 paperback

Contents

List of contributors   *page* vii

Preface   ix

1   Introduction: Bayesian decision theory – foundations and
    problems   1
    Peter Gärdenfors and Nils-Eric Sahlin

Part I: Foundations of Bayesian decision theory

2   Truth and probability   19
    Frank P. Ramsey

3   Individual decision making under uncertainty   48
    R. Duncan Luce and Howard Raiffa

4   The sure-thing principle   80
    Leonard J. Savage

5   Probable knowledge   86
    Richard C. Jeffrey

Part II: Conceptualization of probability and utility

    Introduction   97

6   Bets and beliefs   101
    Henry E. Kyburg

7   Slightly more realistic personal probability   118
    Ian Hacking

8   Risk aversion as a problem of conjoint measurement   136
    Bengt Hansson

Part III: Questionable rules of rationality

Introduction   159

9   Allais' paradox   163
    Leonard J. Savage

10 Sure-thing doubts   166
    Edward F. McClennen

11 Prospect theory: An analysis of decision under risk   183
    Daniel Kahneman and Amos Tversky

12 Generalized expected utility analysis and the nature of
    observed violations of the independence axiom   215
    Mark Machina

Part IV: Unreliable probabilities

Introduction   241

13 Risk, ambiguity, and the Savage axioms   245
    Daniel Ellsberg

14 Self-knowledge, uncertainty, and choice   270
    Frederic Schick

15 On indeterminate probabilities   287
    Isaac Levi

16 Unreliable probabilities, risk taking, and decision making   313
    Peter Gärdenfors and Nils-Eric Sahlin

Part V: Causal decision theory

Introduction   335

17 Counterfactuals and two kinds of expected utility   341
    Allan Gibbard and William L. Harper

18 Causal decision theory   377
    David Lewis

19 Ratifiability and stability   406
    Wlodzimierz Rabinowicz

References   426

Name index   443

Subject index   447

Contributors

*Daniel Ellsberg *The Rand Corporation, Santa Monica*
Peter Gärdenfors *Filosofiska Institutionen, Lunds Universitet*
Allan Gibbard *Department of Philosophy, University of Pittsburgh*
Ian Hacking *Department of Philosophy, University of Toronto*
Bengt Hansson *Filosofiska Institutionen, Lunds Universitet*
William L. Harper *Department of Philosophy, University of Western Ontario*
Richard C. Jeffrey *Department of Philosophy, Princeton University*
Daniel Kahneman *Department of Psychology, University of British Columbia*
Henry E. Kyburg *Department of Philosophy, University of Rochester*
Isaac Levi *Department of Philosophy, Columbia University*
David Lewis *Department of Philosophy, Princeton University*
R. Duncan Luce *Department of Psychology and Social Relations, Harvard University*
Mark Machina *Department of Economics, University of California, San Diego*
Edward F. McClennen *Department of Philosophy, Washington University*
Wlodzimierz Rabinowicz *Filosofiska Institutionen, Uppsala Universitet*
Howard Raiffa *Department of Economics, Harvard University*
*Frank P. Ramsey *King's College, University of Cambridge*
Nils-Eric Sahlin *Filosofiska Institutionen, Lunds Universitet*
*Leonard J. Savage *Department of Statistics, Yale University*
Frederic Schick *Department of Philosophy, Rutgers University*
Amos Tversky *Department of Psychology, Stanford University*

*Asterisk indicates affiliation when article was originally published.

Preface

Since the mid-1950s, decision theory has become a very active research area in economics, statistics, psychology, and philosophy. However, different disciplines deal with different problems, and thus we have a rather heterogeneous field of research. A contribution to decision theory can be anything from a mathematical axiomatization to an experimental report. This does not mean, however, that there is no common theoretical core. The present volume aims at presenting the theoretical foundations of Bayesian decision theory and some alternative theories that have been developed as a consequence of the problems raised by the traditional theory.

We hope that this volume will be useful as a textbook. To this end we have written a self-contained introduction as well as brief presentations of each of the parts. We also hope that it will be of value to researchers interested in the foundations of decision theory. Assuming the reader has some prior knowledge of decision theory (that is, what is given in the introduction), Parts I to V can be read as separate entities.

The volume concludes with a comprehensive bibliography. It mainly consists of works referred to in the text, but we have also added a number of items that we thought would add to its usefulness as a source of reference.

Including the bibliography made it necessary to change some of the original footnotes in the articles. Some cross references have also been added. It should also be noted that some authors have revised their articles for this volume.

We want to thank Ellery Eells, Baruch Fischhoff, Richard Jeffrey, Isaac Levi, Edward McClennen, and Jonathan Sinclair-Wilson for supporting this project in various ways. We also thank Sören Halldén,

Bengt Hansson, and Wlodzimierz Rabinowicz who have helpfully commented on our contributions, and Victoria Höög for assisting us in compiling the indices.

This book is dedicated to those of us who find it difficult to make up our minds.

Lund, Midsummer Day 1986   P.G.
                                                             N.-E.S.

1. Introduction: Bayesian decision theory – foundations and problems
Peter Gärdenfors and Nils-Eric Sahlin

Traditional Bayesian decision theory

We are forced to make decisions numerous times every day. Some of these decisions are made consciously, sometimes well deliberated; some of them are simply the result of habit, be it good or bad. In a sense, our decisions derive from our wants and we make them because we believe that our actions will influence the world in such a direction that our wants are more likely to be fulfilled. Because the outcomes of our decisions have an impact on our everyday lives, we want to make good decisions. This should be the case whether we buy tomatoes in the supermarket, make a major investment in the stock market, or choose a lifelong partner. A key question is, how do we know when our decisions are *rational*? Psychologists keep telling us that, in many situations, we are not particularly good at making rational decisions. This does not mean, however, that we cannot formulate a rational *theory* of how to make good decisions.

This volume contains a number of articles on rational decision making, some of which explicitly formulate recommendations on how to make good decisions, and some of which criticize these recommendations and their theoretical foundations. However, a common core of the articles is that there are two main types of factors determining our decisions. One is our *wants* or *desires*. These determine the *values* or *utilities* of the possible outcomes of our decisions. The other is our *information* or *beliefs* about what the world is like and how our possible actions will influence the world. The beliefs determine the *probabilities* of the possible outcomes. The main aims of a *decision theory* are, first, to provide models for how we handle our wants and our beliefs and, second, to account for how they combine into rational decisions.

In this introduction we begin by presenting the traditional Bayesian decision theory. We also give an outline of the axiomatizations of this theory proposed by Frank P. Ramsey, Leonard J. Savage, and others. In this presentation, we pinpoint the underlying assumptions because these will play an important role in the criticisms of the traditional theory. The validity of these assumptions is crucial for evaluating the *rationality* of different decision theories.

First of all, the traditional notion of a *decision situation* must be introduced. This is best done by an example. Here is Savage's classical omelet:

Your wife has just broken five good eggs into a bowl when you come in and volunteer to finish making the omelet. A sixth egg, which for some reason must either be used for the omelet or wasted altogether, lies unbroken beside the bowl. You must decide what to do with this unbroken egg. Perhaps it is not too great an oversimplification to say that you must decide among three acts only, namely, to break it into the bowl containing the other five, to break it into a saucer for inspection, or to throw it away without inspection. Depending on the state of the egg, each of these three acts will have some consequence of concern to you.... (Savage 1954/1972, pp. 13–14)

In Savage's example, the decision maker has a choice between three different *alternatives* or *acts*. The set of alternatives is the first component in a decision situation. In general, let us denote this set A, and the particular alternatives in this set $a_1, a_2, \ldots$.

A decision maker normally does not have complete control over the factors that determine the outcome of his or her decision; sometimes a chance event influences the outcome of a decision, sometimes part of the control is in the hands of other persons. A second component in a decision situation, the set of possible *states* of the world, is introduced to account for the *uncertainty* as to the consequences of an act. In the omelet example, there are two relevant states of the world: The egg is good and the egg is rotten. The set of all states of the world, relevant to a particular decision situation, will be denoted S, and the particular states in this set $s_1, s_2, \ldots$.

Our acts lead to different *consequences* or *outcomes* depending on which state turns out to be the true state of the world. If we choose to break the egg into the bowl and it turns out to be good, the consequence of this act is a six-egg omelet. If, on the other hand, the egg is rotten, the outcome is different: no omelet and five good eggs destroyed. The outcome that is the result of choosing the alternative a_i when the state of the world turns out to be s_j will be denoted o_{ij}.

A decision situation is thus determined by the sets of alternatives, states, and outcomes. This information can be summarized as a *decision matrix*. The decision matrix corresponding to Savage's problem looks as follows:

	State	
Act	Good	Rotten
Break into bowl	six-egg omelet	no omelet, and five good eggs destroyed
Break into saucer	six-egg omelet and a saucer to wash	five-egg omelet and a saucer to wash
Throw away	five-egg omelet and a good egg destroyed	five-egg omelet

In the general case, a decision matrix can be described in more abstract terms as follows:

	State		
Act	s_1	s_2	$\ldots s_m$
a_1	o_{11}	o_{12}	o_{1m}
a_2	o_{21}	o_{22}	o_{2m}
.	.	.	.
.	.	.	.
.	.	.	.
a_n	o_{n1}	o_{n2}	o_{nm}

A fundamental question for a theory of decision making is, how are the alternatives in a decision situation *evaluated*? This depends, of course, on the desirability of the possible outcomes. It is commonly assumed that the decision maker can numerically evaluate the *utility* of the various outcomes in the decision situation. The intended interpretation of the utility assignment is that it measures the "intrinsic" value of the outcomes for the decision maker, independent of external factors such as the risk involved in the alternative that produces the particular outcomes. We set aside this assumption explicitly for future discussion:

A1 *(Values of outcomes)*: The values of the outcomes in a decision situation are determined by a utility measure which assigns numerical values to the outcomes.

The purpose of utility measure is to reflect not only the ordinal preferences between the possible outcomes, but also the corresponding *numerical* value differences. It is not only of interest to know that I prefer a six-egg omelet to no omelet and five good eggs destroyed, but also exactly how strong this preference is. We will denote the utility value of an outcome o_{ij} by $u(o_{ij})$.

The next problem is how the utilities of the outcomes are used when evaluating the alternatives. In many cases it seems that other factors than the utilities of the outcomes – for example, the risk involved – have an influence on the value of the alternatives. Such "external" factors are discussed in Section 3 of this chapter. However, in the traditional theory we have here another restricting assumption:

> A2 (*Values of alternatives*): When determining the value of a decision alternative, the *only* information about the decision maker's wants and desires that is exploited is the utilities of the possible outcomes of the alternative.

Apart from the values of the outcomes, the other factor that is used when evaluating an alternative is the decision maker's *beliefs* about the possible states of the world. In standard Bayesian theory, it is assumed that the decision maker assigns *probabilities* to states in the set S.

> A3 (*Information about states*): A decision maker's beliefs about the states of the world in a given situation can be represented by a unique probability measure defined over the states.

For each state s_j the decision maker thus assigns a probability value $P(s_j)$ so that the sum of all these values taken over the set S is 1. It should be noted that the probability values are *subjective* probabilities in the sense that they derive totally from the decision maker's personal information about the states.

Because Bayesian decision theory is based on subjective probabilities, it has been criticized for being arbitrary: For any state of the world, the whims of the decision maker can let him pick any probability value between 0 and 1 and thus he can reach almost any kind of decision and still be "rational" according to the principle of maximizing expected utility. However, this criticism does not take the Bayesian theory of learning into account. It is true that the "prior" probability values can be selected arbitrarily, but new information, obtained from experiments or by other means, will adjust the "subjective" probability distribution. One of the fundamental results in Bayesian theory, de Finetti's (1937/1964) representation theorem, entails that even if two decision makers start out from widely different initial distributions – for example, concerning the probabilities of the outcomes of tosses with a particular coin – they will end up arbitrarily close to each other, if given sufficient time to experiment with the coin. Even though de Finetti does not believe that there are such things as "objective" probabilities, one could say that his representation theorem shows that everyone's subjective probability distributions would converge toward an *inter-subjective* probability distribution if given more and more information about what the world actually is like. [For further discussion of this topic, see Jeffrey 1965/1983; Chapter 12, and Kyburg 1968; reprinted as Chapter 6, this volume].

Traditionally, one distinguishes between decision making under *risk* and decision making under *uncertainty*. A decision situation is in the realm of risk if the decision maker has *full* information about the world in the sense that he or she "knows" the exact probabilities of the states, whereas a decision situation is in the realm of uncertainty if he or she has *no* information whatsoever about the states so that the probabilities of these are completely unknown. In our opinion, the use of the term "risk" is misleading, because the riskiness of a decision prospect is dependent not only on the information available to the decision maker but also on several other factors as well (see Sahlin, 1984). We will return to this topic in the last section of this chapter.

Between the two extremes of "risk" and "uncertainty," there is a vast area of different degrees of partial information. In our opinion, most practical decision problems fall within this area. Decision theories focusing on this intermediate area are the topic of Part IV of this volume.

However, as indicated in assumption A3, Bayesian theory *reduces* all problems to decision making under risk by assuming that the decision maker, by processing his or her partial information, can generate a unique probability distribution over the states of the world (see Savage, 1954/1972; reprinted as Chapters 4 and 9 this volume).

Another assumption is that the alternative chosen does not have any influence on which state is the actual state of the world.

A4 (Probabilistic independence): For all states and all alternatives, the probability of the state is independent of the act chosen.

More technically, this assumption can be formulated as requiring that $P(s_j/a_i) = P(s_j)$ for all s_j and all a_i.

Given the utilities of the outcomes and the probabilities of the states, the decision maker can then compute the *expected utility* of the various alternatives. This is defined as follows for any alternative a_i:

$$EU(a_i) = P(s_1) \cdot u(o_{i1}) + P(s_2) \cdot u(o_{i2}) + \cdots + P(s_m) \cdot u(o_{im})$$

We can now formulate the fundamental decision rule of Bayesian theory:

The principle of maximizing expected utility (MEU): In a given decision situation the decision maker should choose the alternative with maximal expected utility (or one of the alternatives with maximal expected utility if there are more than one).

At this point we must introduce a distinction between a *normative* and a *descriptive* decision theory. A normative theory gives rules for what a *rational* decision maker *should* do in various situations. A descriptive theory aims at describing how people *in fact* (rationally or not) make decisions in different situations. The principle of maximizing expected utility can be given both a normative and a descriptive reading. The normative reading of the MEU principle has been prevalent among economists who are interested in making rational decisions in the market (e.g., see Sen, 1977) and among philosophers who are interested in general foundations of rationality. Among psychologists the MEU principle has served as a theory about people's actual behavior (e.g., see Slovic, Lichtenstein, and Fischhoff, 1983, Kahneman, Slovic, and Tversky, 1982, for surveys of this area). In Section 3, we return to some of the problems that the MEU principle has encountered as a descriptive theory. It should be noted, however, that the distinction between a normative and a descriptive interpretation of a decision theory is slippery: A proposed normative theory is tested against our intuitive understanding of rational decisions, and people's actual behavior may be influenced by existing normative theories. For a discussion of the interplay between normative and descriptive interpretations of psychological experiments in general and decision theoretic experiments in particular, cf. Sahlin (1987).

Foundations of Bayesian decision theory

In presenting Bayesian decision theory, we listed some of the assumptions that are necessary for the formulation of the MEU principle. In this section we discuss some results that aim at providing these assumptions with secure foundations. Basically, these results are proved with the aid of systems of *axioms* or *rationality postulates* concerning people's behavior in decision situations. The best known axiom systems for Bayesian decision theory are those formulated by Ramsey (1931/1978) and Savage (1954/1972).

Ramsey's primary interest is to give a foundation to A3, that is, that the beliefs of a decision maker can be represented by a unique (subjective) probability measure. In doing so, he also gives foundations to A1, A2, and A4 (this has not always been observed in the literature). His method consists in an analysis of rational *betting behavior*. The key idea is that the degree of belief the decision maker attaches to a proposition *o*, describing, say, the state that stock prices will fall tomorrow, can be measured by the odds he or she is willing to accept in a bet on *o*.

To do this Ramsey first formulates some axioms which enable him to compare *value distances* between the outcomes in a decision situation. For instance, his fourth axiom says that if the value distance between the outcomes *a* and *b* is the same as the value distance between *c* and *d* and the value distance between *c* and *d* is the same as that between *e* and *f*, then the distance between *a* and *b* is the same as that between *e* and *f*. Note that the formulation of this axiom does not presume that we have assigned any *numbers* to the values of the outcomes, only that we can compare the magnitudes of value distances (the same applies to the other axioms). Ramsey then shows that if the axioms are fulfilled, it is indeed possible to correlate the value of the outcome a with a real number $u(a)$ in such a way that the value distance between the outcomes a and b is represented by the numerical expression $u(a) - u(b)$.

He then uses this way of measuring values of outcomes to derive a way of measuring beliefs in general. The key definition is the following:

> *Ramsey's rule*: If the decision maker is indifferent between the alternative that yields outcome a for certain and the alternative that yields b if p is true and c if p is false, then his or her degree of belief in p is $[u(a) - u(c)]/[u(b) - u(c)]$.

As Ramsey says, "This amounts roughly to defining the degree of belief in p by the odds at which the subject would bet on p, the bet being conducted in terms of values as defined" (Chapter 2, this volume).

On the basis of this and some further definitions, Ramsey is able to prove that the degrees of belief introduced in this way satisfy the fundamental laws of probability, so that they can be represented by a unique probability measure (that is assumption A3 is satisfied). If a person's degrees of belief satisfy the laws of probability, that person's beliefs are said to be *coherent*. Thus Ramsey shows that, given his definition of degrees of belief, a decision maker is coherent if and only if his or her preferences satisfy Ramsey's axioms.

Ramsey also notes that thus anyone who violates the axioms "could have a book made against him by a cunning bettor, and would then stand to lose in any event" (Chapter 2, this volume). This result, named the *Dutch Book Theorem*, can be reformulated as saying that if your beliefs are not coherent, then a smart bookie can wheedle money out of your pocket no matter what happens. [For other versions of this result, see Kemeny (1955) and Shimony (1955).] Although he does not state it explicitly, it is clear that Ramsey's results also yield a foundation for assumptions A1 and A2 as well as the MEU principle. As Ramsey puts it: "I have not worked out the mathematical logic of this in detail, because this would, I think, be rather like working out to seven places of decimals a result only valid to two."

Savage's method of providing foundations for the assumptions A1–A4 is closely related to Ramsey's, and the results he reaches are almost identical (except for some technicalities concerning so-called "null events"). The axioms used by Savage, however, are quite different from those employed by Ramsey. Savage uses as primitive notion a binary relation $\geq$ between pairs of alternatives interpreted to mean "is preferred or indifferent to." All the axioms are formulated as rationality conditions on this relation. One of Savage's axioms, the so-called *sure-thing principle*, has created considerable discussion, as the reader will see in Part III of this volume. To illustrate the content of the axiom, let us consider the following decision situation [which is a variant of the kinds of examples considered by Allais (1953)]:

	s_1	s_2	s_3
a_1	$4,000	$0	$0
a_2	$3,000	$3,000	$0
a_3	$4,000	$0	$3,000
a_4	$3,000	$3,000	$3,000

In this matrix, the outcomes of a_1 and a_2 are the same for state s_3, and the outcomes of a_3 and a_4 are also the same for s_3. Furthermore, a_1 has the same outcomes as a_3, except in s_3, and similarly a_2 has the same outcomes as a_4, except in s_3. In a situation like this, the sure-thing principle requires that if a_1 is preferred to a_2, then a_3 is likewise preferred to a_4, and vice versa (independent of the probabilities of the states). In more general terms, the principle can be formulated as follows:

> *The sure-thing principle*: The choice between two alternatives must be unaffected by the value of outcomes corresponding to states for which both alternatives have the same payoff.

The sure-thing principle corresponds closely to Axiom 8 (Rubin's postulate) in Luce and Raiffa (1957) or Milnor's "column linearity" postulate (see Chapter 3, this volume). On the basis of this and the other axioms, Savage is able to show (in careful mathematical detail), first, that there exists a subjective probability measure P defined over the possible states of the world which represents the decision maker's degrees of belief, and, second, that there exists a utility measure u defined over the possible outcomes such that the expected utilities of the decision alternatives are exactly mirrored by the assumed preference ordering ≥. In other words, Savage proves the following *representation theorem*:

> For any preference ordering ≥ satisfying the axioms, there exist a probability function and a utility measure such that, for all alternatives a_1 and a_2, $a_1 \geq a_2$ if and only if $EU(a_1) \geq EU(a_2)$.

Furthermore, it follows from his results that the probability function is uniquely determined and that the utility function is determined up to a positive linear transformation. If u and u' are two utility functions satisfying the postulates, then there exist real numbers a and b with $a > 0$ such that, for all x, $u(x) = a \cdot u'(x) + b$. An inspection of Savage's arguments shows that his axioms form a foundation for all the assumptions A1–A4 as well as the MEU principle.

A third axiomatization of the MEU principle is given by John von Neumann and Oskar Morgenstern (1944). However, their results do not go as far as Savage's because they assume the existence of *objective* probabilities of the states (that is the probabilities of the possible states are given independently of the decision maker's evaluations). Thus they do not need the assumptions A3 and A4, and they give an axiomatic foundation only for correspondences of A1 and A2 which enables them to derive the MEU principle. In the game theory setting they use as a basis for their axiomatization, the assumption concerning objective probabilities is perhaps reasonable, but in a more general decision theory it is entirely unacceptable, at least from a Bayesian point of view. On the other hand, von Neumann and Morgenstern historically precede Savage, and his derivation of the utility measure is in fact based on their proof techniques (see Savage, 1954/1972, p. 75).

Yet another way of providing theoretical underpinnings of the assumptions A1–A3 as well as a version of the MEU principle has been developed by Richard Jeffrey (1965/1983). However, he rejects A4, the assumption that the probabilities of the states of the world are independent of the alternative chosen. In many cases, it seems quite natural that the acts chosen influence the probabilities of certain states of the world. For example, if the alternatives facing the decision maker are smoking and not smoking and if the relevant states of the world are that he or she will develop lung cancer or not, then certainly his or her decision will affect the probability of this aspect of the world. Savage was aware of this problem concerning the validity of A4, and he suggested another way of representing the states in this decision situation which would make the probability of the states independent of the alternatives chosen (Savage, 1954/1972, p. 15). However, this way of saving A4 is quite artificial and presumes much more about the decision maker's information about the world than Jeffrey's formulation. (For a discussion of this point see Jeffrey, 1977b and Jeffrey, 1965/1983, pp. 21–22.)

These considerations lead Jeffrey to introduce the notion of the *conditional* expected utility of an alternative, which may be defined as follows:

$$CEU(a_i) = P(s_1/a_i) \cdot u(o_{i1}) + P(s_2/a_i) \cdot u(o_{i2}) + \cdots + P(s_m/a_i) \cdot u(o_{im})$$

Jeffrey's decision rule is then the following version of the MEU principle:

The principle of maximizing conditional expected utility (MCEU): In a given decision situation, the decision maker should choose the alternative with maximal conditional expected utility (or one of the alternatives with maximal conditional expected utility if there are more than one).

Among philosophers, it is commonly recognized that high conditional probabilities need not indicate any causal relations. This has led to a different form of criticism of the traditional Bayesian theory. In so-called *causal* decision theory, it is suggested that one should not use the conditional probabilities of the states but rather a probabilistic evaluation of the causal *influence* of an act on a state. Allan Gibbard and William Harper (1978, reprinted as Chapter 17, this volume) analyze this kind of causal relation with the aid of the probabilities of *conditional (subjunctive) statements* of the form "if act a_i were performed, then state s_j would occur." In most cases, the probability of such a statement will be identical with the conditional probability $P(s_j/a_i)$. However, conditional probabilities are sometimes misleading indicators of causal dependence, and they argue that in some interesting decision situations conditional probabilities cannot be used, but one must rely on probabilities of the relevant conditional statements. The most famous such situation is the so-called Newcomb's problem (first presented in Nozick, 1969; see e.g. pp. 335–36, this volume). Causal decision theory is the topic of Part V of this book.

Another difference between Jeffrey on the one hand and Ramsey and Savage on the other is that Jeffrey considers preferences as a relation between *propositions* rather than between outcomes. In his axiomatization, the logical operations of propositions (negation, conjunction, disjunction) replace the work done by the operation of forming bets or gambles in the theories of Ramsey and Savage. Explaining this view, Jeffrey writes:

where acts are characterized with sufficient accuracy by declarative sentences, we can conveniently identify the acts with the propositions that the sentences express. An act is then a proposition which is within the agent's power to make true if he pleases.... (1983, p. 84)

Given the algebra of propositions, some of which describes acts, Jeffrey introduces the notion of the *desirability* of a proposition, which corresponds to the utility measure used in Savage's theory, as well as the probability of the proposition. Jeffrey and Ethan Bolker have developed an axiomatization, again using a preference ordering $\geq$ as primitive, which forms the basis for a representation theorem, where MCEU emerges as the decision rule.

However, on this axiomatization, it is not possible to *separate* completely desirabilities from probabilities in the same way as Ramsey and Savage were able to treat the probability of a state separated from its utility. Furthermore, the relation between the different desirability and probability measures that satisfy a given preference ordering is more complex in the Jeffrey–Bolker case than for Ramsey and Savage, and thus, in a sense, their representation theorems are weaker.

Criticisms of the Bayesian theory and some alternatives

We have now presented some different ways of providing foundations for the rationality of the MEU principle. However, this principle and the assumptions (A1)–(A4) on which it is based are far from being universally accepted. They have been criticized from several points of view, through both normative rationality arguments and more descriptive psychological considerations.

Starting with the rationality arguments, some of the axioms introduced by Ramsey and Savage have been extensively discussed. In particular Savage's sure-thing principle has come under heavy fire. We here present the two most influential criticisms, namely, the Allais paradox and the Ellsberg paradox.

Maurice Allais's (1953) example can be described by the following matrix:

	s_1 $P(s_1) = 0.01$	s_2 $P(s_2) = 0.10$	s_3 $P(s_3) = 0.89$
a_1	$500,000	$500,000	$500,000
a_2	$0	$2,500,000	$500,000
a_3	$500,000	$500,000	$0
a_4	$0	$2,500,000	$0

In this decision situation, most people prefer a_1 to a_2, because they do not find a 10 percent chance of winning a very large fortune in place of receiving a large fortune for certain to be enough compensation for the 1 percent risk of winning no money at all. Also a large majority of people prefer a_4 to a_3, because the chance of winning is nearly the same in both gambles (11 percent vs. 10 percent), so the one with the much larger prize seems better. (For some experimental evidence on the Allais paradox, see Kahneman and Tversky, 1979; reprinted in this volume, Chapter 11.) However, this pattern of preferences clearly violates Savage's sure-thing principle and thus also the MEU recommendation for decision making.

If interpreted as a descriptive principle, it seems very difficult to defend the sure-thing principle. But even if interpreted as a normative principle, it is far from clear that it can derive any support from more fundamental considerations. Edward McClennen (1983, Chapter 10, this volume) provides a careful analysis of the arguments that have been put forward to defend the sure-thing principle, and he concludes that these arguments do not hold water.

The second "paradox" that brings out problems for the MEU principle is due to Daniel Ellsberg (1961, Chapter 13, this volume). He asks us to consider the following decision problem. Imagine an urn known to contain 30 red balls and 60 black or yellow balls, the latter colors in unknown proportion. One ball is to be drawn at random from the urn. A first choice situation consists of two alternatives a_1 and a_2. If you choose a_1, you will receive \$100 if a red ball is drawn and nothing if a black ball or a yellow ball is drawn; if you choose a_2, you will receive \$100 if a black ball is drawn, otherwise nothing. Under the same circumstances, a second choice situation consists of the alternatives a_3 and a_4. If you choose a_3, you will receive \$100 if a red or yellow ball is drawn, otherwise nothing; if you choose a_4 you will receive \$100 if a black or yellow ball is drawn, otherwise nothing. These two decision situations can be summarized by the following matrix:

	Red	Black	Yellow
a_1	\$100	\$0	\$0
a_2	\$0	\$100	\$0
a_3	\$100	\$0	\$100
a_4	\$0	\$100	\$100

The most frequent pattern of response to these two decision situations is that a_1 is preferred to a_2 and a_4 is preferred to a_3. As is easily seen this preference pattern violates the sure-thing principle.

The rationale for these preferences seems to be that there is a difference between the quality of knowledge we have about the states. We *know* that the proportion of red balls in the urn is one third, whereas we are *uncertain* about the proportion of black balls (it can be anything between zero and two thirds). Thus this decision situation falls within the unnamed area between decision making under "risk" and decision making under "uncertainty."

The difference in information about the states is then reflected in the preferences in such a way that the alternative for which the exact probability of winning can be determined is preferred to the alternative where the probability of winning is "ambiguous" (Ellsberg's term). This is in sharp contrast to the assumption A3, that is, that a decision maker's beliefs can be represented by a unique probability measure; the quality of a decision maker's information about the state should thus not affect his or her preferences. Ellsberg's paradox can therefore be seen as an argument against A3.

We next turn to some forms of criticism that have been leveled against A1 and A2. First note that Allais's paradox can, indirectly, be seen as an argument against A2, that is, that the only value judgments relevant to the decision maker are the utilities of the outcomes.

Regarding the notion of utility involved in A1 and A2, it is possible to distinguish two different interpretations (e.g., see Hansson, 1981, and Sen, 1985). One is the *realistic* interpretation where utility is thought of as a sort of measure of subjective *usefulness*. For instance, the usefulness of two umbrellas is less than twice the usefulness of one umbrella. This interpretation has been dominant within economic theory. It was obviously a notion of this kind that Daniel Bernoulli had in mind when he proposed to use utilities instead of monetary values to obtain a solution to the so called St. Petersburg paradox.

The second interpretation, which may be called the *formalistic* interpretation, is the one used in presenting the foundations of Bayesian decision theory. The axiomatizations presented by Ramsey and Savage entail that for any decision behavior that satisfies the axioms, there exists a utility function that assigns values to different outcomes in such a way that the chosen alternative always has the highest expected utility. This interpretation is eloquently defended by John Harsanyi (1977), who writes:

In other words, a Bayesian need not have any special desire to maximize his expected utility *per se*. Rather, he simply wants to act in accordance with a few very important rationality axioms; and he knows that this fact has the inevitable mathematical *implication* of making his behavior equivalent to expected-utility maximization. As long as he obeys these rationality axioms, he simply cannot help acting *as if* he assigned numerical *utilities*, at least implicitly, to alternative possible outcomes of his behavior. . . . (p. 381)

Satisfying the axioms is thus equivalent to maximizing the expected utility in relation to *some* utility function. But is this "derived" utility function the same as the "real" utility function? And how do we know that the axioms are *rationality* axioms?

The difference between these two interpretations is important when considering the relation between utilities and risk behavior. The customary definition, at least in the economic literature, is that a decision maker is called *risk averse* if he or she prefers to have the expected monetary value of a gamble rather than the gamble itself, and *risk prone* if he or she has the opposite preferences. It can be shown that a person is risk averse with respect to a given gamble if and only if his or her utility function as a function of money is concave in the region around the mean value of the gamble. In fact, economic theory (Arrow, 1971;

Pratt, 1964) defines a risk aversion coefficient which can be computed once the utility curve for a given commodity is known. (Essentially, it is the second derivative over the first.) This way of defining risk entails that risk aversion is completely determined by the utility function.

However, intuitively we feel that the utility curve is one thing and risk behavior is another. This distinction cannot be made in the Arrow–Pratt theory of risk because two agents with the same utility functions will, by definition, have the same attitudes toward risk. Furthermore, under the formalistic interpretation of utilities, risk averse behavior will always *violate* some of the axioms and is thus deemed as irrational behavior (see Hansson, Chapter 8, this volume). So if one wants to disentangle the concepts of utility and risk, and allow for the fact that the decision makers' different attitudes toward risk may influence their decisions, then assumption A2 must be given up. (For an alternative analysis of the concept of risk, see Sahlin, 1984.)

Daniel Kahneman and Amos Tversky (1979, Chapter 11, this volume) present some types of psychological experiments, which are relevant for the evaluation of the rationality of assumptions A1 and A2 because the exhibited decision behavior violates the MEU principle. One kind of problem can be illustrated by such examples as the following (based on problems 11 and 12 in Kahneman and Tversky, 1979; Chapter 11, this volume):

> *Situation 1*: In addition to whatever you own, you have been given $1000. Choose between getting another $500 for certain and a 50 percent chance of getting another $1000.

> *Situation 2*: In addition to whatever you own, you have been given $2000. Choose between losing $500 for certain and a 50 percent chance of losing $1000.

Here most subjects chose getting another $500 for certain in the first situation, and they also chose a 50 percent chance of losing $1000 in the second. This pattern of choices is clearly inconsistent with the MEU principle because getting $500 in the first situation gives the same final wealth as losing $500 in the second and a 50 percent chance of winning another $1000 in the first is the same as a 50 percent chance of losing $1000 in the second. According to Bayesian decision theory, the same utility should be assigned to the same wealth regardless of what prior wealth it was reached from.

This example and others indicate that it is not only the utility of the outcomes that determines the choices (as is claimed in assumption A2), but that also a *reference level*, of wealth for example, is an important factor when evaluating the prospects. In situation 1 above, the natural reference level is the decision maker's present wealth increased by the

$1000 he is already guaranteed; in situation 2 it is the present level increased by $2000. This difference of reference level then explains the difference in the subject's decision behavior. An essential feature of the prospect theory suggested by Kahneman and Tversky is that what determines the value of an alternative is its possible *changes* in wealth or welfare, rather than the utility of the final outcomes, where changes are evaluated in relation to a reference level. (For a related discussion of a "relativized" notion of utility, see Hansson, 1975, and Sahlin, 1984.) Because there is no room for the concept of a reference level in traditional Bayesian theory, the introduction of this concept means that other factors than utilities of outcomes determine the value of the alternatives, so that assumption A2 must be given up.

This ends our presentation of Bayesian decision theory and some of the criticism that has been raised against it. We hope that this introduction will make clear the connections between the different articles in the present volume. The articles are divided into five groups:

Part I aims at presenting the foundations of Bayesian decision theory. It contains Ramsey's classic paper, an outline of Savage's theory, and Jeffrey's presentation of his (early) theory.

Part II raises some problems concerning the conceptualization of probability and utility. Among other points subject to criticism are the assumption that a Bayesian can reduce all decision problems to decision making under risk and the traditional notion of risk itself.

Part III focuses on the criticism of the sure-thing principle. Savage gives his defense of the principle, his and others' arguments are scrutinized, and some of the empirical evidence against the principle is presented. Two alternative decision theories, which account for the empirical findings, are introduced.

Part IV deals with decision making with "unreliable" probabilities, the area between the traditional extremes of decision making under risk and decision making under uncertainty. Ellsberg's paradox is a common starting point.

Part V presents causal decision theory. Newcomb's paradox has served as a source for several discussions concerning the role of causation in decision making.

Part I
Foundations of Bayesian decision theory

2. Truth and probability

Frank P. Ramsey

To say of what is that it is not, or of what is not that
it is, is false, while to say of what is that it is and of
what is not that it is not is true.—*Aristotle.*

When several hypotheses are presented to our mind
which we believe to be mutually exclusive and
exhaustive, but about which we know nothing
further, we distribute our belief equally among
them.... This being admitted as an account of the
way in which we *actually do* distribute our belief in
simple cases, the whole of the subsequent theory
follows as a deduction of the way in which we must
distribute it in complex cases *if we would be consistent.*—*W. F. Donkin.*

The object of reasoning is to find out, from the
consideration of what we already know, something
else which we do not know. Consequently, reasoning is good if it be such as to give a true conclusion
from true premises, and not otherwise.—*C. S. Peirce.*

Truth can never be told so as to be understood, and
not be believed.—*W. Blake.*

Foreword

In this essay the Theory of Probability is taken as a branch of logic, the
logic of partial belief and inconclusive argument; but there is no intention of implying that this is the only or even the most important aspect
of the subject. Probability is of fundamental importance not only in

Reprinted from *The Foundations of Mathematics*, ed. by R. B. Braithwaite, Routledge &
Kegan Paul, London, 1931, pp. 156–198, by kind permission of the publisher.

logic but also in statistical and physical science, and we cannot be sure beforehand that the most useful interpretation of it in logic will be appropriate in physics also. Indeed the general difference of opinion between statisticians who for the most part adopt the frequency theory of probability and logicians who mostly reject it renders it likely that the two schools are really discussing different things, and that the word 'probability' is used by logicians in one sense and by statisticians in another. The conclusions we shall come to as to the meaning of probability in logic must not, therefore, be taken as prejudging its meaning in physics.

1. The frequency theory

In the hope of avoiding some purely verbal controversies, I propose to begin by making some admissions in favour of the frequency theory. In the first place this theory must be conceded to have a firm basis in ordinary language, which often uses "probability" practically as a synonym for proportion; for example, if we say that the probability of recovery from smallpox is three-quarters, we mean, I think, simply that that is the proportion of smallpox cases which recover. Secondly, if we start with what is called the calculus of probabilities, regarding it first as a branch of pure mathematics, and then looking round for some interpretation of the formulae which shall show that our axioms are consistent and our subject not entirely useless, then much the simplest and least controversial interpretation of the calculus is one in terms of frequencies. This is true not only of the ordinary mathematics of probability, but also of the symbolic calculus developed by Mr Keynes; for if in his a/h, a and h are taken to be not propositions but propositional functions or class-concepts which define finite classes, and a/h is taken to mean the proportion of members of h which are also members of a, then all his propositions become arithmetical truisms.

Besides these two inevitable admissions, there is a third and more important one, which I am prepared to make temporarily although it does not express my real opinion. It is this. Suppose we start with the mathematical calculus, and ask, not as before what interpretation of it is most convenient to the pure mathematician but what interpretation gives results of greatest value to science in general, then, it may be that the answer is again an interpretation in terms of frequency; that probability as it is used in statistical theories, especially in statistical mechanics – the kind of probability whose logarithm is the entropy – is really a ratio between the numbers of two classes, or the limit of such a ratio. I do not myself believe this, but I am willing for the present to concede to the frequency theory that probability as used in modern science is really the same as frequency.

But, supposing all this admitted, it still remains the case that we have the authority both of ordinary language and of many great thinkers for discussing under the heading of probability what appears to be quite a different subject, the logic of partial belief. It may be that, as some supporters of the frequency theory have maintained, the logic of partial belief will be found in the end to be merely the study of frequencies, either because partial belief is definable as, or by reference to, some sort of frequency, or because it can only be the subject of logical treatment when it is grounded on experienced frequencies. Whether these contentions are valid can, however, only be decided as a result of our investigation into partial belief, so that I propose to ignore the frequency theory for the present and begin an inquiry into the logic of partial belief. In this, I think, it will be most convenient if, instead of straight away developing my own theory, I begin by examining the views of Mr Keynes, which are so well known and in essentials so widely accepted that readers probably feel that there is no ground for re-opening the subject *de novo* until they have been disposed of.

2. Mr Keynes' theory

Mr Keynes[1] starts from the supposition that we make probable inferences for which we claim objective validity; we proceed from full belief in one proposition to partial belief in another, and we claim that this procedure is objectively right, so that if another man in similar circumstances entertained a different degree of belief, he would be wrong in doing so. Mr Keynes accounts for this by supposing that between any two propositions, taken as premiss [sic.] and conclusion, there holds one and only one relation of a certain sort called probability relations; and that if, in any given case, the relation is that of degree a, from full belief in the premiss, we should, if we were rational, proceed to a belief of degree a in the conclusion.

Before criticising this view, I may perhaps be allowed to point out an obvious and easily corrected defect in the statement of it. When it is said that the degree of the probability relation is the same as the degree of belief which it justifies, it seems to be presupposed that both probability relations, on the one hand, and degrees of belief on the other can be naturally expressed in terms of numbers, and then that the number expressing or measuring the probability relation is the same as that expressing the appropriate degree of belief. But if, as Mr Keynes holds, these things are not always expressible by numbers, then we cannot give his statement that the degree of the one is the same as the

[1] J. M. Keynes, *A Treatise on Probability* (1921).

degree of the other such a simple interpretation, but must suppose him to mean only that there is a one-one correspondence between probability relations and the degrees of belief which they justify. This correspondence must clearly preserve the relations of greater and less, and so make the manifold of probability relations and that of degrees of belief similar in Mr Russell's sense. I think it is a pity that Mr Keynes did not see this clearly, because the exactitude of this correspondence would have provided quite as worthy material for his scepticism as did the numerical measurement of probability relations. Indeed some of his arguments against their numerical measurement appear to apply quite equally well against their exact correspondence with degrees of belief; for instance, he argues that if rates of insurance correspond to subjective, i.e. actual, degrees of belief, these are not rationally determined, and we cannot infer that probability relations can be similarly measured. It might be argued that the true conclusion in such a case was not that, as Mr Keynes thinks, to the non-numerical probability relation corresponds a non-numerical degree of rational belief, but that degrees of belief, which were always numerical, did not correspond one to one with the probability relations justifying them. For it is, I suppose, conceivable that degrees of belief could be measured by a psychogalvanometer or some such instrument, and Mr Keynes would hardly wish it to follow that probability relations could all be derivatively measured with the measures of the beliefs which they justify.

But let us now return to a more fundamental criticism of Mr Keynes' views, which is the obvious one that there really do not seem to be any such things as the probability relations he describes. He supposes that, at any rate in certain cases, they can be perceived; but speaking for myself I feel confident that this is not true. I do not perceive them, and if I am to be persuaded that they exist it must be by argument; moreover I shrewdly suspect that others do not perceive them either, because they are able to come to so very little agreement as to which of them relates any two given propositions. All we appear to know about them are certain general propositions, the laws of addition and multiplication; it is as if everyone knew the laws of geometry but no one could tell whether any given object were round or square; and I find it hard to imagine how so large a body of general knowledge can be combined with so slender a stock of particular facts. It is true that about some particular cases there is agreement, but these somehow paradoxically are always immensely complicated; we all agree that the probability of a coin coming down heads is ½, but we can none of us say exactly what is the evidence which forms the other term for the probability relation about which we are then judging. If, on the other hand, we take the simplest possible pairs of propositions such as "This is red" and "That is blue" or "This is red" and "That is red", whose logical relations should surely be easiest to see, no one, I think, pretends to be sure what is the probability relation which connects them. Or, perhaps, they may claim to see the relation but they will not be able to say anything about it with certainty, to state if it is more or less than ⅓, or so on. They may, of course, say that it is incomparable with any numerical relation, but a relation about which so little can be truly said will be of little scientific use and it will be hard to convince a sceptic of its existence. Besides this view is really rather paradoxical; for any believer in induction must admit that between "This is red" as conclusion and "This is round", together with a billion propositions of the form "a is round and red" as evidence, there is a finite probability relation; and it is hard to suppose that as we accumulate instances there is suddenly a point, say after 233 instances, at which the probability relation becomes finite and so comparable with some numerical relations.

It seems to me that if we take the two propositions "a is red", "b is red", we cannot really discern more than four simple logical relations between them; namely identity of form, identity of predicate, diversity of subject, and logical independence of import. If anyone were to ask me what probability one gave to the other, I should not try to answer by contemplating the propositions and trying to discern a logical relation between them, I should, rather, try to imagine that one of them was all that I knew, and to guess what degree of confidence I should then have in the other. If I were able to do this, I might no doubt still not be content with it but might say "This is what I should think, but, of course, I am only a fool" and proceed to consider what a wise man would think and call that the degree of probability. This kind of self-criticism I shall discuss later when developing my own theory: all that I want to remark here is that no one estimating a degree of probability simply contemplates the two propositions supposed to be related by it; he always considers *inter alia* his own actual or hypothetical degree of belief. This remark seems to me to be borne out by observation of my own behaviour; and to be the only way of accounting for the fact that we can all give estimates of probability in cases taken from actual life, but are quite unable to do so in the logically simplest cases in which, were probability a logical relation, it would be easiest to discern.

Another argument against Mr Keynes' theory can, I think, be drawn from his inability to adhere to it consistently even in discussing first principles. There is a passage in his chapter on the measurement of probabilities which reads as follows:

Probability is, *vide* Chapter II (§12), relative in a sense to the principles of *human* reason. The degree of probability, which it is rational for *us* to entertain, does not presume perfect logical insight, and is relative in part to the secondary propositions which we in fact know; and it is not dependent upon whether more perfect logical insight is or is not conceivable. It is the degree of probability to which those logical processes lead, of which our minds are capable; or, in the language of Chapter II, which those secondary propositions justify, which we in fact know. If we do not take this view of probability, if we do not limit it in this way and make it, to this extent, relative to human powers, we are altogether adrift in the unknown; for we cannot ever know what degree of probability would be justified by the perception of logical relations which we are, and must always be, incapable of comprehending. (p. 32, his italics)

This passage seems to me quite unreconcilable with the view which Mr Keynes adopts everywhere except in this and another similar passage. For he generally holds that the degree of belief which we are justified in placing in the conclusion of an argument is determined by what relation of probability unites that conclusion to our premisses. There is only one such relation and consequently only one relevant true secondary proposition, which, of course, we may or may not know, but which is necessarily independent of the human mind. If we do not know it, we do not know it and cannot tell how far we ought to believe the conclusion. But often, he supposes, we do know it; probability relations are not ones which we are incapable of comprehending. But on this view of the matter the passage quoted above has no meaning: the relations which justify probable beliefs are probability relations, and it is nonsense to speak of them being justified by logical relations which we are, and must always be, incapable of comprehending.

The significance of the passage for our present purpose lies in the fact that it seems to presuppose a different view of probability, in which indefinable probability relations play no part, but in which the degree of rational belief depends on a variety of logical relations. For instance, there might be between the premiss and conclusion the relation that the premiss was the logical product of a thousand instances of a generalization of which the conclusion was one other instance, and this relation, which is not an indefinable probability relation but definable in terms of ordinary logic and so easily recognizable, might justify a certain degree of belief in the conclusion on the part of one who believed the premiss. We should thus have a variety of ordinary logical relations justifying the same or different degrees of belief. To say that the probability of a given h was such-and-such would mean that between a and h was some relation justifying such-and-such a degree of belief. And on this view it would be a real point that the relation in question must not be one which the human mind is incapable of comprehending.

This second view of probability as depending on logical relations but not itself a new logical relation seems to me more plausible than Mr Keynes' usual theory; but this does not mean that I feel at all inclined to agree with it. It requires the somewhat obscure idea of a logical relation justifying a degree of belief, which I should not like to accept as indefinable because it does not seem to be at all a clear or simple notion. Also it is hard to say what logical relations justify what degrees of belief, and why; any decision as to this would be arbitrary, and would lead to a logic of probability consisting of a host of so-called "necessary" facts, like formal logic on Mr Chadwick's (1927) view of logical constants. Whereas I think it far better to seek an explanation of this "necessity" after the model of the work of Mr Wittgenstein, which enables us to see clearly in what precise sense and why logical propositions are necessary, and in a general way why the system of formal logic consists of the propositions it does consist of, and what is their common characteristic. Just as natural science tries to explain and account for the facts of nature, so philosophy should try, in a sense, to explain and account for the facts of logic; a task ignored by the philosophy which dismisses these facts as being unaccountably and in an indefinable sense "necessary".

Here I propose to conclude this criticism of Mr Keynes' theory, not because there are not other respects in which it seems open to objection, but because I hope that what I have already said is enough to show that it is not so completely satisfactory as to render futile any attempt to treat the subject from a rather different point of view.

3. Degrees of belief

The subject of our inquiry is the logic of partial belief, and I do not think we can carry it far unless we have at least an approximate notion of what partial belief is, and how, if at all, it can be measured. It will not be very enlightening to be told that in such circumstances it would be rational to believe a proposition to the extent of ⅔ unless we know what sort of a belief in it that means. We must therefore try to develop a purely psychological method of measuring belief. It is not enough to measure probability; in order to apportion correctly our belief to the probability we must also be able to measure our belief.

It is a common view that belief and other psychological variables are not measurable, and if this is true our inquiry will be vain; and so will the whole theory of probability conceived as a logic of partial belief; for if the phrase "a belief two-thirds of certainty" is meaningless, a calculus whose sole object is to enjoin such beliefs will be meaningless also. Therefore unless we are prepared to give up the whole thing as a bad job we are bound to hold that beliefs can to some extent be measured. If we were to follow the analogy of Mr Keynes' treatment of probabilities we should say that some beliefs were measurable and some not; but this does not seem to me likely to be a correct account of the matter: I do not see how we can sharply divide beliefs into those which have a position in the numerical scale and those which have not. But I think beliefs do differ in measurability in the following two ways. First, some beliefs can be measured more accurately than others; and, secondly, the measurement of beliefs is almost certainly an ambiguous process leading to a variable answer depending on how exactly the measurement is conducted. The degree of a belief is in this respect like the time interval between two events; before Einstein it was supposed that all the ordinary ways of measuring a time interval would lead to the same result if properly performed. Einstein showed that this was not the case; and time interval can no longer be regarded as an exact notion, but must be discarded in all precise investigations. Nevertheless, time interval and the Newtonian system are sufficiently accurate for many purposes and easier to apply.

I shall try to argue later that the degree of a belief is just like a time interval; it has no precise meaning unless we specify more exactly how it is to be measured. But for many purposes we can assume that the alternative ways of measuring it lead to the same result, although this is only approximately true. The resulting discrepancies are more glaring in connection with some beliefs than with others, and these therefore appear less measurable. Both these types of deficiency in measurability, due respectively to the difficulty in getting an exact enough measurement and to an important ambiguity in the definition of the measurement process, occur also in physics and so are not difficulties peculiar to our problem; what is peculiar is that it is difficult to form any idea of how the measurement is to be conducted, how a unit is to be obtained, and so on.

Let us then consider what is implied in the measurement of beliefs. A satisfactory system must in the first place assign to any belief a magnitude or degree having a definite position in an order of magnitudes; beliefs which are of the same degree as the same belief must be of the same degree as one another, and so on. Of course this cannot be accomplished without introducing a certain amount of hypothesis or fiction. Even in physics we cannot maintain that things that are equal to the same thing are equal to one another unless we take "equal" not as meaning "sensibly equal" but a fictitious or hypothetical relation. I do not want to discuss the metaphysics or epistemology of this process, but merely to remark that if it is allowable in physics it is allowable in psychology also. The logical simplicity characteristic of the relations dealt with in a science is never attained by nature alone without any admixture of fiction.

But to construct such an ordered series of degrees is not the whole of our task; we have also to assign numbers to these degrees in some intelligible manner. We can of course easily explain that we denote full belief by 1, full belief in the contradictory by 0, and equal beliefs in the proposition and its contradictory by ½. But it is not so easy to say what is meant by a belief ⅔ of certainty, or a belief in the proposition being twice as strong as that in its contradictory. This is the harder part of the task, but it is absolutely necessary; for we do calculate numerical probabilities, and if they are to correspond to degrees of belief we must discover some definite way of attaching numbers to degrees of belief. In physics we often attach numbers by discovering a physical process of addition (Campbell, 1920, p. 277): the measure-numbers of lengths are not assigned arbitrarily subject only to the proviso that the greater length shall have the greater measure; we determine them further by deciding on a physical meaning for addition; the length got by putting together two given lengths must have for its measure the sum of their measures. A system of measurement in which there is nothing corresponding to this is immediately recognized as arbitrary, for instance Mohs' scale of hardness (Campbell, 1920, p. 271) in which 10 is arbitrarily assigned to diamond, the hardest known material, 9 to the next hardest, and so on. We have therefore to find a process of addition for degrees of belief, or some substitute for this which will be equally adequate to determine a numerical scale.

Such is our problem; how are we to solve it? There are, I think, two ways in which we can begin. We can, in the first place, suppose that the degree of a belief is something perceptible by its owner; for instance that beliefs differ in the intensity of a feeling by which they are accompanied, which might be called a belief-feeling or feeling of conviction, and that by the degree of belief we mean the intensity of this feeling. This view would be very inconvenient, for it is not easy to ascribe numbers to the intensities for feelings; but apart from this it seems to me observably false, for the beliefs which we hold most strongly are often accompanied by practically no feeling at all; no one feels strongly about things he takes for granted.

We are driven therefore to the second supposition that the degree of a belief is a causal property of it, which we can express vaguely as the extent to which we are prepared to act on it. This is a generalization of the well-known view, that the differentia of belief lies in its causal efficacy, which is discussed by Mr Russell in his *Analysis of Mind*. He there dismisses it for two reasons, one of which seems entirely to miss the point. He argues that in the course of trains of thought we believe many things which do not lead to action. This objection is however beside the mark, because it is not asserted that a belief is an idea which does actually lead to action, but one which would lead to action in suitable circumstances; just as a lump of arsenic is called poisonous not because it actually has killed or will kill anyone, but because it would kill anyone if he ate it. Mr Russell's second argument is, however, more formidable. He points out that it is not possible to suppose that beliefs differ from other ideas only in their effects, for if they were otherwise identical their effects would be identical also. This is perfectly true, but it may still remain the case that the nature of the difference between the causes is entirely unknown or very vaguely known, and that what we want to talk about is the difference between the effects, which is readily observable and important.

As soon as we regard belief quantitatively, this seems to me the only view we can take of it. It could well be held that the difference between believing and not believing lies in the presence or absence of intro-spectible feelings. But when we seek to know what is the difference between believing more firmly and believing less firmly, we can no longer regard it as consisting in having more or less of certain observable feelings; at least I personally cannot recognize any such feelings. The difference seems to me to lie in how far we should act on these beliefs: this may depend on the degree of some feeling or feelings, but I do not know exactly what feelings and I do not see that it is indispensable that we should know. Just the same thing is found in physics; men found that a wire connecting plates of zinc and copper standing in acid deflected a magnetic needle in its neighbourhood. According as the needle was more or less deflected the wire was said to carry a larger or a smaller current. The nature of this 'current' could only be conjectured: what were observed and measured were simply its effects.

It will no doubt be objected that we know how strongly we believe things, and that we can only know this if we can measure our belief by introspection. This does not seem to me necessarily true; in many cases, I think, our judgment about the strength of our belief is really about how we should act in hypothetical circumstances. It will be answered that we can only tell how we should act by observing the present belief-feeling which determines how we should act; but again I doubt the cogency of the argument. It is possible that what determines how we should act determines us also directly or indirectly to have a correct opinion as to how we should act, without its ever coming into consciousness.

Suppose, however, I am wrong about this and that we can decide by introspection the nature of belief, and measure its degree; still, I shall argue, the kind of measurement of belief with which probability is concerned is not this kind but is a measurement of belief *qua* basis of action. This can I think be shown in two ways. First, by considering the scale of probabilities between 0 and 1, and the sort of way we use it, we shall find that it is very appropriate to the measurement of belief as a basis of action, but in no way related to the measurement of an introspected feeling. For the units in terms of which such feelings or sensations are measured are always, I think, differences which are just perceptible: there is no other way of obtaining units. But I see no ground for supposing that the interval between a belief of degree $\frac{1}{3}$ and one of degree $\frac{1}{2}$ consists of as many just perceptible changes as does that between one of $\frac{2}{3}$ and one of $\frac{5}{6}$, or that a scale based on just perceptible differences would have any simple relation to the theory of probability. On the other hand the probability of $\frac{1}{3}$ is clearly related to the kind of belief which would lead to a bet of 2 to 1, and it will be shown below how to generalize this relation so as to apply to action in general. Secondly, the quantitative aspects of beliefs as the basis of action are evidently more important than the intensities of belief-feelings. The latter are no doubt interesting, but may be very variable from individual to individual, and their practical interest is entirely due to their position as the hypothetical causes of beliefs *qua* bases of action.

It is possible that some one will say that the extent to which we should act on a belief in suitable circumstances is a hypothetical thing, and therefore not capable of measurement. But to say this is merely to reveal ignorance of the physical sciences which constantly deal with and measure hypothetical quantities; for instance, the electric intensity at a given point is the force which would act on a unit charge if it were placed at the point.

Let us now try to find a method of measuring beliefs as bases of possible actions. It is clear that we are concerned with dispositional rather than with actualized beliefs; that is to say, not with beliefs at the moment when we are thinking of them, but with beliefs like my belief that the earth is round, which I rarely think of, but which would guide my action in any case to which it was relevant.

The old-established way of measuring a person's belief is to propose a bet, and see what are the lowest odds which he will accept. This method I regard as fundamentally sound; but it suffers from being insufficiently general, and from being necessarily inexact. It is inexact partly because of the diminishing marginal utility of money, partly because the person may have a special eagerness or reluctance to bet, because he either enjoys or dislikes excitement or for any other reason, e.g. to make a book. The difficulty is like that of separating two different co-operating forces. Besides, the proposal of a bet may inevitably alter his state of opinion, just as we could not always measure electric intensity by actually introducing a charge and seeing what force it was subject to, because the introduction of the charge would change the distribution to be measured.

In order therefore to construct a theory of quantities of belief which shall be both general and more exact, I propose to take as a basis a general psychological theory, which is now universally discarded, but nevertheless comes, I think, fairly close to the truth in the sort of cases with which we are most concerned. I mean the theory that we act in the way we think most likely to realize the objects of our desires, so that a person's actions are completely determined by his desires and opinions. This theory cannot be made adequate to all the facts, but it seems to me a useful approximation to the truth particularly in the case of our self-conscious or professional life, and it is presupposed in a great deal of our thought. It is a simple theory and one which many psychologists would obviously like to preserve by introducing unconscious desires and unconscious opinions in order to bring it more into harmony with the facts. How far such fictions can achieve the required result I do not attempt to judge: I only claim for what follows approximate truth, or truth in relation to this artificial system of psychology, which like Newtonian mechanics can, I think, still be profitably used even though it is known to be false.

It must be observed that this theory is not to be identified with the psychology of the Utilitarians, in which pleasure had a dominating position. The theory I propose to adopt is that we seek things which we want, which may be our own or other people's pleasure, or anything else whatever, and our actions are such as we think most likely to realize these goods. But this is not a precise statement, for a precise statement of the theory can only be made after we have introduced the notion of quantity of belief.

Let us call the things a person ultimately desires 'goods', and let us at first assume that they are numerically measurable and additive. That is to say that if he prefers for its own sake an hour's swimming to an hour's reading, he will prefer two hours' swimming to one hour's swimming and one hour's reading. This is of course absurd in the given case but this may only be because swimming and reading are not ultimate goods, and because we cannot imagine a second hour's swimming precisely similar to the first, owing to fatigue, etc.

Let us begin by supposing that our subject has no doubts about anything, but certain opinions about all propositions. Then we can say that he will always choose the course of action which will lead in his opinion to the greatest sum of good.

It should be emphasized that in this essay good and bad are never to be understood in any ethical sense but simply as denoting that to which a given person feels desire and aversion.

The question then arises how we are to modify this simple system to take account of varying degrees of certainty in his beliefs. I suggest that we introduce as a law of psychology that his behaviour is governed by what is called the mathematical expectation; that is to say that, if p is a proposition about which he is doubtful, any goods or bads for whose realization p is in his view a necessary and sufficient condition enter into his calculations multiplied by the same fraction, which is called the "degree of his belief in p". We thus define degree of belief in a way which presupposes the use of the mathematical expectation.

We can put this in a different way. Suppose his degree of belief in p is m/n; then his action is such as he would choose it to be if he had to repeat it exactly n times, in m of which p was true, and in the others false. [Here it may be necessary to suppose that in each of the n times he had no memory of the previous ones.]

This can also be taken as a definition of the degree of belief, and can easily be seen to be equivalent to the previous definition. Let us give an instance of the sort of case which might occur. I am at a cross-roads and do not know the way; but I rather think one of the two ways is right. I propose therefore to go that way but keep my eyes open for someone to ask; if now I see someone half a mile away over the fields, whether I turn aside to ask him will depend on the relative inconvenience of going out of my way to cross the fields or of continuing on the wrong road if it is the wrong road. But it will also depend on how confident I am that I am right; and clearly the more confident I am of this the less distance I should be willing to go from the road to check my opinion. I propose therefore to use the distance I would be prepared to go to ask, as a measure of the confidence of my opinion; and what I have said above explains how this is to be done. We can set it out as follows: suppose the disadvantage of going x yards to ask is $f(x)$, the advantage of arriving at the right destination is r, that of arriving at the wrong one w. Then if I should just be willing to go a distance d to ask, the degree of my belief that I am on the right road is given by

$$p = 1 - \frac{f(d)}{r - w}.$$

For such an action is one it would just pay me to take, if I had to act in the same way n times, in np of which I was on the right way but in the others not.

For the total good resulting from not asking each time

$$= npr + n(1 - p)w$$
$$= nw + np(r - w),$$

that resulting from asking at distance x each time

$$= nr - nf(x), \qquad \text{[I now always go right.]}$$

This is greater than the preceding expression, provided

$$f(x) < (r - w)(1 - p),$$

$\therefore$ the critical distance d is connected with p, the degree of belief, by the relation $f(d) = (r - w)(1 - p)$

$$\text{or } p = 1 - \frac{f(d)}{r - w}$$

as asserted above.

It is easy to see that this way of measuring beliefs gives results agreeing with ordinary ideas; at any rate to the extent that full belief is denoted by 1, full belief in the contradictory by 0, and equal belief in the two by ½. Further, it allows validity to betting as means of measuring beliefs. By proposing a bet on p we give the subject a possible course of action from which so much extra good will result to him if p is true and so much extra bad if p is false. Supposing the bet to be in goods and bads instead of in money, he will take a bet at any better odds than those corresponding to his state of belief; in fact his state of belief is measured by the odds he will just take; but this is vitiated, as already explained, by love or hatred of excitement, and by the fact that the bet is in money and not in goods and bads. Since it is universally agreed that money has a diminishing marginal utility, if money bets are to be used, it is evident that they should be for as small stakes as possible. But then again the measurement is spoiled by introducing the new factor of reluctance to bother about trifles.

Let us now discard the assumption that goods are additive and immediately measurable, and try to work out a system with as few assumptions as possible. To begin with we shall suppose, as before, that our subject has certain beliefs about everything; then he will act so that what he believes to be the total consequences of his action will be the best possible. If then we had the power of the Almighty, and could persuade our subject of our power, we could, by offering him options, discover how he placed in order of merit all possible courses of the world. In this way all possible worlds would be put in an order of value, but we should have no definite way of representing them by numbers. There would be no meaning in the assertion that the difference in value between α and β was equal to that between γ and δ. [Here and elsewhere we use Greek letters to represent the different possible totalities of events between which our subject chooses – the ultimate organic unities.]

Suppose next that the subject is capable of doubt; then we could test his degree of belief in different propositions by making him offers of the following kind. Would you rather have world α in any event; or world β if p is true, and world γ if p is false? If, then, he were certain that p was true, he would simply compare α and β and choose between them as if no conditions were attached; but if he were doubtful his choice would not be decided so simply. I propose to lay down axioms and definitions concerning the principles governing choices of this kind. This is, of course, a very schematic version of the situation in real life, but it is, I think, easier to consider it in this form.

There is first a difficulty which must be dealt with; the propositions like p in the above case which are used as conditions in the options offered may be such that their truth or falsity is an object of desire to the subject. This will be found to complicate the problem, and we have to assume that there are propositions for which this is not the case, which we shall call ethically neutral. More precisely an atomic proposition p is called ethically neutral if two possible worlds differing only in regard to the truth of p are always of equal value; and a non-atomic proposition p is called ethically neutral if all its atomic truth-arguments[2] are ethically neutral.

We begin by defining belief of degree ½ in an ethically neutral proposition. The subject is said to have belief of degree ½ in such a proposition p if he has no preference between the options (1) α if p is true, β if p is false, and (2) α if p is false, β if p is true, but has a preference between α and β simply. We suppose by an axiom that if this is true of any one pair α, β it is true of all such pairs.[3] This comes roughly to defining belief of degree ½ as such a degree of belief as leads to indifference between betting one way and betting the other for the same stakes.

Belief of degree ½ as thus defined can be used to measure values numerically in the following way. We have to explain what is meant by the difference in value between α and β being equal to that between γ and δ; and we define this to mean that, if p is an ethically neutral proposition believed to degree ½, the subject has no preference between the options (1) α if p is true, δ if p is false, and (2) β if p is true, γ if p is false.

This definition can form the basis of a system of measuring values in the following way:

Let us call any set of all worlds equally preferable to a given world a value: we suppose that if world α is preferable to β any world with the same value as α is preferable to any world with the same value as β and shall say that the value of α is greater than that of β. This relation 'greater than' orders values in a series. We shall use α henceforth both for the world and its value.

Axioms

1. There is an ethically neutral proposition p believed to degree ½.
2. If p, q are such propositions and the option

   α if p, δ if not-p is equivalent to β if p, γ if not-p

   then α if q, δ if not-q is equivalent to β if q, γ if not-q.

---

[2] I assume here Wittgenstein's theory of propositions; it would probably be possible to give an equivalent definition in terms of any other theory.

[3] α and β must be supposed so far undefined as to be compatible with both p and not-p.

Def. In the above case we say $\alpha\beta = \gamma\delta$.

Theorems. If $\alpha\beta = \gamma\delta$,

then $\beta\alpha = \delta\gamma$, $\alpha\gamma = \beta\delta$, $\gamma\alpha = \delta\beta$.

2a. If $\alpha\beta = \gamma\delta$, then $\alpha > \beta$ is equivalent to $\gamma > \delta$

and $\alpha = \beta$ is equivalent to $\gamma = \delta$.

3. If option A is equivalent to option B and B to C then A to C.

Theorem. If $\alpha\beta = \gamma\delta$ and $\beta\eta = \zeta\gamma$,

then $\alpha\eta = \zeta\delta$.

4. If $\alpha\beta = \gamma\delta$, $\gamma\delta = \eta\zeta$, then $\alpha\beta = \eta\zeta$.

5. (α, β, γ). E! (ιx) $(\alpha x = \beta\gamma)$.

6. (α, β). E! (ιx) $(\alpha x = x\beta)$.

7. Axiom of continuity: – Any progression has a limit (ordinal).

8. Axiom of Archimedes.

These axioms enable the values to be correlated one-one with real numbers so that if α^1 corresponds to α, etc.

$$\alpha\beta = \gamma\delta \; . \; \equiv \; . \; \alpha^1 - \beta^1 = \gamma^1 - \delta^1.$$

Henceforth we use α for the correlated real number α^1 also.

Having thus defined a way of measuring value we can now derive a way of measuring belief in general. If the option of α for certain is indifferent with that of β if p is true and γ if p is false,[4] we can define the subject's degree of belief in p as the ratio of the difference between α and γ to that between β and γ; which we must suppose the same for all α's, β's and γ's that satisfy the conditions. This amounts roughly to defining the degree of belief in p by the odds at which the subject would bet on p, the bet being conducted in terms of differences of value as defined. The definition only applies to partial belief and does not include certain beliefs; for belief of degree 1 in p, α for certain is indifferent with α if p and any β if not-p.

We are also able to define a very useful new idea – "the degree of belief in p given q". This does not mean the degree of belief in "If p then q", or that in "p entails q", or that which the subject would have in p if he knew q, or that which the subject would have in p if he knew q, or that which he ought to have. It roughly expresses the odds at which he would now bet on p, the bet only to be valid if q is true. Such conditional bets were often made in the eighteenth century.

The degree of belief in p given q is measured thus. Suppose the subject indifferent between the options (1) α if q true, β if q false, (2) γ

---

[4] Here β must include the truth of p, γ its falsity; p need no longer be ethically neutral. But we have to assume that there is a world with an assigned value in which p is true, and one in which p is false.

if p true and q true, δ if p false and q true, β if q false. Then the degree
of his belief in p given q is the ratio of the difference between α and δ to
that between γ and δ, which we must suppose the same for any α, β,
γ, δ which satisfy the given conditions. This is not the same as the
degree to which he would believe p, if he believed q for certain; for
knowledge of q might for psychological reasons profoundly alter his
whole system of beliefs.

Each of our definitions has been accompanied by an axiom of consistency, and in so far as this is false, the notion of the corresponding
degree of belief becomes invalid. This bears some analogy to the
situation in regard to simultaneity discussed above.

I have not worked out the mathematical logic of this in detail,
because this would, I think, be rather like working out to seven places
of decimals a result only valid to two. My logic cannot be regarded as
giving more than the sort of way it might work.

From these definitions and axioms it is possible to prove the fundamental laws of probable belief (degrees of belief lie between 0 and 1):

1. Degree of belief in p + degree of belief in $\bar{p}$ = 1.
2. Degree of belief in p given q + degree of belief in $\bar{p}$ given q = 1.
3. Degree of belief in (p and q) = degree of belief in p × degree of
   belief in q given p.
4. Degree of belief in (p and q) + degree of belief in (p and $\bar{q}$) =
   degree of belief in p.

The first two are immediate. (3) is proved as follows.

Let degree of belief in $p = x$, that in q given $p = y$.

Then ξ for certain $\equiv \xi + (1 - x)t$ if p true, $\xi - xt$ if p false, for any t.

$$\xi + (1 - x)\, t \text{ if } p \text{ true} \equiv$$

$$\begin{cases} \xi + (1 - x)\, t + (1 - y)\, u \text{ if } ``p \text{ and } q"\text{true,} \\ \xi + (1 - x)\, t - yu \text{ if } p \text{ true } q \text{ false;} \quad \text{for any } u. \end{cases}$$

Choose u so that $\xi + (1 - x)\, t - yu = \xi - xt$,

i.e. let $u = t/y \ (y \neq 0)$

Then ξ for certain $\equiv$

$$\begin{cases} \xi + (1 - x)\, t + (1 - y)\, t/y \text{ if } p \text{ and } q \text{ true} \\ \xi - xt \text{ otherwise,} \end{cases}$$

$\therefore$ degree of belief in $``p$ and $q" = \dfrac{xt}{t + (1 - y)t/y} = xy.\,(t \neq 0)$

If $y = 0$, take $t = 0$.

Then ξ for certain $\equiv \xi$ if p true, ξ if p false

$\equiv \xi + u$ if p true, q true; ξ if p false, q false;
ξ if p false

$\equiv \xi + u$, pq true; ξ, pq false

$\therefore$ degree of belief in $pq = 0$.

(4) follows from (2), (3) as follows:

Degree of belief in pq = that in $p \times$ that in q given p, by (3). Similarly degree of belief in $p\bar{q}$ = that in $p \times$ that in $\bar{q}$ given p $\therefore$ sum = degree of belief in p, by (2).

These are the laws of probability, which we have proved to be necessarily true of any consistent set of degrees of belief. Any definite set of degrees of belief which broke them would be inconsistent in the sense that it violated the laws of preference between options, such as that preferability is a transitive asymmetrical relation, and that if α is preferable to β, β for certain cannot be preferable to α if p, β if not-p. If anyone's mental condition violated these laws, his choice would depend on the precise form in which the options were offered him, which would be absurd. He could have a book made against him by a cunning better and would then stand to lose in any event.

We find, therefore, that a precise account of the nature of partial belief reveals that the laws of probability are laws of consistency, an extension to partial beliefs of formal logic, the logic of consistency. They do not depend for their meaning on any degree of belief in a proposition being uniquely determined as the rational one; they merely distinguish those sets of beliefs which obey them as consistent ones.

Having any definite degree of belief implies a certain measure of consistency, namely willingness to bet on a given proposition at the same odds for any stake, the stakes being measured in terms of ultimate values. Having degrees of belief obeying the laws of probability implies a further measure of consistency, namely such a consistency between the odds acceptable on different propositions as shall prevent a book being made against you.

Some concluding remarks on this section may not be out of place. First, it is based fundamentally on betting, but this will not seem unreasonable when it is seen that all our lives we are in a sense betting. Whenever we go to the station we are betting that a train will really run, and if we had not a sufficient degree of belief in this we should decline the bet and stay at home. The options God gives us are always conditional on our guessing whether a certain proposition is true. Secondly, it is based throughout on the idea of mathematical expectation; the dissatisfaction often felt with this idea is due mainly to the inaccurate measurement of goods. Clearly mathematical expectations in terms of money are not proper guides to conduct. It should be remembered, in judging my system, that in it value is actually defined by means of mathematical expectation in the case of beliefs of degree ½, and so may be expected to be scaled suitably for the valid application of the mathematical expectation in the case of other degrees of belief also.

Thirdly, nothing has been said about degrees of belief when the number of alternatives is infinite. About this I have nothing useful to say, except that I doubt if the mind is capable of contemplating more than a finite number of alternatives. It can consider questions to which an infinite number of answers are possible, but in order to consider the answers it must lump them into a finite number of groups. The difficulty becomes practically relevant when discussing induction, but even then there seems to me no need to introduce it. We can discuss whether past experience gives a high probability to the sun's rising to-morrow without bothering about what probability it gives to the sun's rising each morning for evermore. For this reason I cannot but feel that Mr Ritchie's discussion of the problem[5] is unsatisfactory; it is true that we can agree that inductive generalizations need have no finite probability, but particular expectations entertained on inductive grounds undoubtedly do have a high numerical probability in the minds of all of us. We all are more certain that the sun will rise to-morrow than that I shall not throw 12 with two dice first time, i.e. we have a belief of higher degree than $35/36$ in it. If induction ever needs a logical justification it is in connection with the probability of an event like this.

4. The logic of consistency

We may agree that in some sense it is the business of logic to tell us what we ought to think; but the interpretation of this statement raises considerable difficulties. It may be said that we ought to think what is true, but in that sense we are told what to think by the whole of science and not merely by logic. Nor, in this sense, can any justification be found for partial belief; the ideally best thing is that we should have beliefs of degree 1 in all true propositions and beliefs of degree 0

[5] A. D. Ritchie (1926, p. 318). 'The conclusion of the foregoing discussion may be simply put. If the problem of induction be stated to be "How can inductive generalizations acquire a large numerical probability?" then this is a pseudo-problem, because the answer is "They cannot". This answer is not, however, a denial of the validity of induction but is a direct consequence of the nature of probability. It still leaves untouched the real problem of induction which is "How can the probability of an induction be increased?" and it leaves standing the whole of Keynes' discussion on this point.'

in all false propositions. But this is too high a standard to expect of mortal men, and we must agree that some degree of doubt or even of error may be humanly speaking justified.

Many logicians, I suppose, would accept as an account of their science the opening words of Mr Keynes' *Treatise on Probability*: "Part of our knowledge we obtain direct; and part by argument. The Theory of Probability is concerned with that part which we obtain by argument, and it treats of the different degrees in which the results so obtained are conclusive or inconclusive." Where Mr Keynes says "the Theory of Probability", others would say Logic. It is held, that is to say, that our opinions can be divided into those we hold immediately as a result of perception or memory, and those which we derive from the former by argument. It is the business of Logic to accept the former class and criticise merely the derivation of the second class from them.

Logic as the science of argument and inference is traditionally and rightly divided into deductive and inductive; but the difference and relation between these two divisions of the subject can be conceived in extremely different ways. According to Mr Keynes valid deductive and inductive arguments are fundamentally alike; both are justified by logical relations between premiss and conclusion which differ only in degree. This position, as I have already explained, I cannot accept. I do not see what these inconclusive logical relations can be or how they can justify partial beliefs. In the case of conclusive logical arguments I can accept the account of their validity which has been given by many authorities, and can be found substantially the same in Kant, De Morgan, Peirce and Wittgenstein. All these authors agree that the conclusion of a formally valid argument is contained in its premisses; that to deny the conclusion while accepting the premisses would be self-contradictory; that a formal deduction does not increase our knowledge, but only brings out clearly what we already know in another form; and that we are bound to accept its validity on pain of being inconsistent with ourselves. The logical relation which justifies the inference is that the sense or import of the conclusion is contained in that of the premisses.

But in the case of an inductive argument this does not happen in the least; it is impossible to represent it as resembling a deductive argument and merely weaker in degree; it is absurd to say that the sense of the conclusion is partially contained in that of the premisses. We could accept the premisses and utterly reject the conclusion without any sort of inconsistency or contradiction.

It seems to me, therefore, that we can divide arguments into two radically different kinds, which we can distinguish in the words of Peirce (1923, p. 92) as (1) "explicative, analytic, or deductive" and (2) "amplifiative, synthetic, or (loosely speaking) inductive". Arguments of the second type are from an important point of view much closer to memories and perceptions than to deductive arguments. We can regard perception, memory and induction as the three fundamental ways of acquiring knowledge; deduction on the other hand is merely a method of arranging our knowledge and eliminating inconsistencies or contradictions.

Logic must then fall very definitely into two parts: (excluding analytic logic, the theory of terms and propositions) we have the lesser logic, which is the logic of consistency, or formal logic; and the larger logic, which is the logic of discovery, or inductive logic.

What we have now to observe is that this distinction in no way coincides with the distinction between certain and partial beliefs; we have seen that there is a theory of consistency in partial beliefs just as much as of consistency in certain beliefs, although for various reasons the former is not so important as the latter. The theory of probability is in fact a generalization of formal logic; but in the process of generalization one of the most important aspects of formal logic is destroyed. If p and $\bar{q}$ are inconsistent so that q follows logically from p, that p implies q is what is called by Wittgenstein a "tautology" and can be regarded as a degenerate case of a true proposition not involving the idea of consistency. This enables us to regard (not altogether correctly) formal logic including mathematics as an objective science consisting of objectively necessary propositions. It thus gives us not merely the ἀνάγκη λέγειν, that if we assert p we are bound in consistency to assert q also, but also the ἀνάγκη εἶναι, that if p is true, so must q be. But when we extend formal logic to include partial beliefs this direct objective interpretation is lost; if we believe pq to the extent of $\frac{1}{3}$, and $p\bar{q}$ to the extent of $\frac{1}{3}$, we are bound in consistency to believe $\bar{p}$ also to the extent of $\frac{1}{3}$. This is the ἀνάγκη λέγειν; but we cannot say that if pq is $\frac{1}{3}$ true and $p\bar{q}$ $\frac{1}{3}$ true, $\bar{p}$ also must be $\frac{1}{3}$ true, for such a statement would be sheer nonsense. There is no corresponding ἀνάγκη εἶναι. Hence, unlike the calculus of consistent full belief, the calculus of objective partial belief cannot be immediately interpreted as a body of objective tautology.

This is, however, possible in a roundabout way; we saw at the beginning of this essay that the calculus of probabilities could be interpreted in terms of class-ratios; we have now found that it can also be interpreted as a calculus of consistent partial belief. It is natural, therefore, that we should expect some intimate connection between these two interpretations, some explanation of the possibility of applying the same mathematical calculus to two such different sets of phenomena. Nor is an explanation difficult to find; there are many connections between partial beliefs and frequencies. For instance, experienced frequencies often lead to corresponding partial beliefs, and partial beliefs lead to the expectation of corresponding frequencies in accordance with Bernoulli's Theorem. But neither of these is exactly the connection we want; a partial belief cannot in general be connected uniquely with any actual frequency, for the connection is always made by taking the proposition in question as an instance of a propositional function. What propositional function we choose is to some extent arbitrary and the corresponding frequency will vary considerably with our choice. The pretensions of some exponents of the frequency theory that partial belief means full belief in a frequency proposition cannot be sustained. But we found that the very idea of partial belief involves reference to a hypothetical or ideal frequency; supposing goods to be additive, belief of degree m/n is the sort of belief which leads to the action which would be best if repeated n times in m of which the proposition is true; or we can say more briefly that it is the kind of belief most appropriate to a number of hypothetical occasions other-wise identical in a proportion m/n of which the proposition in question is true. It is this connection between partial belief and frequency which enables us to use the calculus of frequencies as a calculus of consistent partial belief. And in a sense we may say that the two interpretations are the objective and subjective aspects of the same inner meaning, just as formal logic can be interpreted objectively as a body of tautology and subjectively as the laws of consistent thought.

We shall, I think, find that this view of the calculus of probability removes various difficulties that have hitherto been found perplexing. In the first place it gives us a clear justification for the axioms of the calculus, which on such a system as Mr Keynes' is entirely wanting. For now it is easily seen that if partial beliefs are consistent they will obey these axioms, but it is utterly obscure why Mr Keynes' mysterious logical relations should obey them.[6] We should be so curiously ignorant of the instances of these relations, and so curiously knowl-edgeable about their general laws.

Secondly, the Principle of Indifference can now be altogether dispensed with; we do not regard it as belonging to formal logic to say what should be a man's expectation of drawing a white or a black ball from an urn; his original expectations may within the limits of consis-tency be any he likes; all we have to point out is that if he has certain expectations he is bound in consistency to have certain others. This is simply bringing probability into line with ordinary formal logic, which does not criticize premises but merely declares that certain

---

[6] It appears in Mr Keynes' system as if the principal axioms – the laws of addition and multiplication – were nothing but definitions. This is merely a logical mistake; his definitions are formally invalid unless corresponding axioms are presupposed. Thus his definition of multiplication presupposes the law that if the probability of a given bh is equal to that of c given dk, and the probability of b given h is equal to that of d given k, then will the probabilities of ab given h and of cd given k be equal.

conclusions are the only ones consistent with them. To be able to turn the Principle of Indifference out of formal logic is a great advantage; for it is fairly clearly impossible to lay down purely logical conditions for its validity, as is attempted by Mr Keynes. I do not want to discuss this question in detail, because it leads to hair-splitting and arbitrary distinctions which could be discussed for ever. But anyone who tries to decide by Mr Keynes' methods what are the proper alternatives to regard as equally probable in molecular mechanics, e.g. in Gibbs' phase-space, will soon be convinced that it is a matter of physics rather than pure logic. By using the multiplication formula, as it is used in inverse probability, we can on Mr Keynes' theory reduce all probabilities to quotients of *a priori* probabilities; it is therefore in regard to these latter that the Principle of Indifference is of primary importance; but here the question is obviously not one of formal logic. How can we on merely logical grounds divide the spectrum into equally probable bands?

A third difficulty which is removed by our theory is the one which is presented to Mr Keynes' theory by the following case. I think I perceive or remember something but am not sure; this would seem to give me some ground for believing it, contrary to Mr Keynes' theory, by which the degree of belief in it which it would be rational for me to have is that given by the probability relation between the proposition in question and the things I know for certain. He cannot justify a probable belief founded not on argument but on direct inspection. In our view there would be nothing contrary to formal logic in such a belief; whether it would be reasonable would depend on what I have called the larger logic which will be the subject of the next section; we shall there see that there is no objection to such a possibility, with which Mr Keynes' method of justifying probable belief solely by relation to certain knowledge is quite unable to cope.

5. The logic of truth

The validity of the distinction between the logic of consistency and the logic of truth has been often disputed; it has been contended on the one hand that logical consistency is only a kind of factual consistency; that if a belief in p is inconsistent with one in q, that simply means that p and q are not both true, and that this is a necessary or logical fact. I believe myself that this difficulty can be met by Wittgenstein's theory of tautology, according to which if a belief in p is inconsistent with one in q, that p and q are not both true is not a fact but a tautology. But I do not propose to discuss this question further here.

From the other side it is contended that formal logic or the logic of consistency is the whole of logic, and inductive logic either nonsense or part of natural science. This contention, which would I suppose be made by Wittgenstein, I feel more difficulty in meeting. But I think it would be a pity, out of deference to authority, to give up trying to say anything useful about induction.

Let us therefore go back to the general conception of logic as the science of rational thought. We found that the most generally accepted parts of logic, namely, formal logic, mathematics and the calculus of probabilities, are all concerned simply to ensure that our beliefs are not self-contradictory. We put before ourselves the standard of consistency and construct these elaborate rules to ensure its observance. But this is obviously not enough; we want our beliefs to be consistent not merely with one another but also with the facts[7]: nor is it even clear that consistency is always advantageous; it may well be better to be sometimes right than never right. Nor when we wish to be consistent are we always able to be: there are mathematical propositions whose truth or falsity cannot as yet be decided. Yet it may humanly speaking be right to entertain a certain degree of belief in them on inductive or other grounds: a logic which proposes to justify such a degree of belief must be prepared actually to go against formal logic; for to a formal truth formal logic can only assign a belief of degree 1. We could prove in Mr Keynes' system that its probability is 1 on any evidence. This point seems to me to show particularly clearly that human logic or the logic of truth, which tells men how they should think, is not merely independent of but sometimes actually incompatible with formal logic.

In spite of this nearly all philosophical thought about human logic and especially induction has tried to reduce it in some way to formal logic. Not that it is supposed, except by a very few, that consistency will of itself lead to truth; but consistency combined with observation and memory is frequently credited with this power.

Since an observation changes (in degree at least) my opinion about the fact observed, some of my degrees of belief after the observation are necessarily inconsistent with those I had before. We have therefore to explain how exactly the observation should modify my degrees of belief; obviously if p is the fact observed, my degree of belief in q after the observation should be equal to my degree of belief in q given p before, or by the multiplication law to the quotient of my degree of belief in pq by my degree of belief in p. When my degrees of belief change in this way we can say that they have been changed consistently by my observation.

By using this definition, or on Mr Keynes' system simply by using

---

[7] Cf. Kant: 'Denn obgleich eine Erkenntnis der logischen Form völlig gemäss sein möchte, dass ist sich selbst nicht widerspräche, so kann sie doch noch immer dem Gegenstande widersprechen.' *Kritik der reinen Vernunft*, First Edition, p. 59.

the multiplication law, we can take my present degrees of belief, and by considering the totality of my observations, discover from what initial degrees of belief my present ones would have arisen by this process of consistent change. My present degrees of belief can then be considered logically justified if the corresponding initial degrees of belief are logically justified. But to ask what initial degrees of belief are justified, or in Mr Keynes' system what are the absolutely *a priori* probabilities, seems to me a meaningless question; and even if it had a meaning I do not see how it could be answered.

If we actually applied this process to a human being, found out, that is to say, on what *a priori* probabilities his present opinions could be based, we should obviously find them to be ones determined by natural selection, with a general tendency to give a higher probability to the simpler alternatives. But, as I say, I cannot see what could be meant by asking whether these degrees of belief were logically justified. Obviously the best thing would be to know for certain in advance what was true and what false, and therefore if any one system of initial beliefs is to receive the philosopher's approbation it should be this one. But clearly this would not be accepted by thinkers of the school I am criticising. Another alternative is to apportion initial probabilities on the purely formal system expounded by Wittgenstein, but as this gives no justification for induction it cannot give us the human logic which we are looking for.

Let us therefore try to get an idea of a human logic which shall not attempt to be reducible to formal logic. Logic, we may agree, is concerned not with what men actually believe, but what they ought to believe, or what it would be reasonable to believe. What then, we must ask, is meant by saying that it is reasonable for a man to have such and such a degree of belief in a proposition? Let us consider possible alternatives.

First, it sometimes means something explicable in terms of formal logic; this possibility for reasons already explained we may dismiss. Secondly, it sometimes means simply that were I in his place (and not e.g. drunk) I should have such a degree of belief. Thirdly, it sometimes means that if his mind worked according to certain rules, which we may roughly call "scientific method", he would have such a degree of belief. But fourthly it need mean none of these things; for men have not always believed in scientific method, and just as we ask "But am I necessarily reasonable", we can also ask "But is the scientist necessarily reasonable?" In this ultimate meaning it seems to me that we can identify reasonable opinion with the opinion of an ideal person in similar circumstances. What, however, would this ideal person's opinion be? As has previously been remarked, the highest ideal would be always to have a true opinion and be certain of it; but this ideal is more suited to God than to man.[8]

We have therefore to consider the human mind and what is the most we can ask of it.[9] The human mind works essentially according to general rules or habits; a process of thought not proceeding according to some rule would simply be a random sequence of ideas; whenever we infer A from B we do so in virtue of some relation between them. We can therefore state the problem of the ideal as "What habits in a general sense would it be best for the human mind to have?" This is a large and vague question which could hardly be answered unless the possibilities were first limited by a fairly definite conception of human nature. We could imagine some very useful habits unlike those possessed by any men. [It must be explained that I use habit in the most general possible sense to mean simply rule or law of behaviour, including instinct: I do not wish to distinguish acquired rules or habits in the narrow sense from innate rules or instincts, but propose to call them all habits alike.] A completely general criticism of the human mind is therefore bound to be vague and futile, but something useful can be said if we limit the subject in the following way.

Let us take a habit of forming opinion in a certain way; e.g. the habit of proceeding from the opinion that a toadstool is yellow to the opinion that it is unwholesome. Then we can accept the fact that the person has a habit of this sort, and ask merely what degree of opinion that the toadstool is unwholesome it would be best for him to entertain when he see it; i.e. granting that he is going to think always in the same way about all yellow toadstools, we can ask what degree of confidence it

---

[8] [Earlier draft of matter of preceding paragraph in some ways better – F.P.R.

What is meant by saying that a degree of belief is reasonable? First and often that it is what I should entertain if I had the opinions of the person in question at the time but was otherwise as I am now, e.g. not drunk. But sometimes we go beyond this and ask: "Am I reasonable?" This may mean, do I conform to certain enumerable standards which we call scientific method, and which we value on account of those who practise them and the success they achieve. In this sense to be reasonable means to think like a scientist, or to be guided only by ratiocination and induction or something of the sort (i.e. reasonable means reflective). Thirdly, we may go to the root of why we admire the scientist and criticize not primarily an individual opinion but a mental habit as being conducive or otherwise to the discovery of truth or to entertaining such degrees of belief as will be most useful. (To include habits of doubt or partial belief.) Then we can criticize an opinion according to the habit which produced it. This is clearly right because it all depends on this habit; it would not be reasonable to get the right conclusion to a syllogism by remembering vaguely that you leave out a term which is common to both premises.

We use reasonable in sense 1 when we say of an argument of a scientist this does not seem to me reasonable; in sense 2 when we *contrast* reason and superstition or instinct; in sense 3 when we *estimate* the value of new methods of thought such as soothsaying.]

[9] What follows to the end of the section is almost entirely based on the writings of C. S. Peirce. [Especially his "Illustrations of the Logic of Science", *Popular Science Monthly*, 1877 and 1878, reprinted in *Chance Love and Logic* (1923).]

would be best for him to have that they are unwholesome. And the answer is that it will in general be best for his degree of belief that a yellow toadstool is unwholesome to be equal to the proportion of yellow toadstools which are in fact unwholesome. (This follows from the meaning of degree of belief.) This conclusion is necessarily vague in regard to the spatio-temporal range of toadstools which it includes, but hardly vaguer than the question which it answers. (Cf. density at a point of gas composed of molecules.)

Let us put it in another way: whenever I make an inference, I do so according to some rule or habit. An inference is not completely given when we are given the premiss and conclusion; we require also to be given the relation between them in virtue of which the inference is made. The mind works by general laws; therefore if it infers q from p, this will generally be because q is an instance of a function ϕx and p the corresponding instance of a function ψx such that the mind would always infer ϕx from ψx. When therefore we criticize not opinions but the processes by which they are formed, the rule of the inference determines for us a range to which the frequency theory can be applied. The rule of the inference may be narrow, as when seeing lightning I expect thunder, or wide, as when considering 99 instances of a generalization which I have observed to be true I conclude that the 100th is true also. In the first case the habit which determines the process is "After lightning expect thunder"; the degree of expectation which it would be best for this habit to produce is equal to the proportion of cases of lightning which are actually followed by thunder. In the second case the habit is the more general one of inferring from 99 observed instances of a certain sort of generalization that the 100th instance is true also; the degree of belief it would be best for this habit to produce is equal to the proportion of all cases of 99 instances of a generalization being true, in which the 100th is true also.

Thus given a single opinion, we can only praise or blame it on the ground of truth or falsity: given a habit of a certain form, we can praise or blame it accordingly as the degree of belief it produces is near or far from the actual proportion in which the habit leads to truth. We can then praise or blame opinions derivatively from our praise or blame of the habits that produce them.

This account can be applied not only to habits of inference but also to habits of observation and memory; when we have a certain feeling in connection with an image we think the image represents something which actually happened to us, but we may not be sure about it; the degree of direct confidence in our memory varies. If we ask what is the best degree of confidence to place in a certain specific memory feeling, the answer must depend on how often when that feeling occurs the event whose image it attaches to has actually taken place.

Among the habits of the human mind a position of peculiar importance is occupied by induction. Since the time of Hume a great deal has been written about the justification for inductive inference. Hume showed that it could not be reduced to deductive inference or justified by formal logic. So far as it goes his demonstration seems to me final; and the suggestion of Mr Keynes that it can be got round by regarding induction as a form of probable inference cannot in my view be maintained. But to suppose that the situation which results from this is a scandal to philosophy is, I think, a mistake.

We are all convinced by inductive arguments, and our conviction is reasonable because the world is so constituted that inductive arguments lead on the whole to true opinions. We are not, therefore, able to help trusting induction, nor if we could help it do we see any reason why we should, because we believe it to be a reliable process. It is true that if any one has not the habit of induction, we cannot prove to him that he is wrong; but there is nothing peculiar in that. If a man doubts his memory or his perception we cannot prove to him that they are trustworthy; to ask for such a thing to be proved is to cry for the moon, and the same is true of induction. It is one of the ultimate sources of knowledge just as memory is: no one regards it as a scandal to philosophy that there is no proof that the world did not begin two minutes ago and that all our memories are not illusory.

We all agree that a man who did not make inductions would be unreasonable: the question is only what this means. In my view it does not mean that the man would in any way sin against formal logic or formal probability; but that he had not got a very useful habit, without which he would be very much worse off, in the sense of being much less likely[10] to have true opinions.

This is a kind of pragmatism: we judge mental habits by whether they work, i.e. whether the opinions they lead to are for the most part true, or more often true than those which alternative habits would lead to.

Induction is such a useful habit, and so to adopt it is reasonable. All that philosophy can do is to analyse it, determine the degree of its utility, and find on what characteristics of nature this depends. An indispensable means for investigating these problems is induction itself, without which we should be helpless. In this circle lies nothing vicious. It is only through memory that we can determine the degree of accuracy of memory; for if we make experiments to determine this effect, they will be useless unless we remember them.

Let us consider in the light of the preceding discussion what sort of

---

[10] "Likely" here simply means that I am not sure of this, but only have a certain degree of belief in it.

subject is inductive or human logic – the logic of truth. Its business is to consider methods of thought, and discover what degree of confidence should be placed in them, i.e. in what proportion of cases they lead to truth. In this investigation it can only be distinguished from the natural sciences by the greater generality of its problems. It has to consider the relative validity of different types of scientific procedure, such as the search for a causal law by Mill's Methods, and the modern mathematical methods like the *a priori* arguments used in discovering the Theory of Relativity. The proper plan of such a subject is to be found in Mill[11]; I do not mean the details of his Methods or even his use of the Law of Causality. But his way of treating the subject as a body of inductions about inductions, the Law of Causality governing lesser laws and being itself proved by induction by simple enumeration. The different scientific methods that can be used are in the last resort judged by induction by simple enumeration; we choose the simplest law that fits the facts, but unless we found that laws so obtained also fitted facts other than those they were made to fit, we should discard this procedure for some other.

[11] Cf. also the account of "general rules" in the Chapter "Of Unphilosophical Probability" in Hume's *Treatise*.

3. Individual decision making under uncertainty

R. Duncan Luce and Howard Raiffa

1. Introduction and statement of problem

Possibly the best way to begin this chapter is to reread section 2. [cross-references to Luce and Raiffa, 1957], where we discussed the classification of decision making according to whether it is by a group or an individual and according to whether it is being carried out under conditions of certainty, risk, or uncertainty. For the ten intervening chapters we have been concerned with individual decision making in a very particular context of uncertainty known as a game. In a game the uncertainty is due entirely to the unknown decisions of the other players, and, in the model, the degree of uncertainty is reduced through the assumption that each player knows the desires of the other players and the assumption that they will each take whatever actions appear to gain their ends. Traditionally, the game model is not called decision making under uncertainty; that title is reserved for another special class of problems which lie in the domain of uncertainty. These problems, which we shall discuss presently, have for the most part grown up and been examined in the statistical literature, for they are very much involved in an understanding of experimental evidence and in drawing appropriate inferences from data.

The gist of the problem is simple to state. A choice must be made from among a set of acts $A_1, A_2, \ldots, A_m$, but the relative desirability of each act depends upon which "state of nature" prevails, either $s_1, s_2, \ldots, s_n$. The term "state of nature" will be more fully explicated later, but we hope the idea is intuitively clear. As the decision maker, we are aware that one of several possible things is true; which one it is is

Reprinted from *Games and Decisions*, John Wiley & Sons, Inc., 1957, pp. 275–306, by kind permission of the authors and the publisher.

Table 3.1

Act	State	
	Good	Rotten
Break into bowl	Six-egg omelet	No omelet, and five good eggs destroyed
Break into saucer	Six-egg omelet and a saucer to wash	Five-egg omelet and a saucer to wash
Throw away	Five-egg omelet, and one good egg destroyed	Five-egg omelet

relevant to our choice, but we do not even know the relative probabilities of their truth – or, indeed, if it is even meaningful to talk about probabilities – let alone which one obtains. A simple example will illustrate the dilemma; this one is due to Savage (1954):

Your wife has just broken five good eggs into a bowl when you come in and volunteer to finish making the omelet. A sixth egg, which for some reason must be either used for the omelet or wasted altogether, lies unbroken beside the bowl. You must decide what to do with this unbroken egg. Perhaps it is not too great an oversimplification to say that you must decide among three acts only, namely, to break it into the bowl containing the other five, to break it into a saucer for inspection, or to throw it away without inspection. Depending on the state of the egg, each of these three acts will have some consequence of concern to you, say that indicated by Table 3.1.

In general, to each pair (A_i, s_j), consisting of an act and a state, there will be a consequence or outcome. We assume that our subject's preferences among these outcomes, and among hypothetical lotteries with these outcomes as prizes, are consistent in the sense that they may be summarized by means of a utility function [see Luce & Raiffa (1957), Chapter 2]. If we arbitrarily choose some specific utility function, in other words, choose the origin and a unit of measurement, then we can summarize the decision problem under uncertainty (d. p. u. u.) as in Table 3.2. Here u_{ij} is the utility associated to the consequence of the pair (A_i, s_j). So the problem reduces to: Given an m by n array of numbers u_{ij}, to choose a row (act) which is optimal in some sense – or, more generally, to rank the rows (acts) according to some optimality criterion.

Somewhat more must be said about the states of nature. With respect to any decision problem, the set of "states of nature" is assumed to form a mutually exclusive and exhaustive listing of those aspects of nature which are relevant to this particular choice problem

Table 3.2

Acts	s_1	s_2	$\ldots$	States s_j	$\ldots$	s_n
A_1	u_{11}	u_{12}	$\ldots$	u_{1j}	$\ldots$	u_{1n}
A_2	u_{21}	u_{22}	$\ldots$	u_{2j}	$\ldots$	u_{2n}
.	.	.		.		.
.	.	.		.		.
.	.	.		.		.
A_i	u_{i1}	u_{i2}	$\ldots$	u_{ij}	$\ldots$	u_{in}
.	.	.		.		.
.	.	.		.		.
.	.	.		.		.
A_m	u_{m1}	u_{m2}	$\ldots$	u_{mj}	$\ldots$	u_{mn}

and about which the decision maker is uncertain. Although this characterization is quite vague, often there is a natural enumeration of the possible, pertinent, states of the world in particular contexts. We assume that there is a "true" state of the world which is unknown to the decision maker at the time of choice.

One extreme possibility we know how to treat – namely, risk. In that case a probability distribution over the set of states is known – or, better yet, the decision maker deems it suitable to act as if it were known. For example, suppose in the omelet problem described above, the husband – a scientifically minded farmer – "knows" that in a random sample of six eggs the conditional probability of the sixth egg's being rotten when the other five are good is 0.008. Thus, he may view breaking the sixth egg into the bowl as the lottery: 0.992 probability of the six-egg-omelet prize and 0.008 probability of the no-omelet-and-five-good-eggs-destroyed prize. In other words, an *a priori* probability distribution over the states "good" and "rotten" allows one to structure the problem as one of decision making under risk – as a choice among lotteries.

In general, if an *a priori* probability distribution over the states of nature exists, or is assumed as meaningful by the decision maker, then the problem can be transformed into the domain of decision making under risk. In particular, if the probabilities of states $s_1, s_2, \ldots, s_n$ are $p_1, p_2, \ldots, p_n,$ respectively, $\left(\text{where } \sum_{j=1}^{n} p_j = 1, \; p_j \geq 0\right),$ then the utility index for act A_i is its expected utility, i.e., $u_{i1}p_1 + u_{i2}p_2 + \cdots + u_{in}p_n.$ The act having the maximum utility index is chosen, and we say that this act is "best against the given *a priori* probability distribution." (Equivalently, we can think of the decision problem as a game: the decision maker is player 1 who has strategies $A_1, A_2, \ldots, A_m$; "nature" is player 2 who has strategies $s_1, s_2, \ldots, s_n$; the payoff to 1 for the strategy pair (A_i, s_j) is u_{ij}; and, if 1 knows that 2 is employing the mixed strategy $(p_1s_1, p_2s_2, \ldots, p_ns_n)$, 1 should adopt a strategy (act) which is best against this mixed strategy, i.e., against the given *a priori* probability distribution.)

Thus, one extreme assumption leads us to a problem we have already examined in detail. Let us, therefore, turn to the other extreme in which we assume that the decision maker is "completely ignorant" as to which state of nature prevails. This phrase "completely ignorant" is vague, we know, and it has led to much philosophical controversy. The vagueness will be considerably diminished when later we attempt to cope axiomatically with decision making under uncertainty; however, perhaps it can now be reduced some by an illustration. Let us again examine the omelet problem, but with the cast changed. Instead of a scientific farmer, suppose the omelet is completed by a city boy unaccustomed to the ways of eggs. Furthermore, assume that the five eggs already broken were white, whereas the sixth is speckled brown and (to the city boy!) of unusual size. He doesn't have the faintest idea what to expect, having had no previous experience in matters of this kind. Nonetheless, he must make a decision, which leads to the question of criteria for decision making when the states are completely uncertain.

2. Some decision criteria

We shall now list, but only partially discuss, certain criteria which have been offered to resolve the decision problem under uncertainty, which we shall abbreviate as d. p. u. u. A criterion is well-defined if and only if it prescribes a precise algorithm which, for any d. p. u. u., unambiguously selects the act(s) which is (are) tautologically termed "optimal according to the criterion."

In each of the following criteria we shall suppose that we are given a d. p. u. u. having acts $A_1, A_2, \ldots, A_m$, states $s_1, s_2, \ldots, s_n$, and utility payoffs u_{ij}, $i = 1, \ldots, m$ and $j = 1, \ldots, n$.

The maximin criterion

To each act assign its security level as an index. Thus, the index for A_i is the minimum of the numbers $u_{i1}, u_{i2}, \ldots, u_{in}$. Choose that act whose associated index is maximum – i.e., choose the act which maximizes the minimum payoff. Thus, each act is appraised by looking at the worst state for that act, and the "optimal choice" is the one with the best worst state.

We have seen in the theory of games that the optimal security level often can be raised by allowing randomizations over acts. Consider, for example:

$$\begin{array}{c} \\ A_1 \\ A_2 \end{array}\begin{array}{cc} s_1 & s_2 \\ \left[\begin{array}{cc} 0 & 1 \\ 1 & 0 \end{array}\right]. \end{array}$$

In this case, the security level for each act is 0, but if we permit randomization between A_1 and A_2 the security level can be raised to ½ by using $(½A_1, ½A_2)$. This is the hedging principle discussed in section 4.7. It is suggested that the reader review section 4.10, which dealt with the appropriateness and interpretation of a randomized strategy (act).

The maximin principle can be given another interpretation which, although often misleading in our opinion, is sufficiently prevalent to warrant some comment. According to this view the decision problem is a two-person zero-sum game where the decision maker plays against a diabolical Miss Nature.[1] The maximin strategy is then a best retort against nature's minimax strategy, i.e., against the "least favorable" *a priori* distribution nature can employ. We recall that in a two-person zero-sum game the maximin strategy makes good sense from various points of view: it maximizes 1's security level; and it is good against player 2's minimax strategy, which there is reason to suspect 2 will employ since it optimizes his security level and, in turn, it is good against 1's maximin strategy. In a game against nature, however, such a cyclical reinforcing effect is completely lacking.

Nonetheless, just because a close conceptual parallelism between a d. p. u. u. and a zero-sum game is lacking, it does not follow that the maximin procedure is not a wise criterion to adopt. It has the merit that it is extremely conservative in a context where conservatism *might* make good sense. We will have more to say about this later.

(It is customary in the literature to consider negative utility, disutility, or loss, as an index appraising consequences. With that orientation the decision maker, therefore, attempts to minimize the maximum loss he runs from adopting an act – i.e., he "minimaxes" instead of "maximining." Consequently, the principle described above is usually called the *minimax principle*.)

The following simple example exhibits a possible objection to the maximin principle:

---

[1] In a recent lecture to statisticians one of the authors spoke of "diabolical Mr Nature." The audience reaction was so antagonistic that we have elected the path of least resistance.

$$\begin{array}{c} & s_1 & s_2 \\ A_1 & \begin{bmatrix} 0 & 100 \\ A_2 & 1 & 1 \end{bmatrix}. \end{array}$$

Since A_1 and A_2 have security levels of 0 and 1 respectively, A_2 is preferred to A_1 relative to the maximin criterion. This remains true even if randomized acts are considered. Some consider this unreasonable, and to emphasize their objection they point out that this criterion would still select A_2 even if the 1 were reduced to 0.00001 and the 100 increased to 10^6. These critics agree that act A_2 is reasonable *if* player 2 is a conscious adversary of 1, for then 2 should choose s_1, and A_2 is best against s_1; but, they emphasize, nature does not behave in that way, and if we are completely ignorant about the true state of nature, then they claim A_1 is manifestly better.

The minimax risk criterion

This has been suggested by Savage (1951) as an improvement over the maximin (utility) criterion. This criterion can be suggested by continuing the analysis of the above d. p. u. u. If s_1 is the true state, then we have no "risk" or "regret" if we choose A_2, but some "risk" if we choose A_1; if s_2 is the true state, then we have no risk if we choose A_1 and a good deal of risk if we choose A_2. Schematically:

$$\begin{array}{cc} \text{Utility Payoffs} & \text{"Risk" Payoffs} \\ \begin{array}{c} & s_1 & s_2 \\ A_1 & \begin{bmatrix} 1 & 100 \\ A_2 & 1 & 1 \end{bmatrix} \end{array} & \rightarrow & \begin{array}{c} & s_1 & s_2 \\ A_1 & \begin{bmatrix} 1 & 0 \\ A_2 & 0 & 99 \end{bmatrix}. \end{array} \end{array}$$

In terms of "risk" payoffs, A_1 has a possible maximum risk of 1, whereas A_2 has a possible maximum risk of 99. Consequently, A_1 minimizes the maximum risk. However, if randomization is permitted, neither A_1 nor A_2 is optimal.

The general procedure goes as follows:

1. To a d. p. u. u. with utility entries u_{ij}, associate a new table with risk payoffs r_{ij}, where r_{ij} is defined as the amount that has to be added to u_{ij} to equal the maximum utility payoff in the jth column.
2. Choose that act which minimizes the maximum risk index for each act.

To illustrate the "reasonableness" of a criterion based upon risk payoffs rather than utility payoffs, consider some d. p. u. u. with money payoffs and a decision maker whose utility function is linear with money. Now suppose this d. p. u. u. is modified by giving a $10 bonus to the decision maker, regardless of his choice, provided a particular state, say s_3, turns out to be the true state. This bonus, so it is argued, cannot alter the strategic aspects of the decision problem, hence the preference pattern among acts should be identical for both the original and the modified problem. This amounts to saying that adding a constant to any column of the payoff array should not change the preference ordering of acts. In particular, then, the arrays

$$\begin{bmatrix} 0 & 100 \\ 1 & 1 \end{bmatrix} \quad \text{and} \quad \begin{bmatrix} 0 + a & 100 + b \\ 1 + a & 1 + b \end{bmatrix}$$

should be strategically equivalent for any a and b. By setting a equal to -1 and b equal to -100, we get

$$\begin{bmatrix} -1 & 0 \\ 0 & -99 \end{bmatrix},$$

which is the negative of the risk payoff array. Therefore, the maximin criterion for this payoff array is the same as the minimax criterion for the risk array.

In criticism of this proposal, we quote from Chernoff (1954):

Unfortunately, the minimax regret [risk] criterion has several drawbacks. First, it has never been clearly demonstrated that differences in utility do in fact measure what one may call regret [risk]. In other words, it is not clear that the "regret" of going from a state of utility 5 to a state of utility 3 is equivalent in some sense to that of going from a state of utility 11 to one of utility 9. Secondly, one may construct examples where an arbitrarily small advantage in one state of nature outweighs a considerable advantage in another state. Such examples tend to produce the same feelings of uneasiness which led many to object to the [maximin utility] criterion.

A third objection which the author considers very serious is the following. In some examples the minimax regret criterion may select a strategy [act] A_3 among the available strategies[2] A_1, A_2, A_3, and A_4. On the other hand, if for some reason A_4 is made unavailable, the minimax regret criterion will select A_2 among A_1, A_2, and A_3. The author feels that for a reasonable criterion the presence of an undesirable strategy A_4 should not have an influence on the choice among the remaining strategies.

Chernoff's third objection to the minimax risk principle is a variation on our old theme of the "independence of irrelevant alternatives." There is an obvious modification of the minimax risk principle which copes with the problem of non-independence of irrelevant alternatives – but, unfortunately, it has its own, more serious fault. Roughly, the idea is: instead of comparing an act with all others to ascertain the risk,

---

[2] Chernoff uses letters d_1, d_2, d_3 and d_4

which introduces the difficulties when new acts are added, simply make paired comparisons between acts. *Relative to the universe of any two acts*, and for each state, determine the risk of taking each act. Of the two acts, choose the one whose maximum risk is least. An optimal act is then defined as one which is preferred or indifferent, when compared in this way, to every other act. This procedure is unsatisfactory because there are d. p. u. u.'s in which intransitivities occur, and so for these cases it fails to lead to an unambiguous optimal act. An example is the d. p. u. u.

$$
\begin{array}{c}
 & \begin{array}{ccc} s_1 & s_2 & s_3 \end{array} \\
\begin{array}{c} A_1 \\ A_2 \\ A_3 \end{array} &
\left[\begin{array}{ccc}
10 & 5 & 1 \\
0 & 10 & 4 \\
5 & 2 & 10
\end{array}\right]
\end{array}
\qquad \text{(payoff in utility units)}.
$$

The procedure outlined yields the following:

1. A_1 over A_2 for: A_1 has a maximum risk of 5 (from s_2) whereas A_2 has a maximum risk of 10 (from s_1).
2. A_2 over A_3 for: A_2 has a maximum risk of 6 (from s_3) whereas A_3 has a maximum risk of 8 (from s_2).
3. A_3 over A_1 for: A_3 has a maximum risk of 5 (from s_1) whereas A_1 has a maximum risk of 9 (from s_3).

Consequently, none of the three acts can be optimal since each is less preferred (in a paired comparison) than one of the others.

This same example also illustrates Chernoff's third objection to the minimax risk criterion. Restricting ourselves to acts A_2 and A_3, that criterion selects A_2 as optimal and A_3 *as non-optimal*. When A_1 is added, the risk matrix is

$$
\begin{array}{c}
 & \begin{array}{ccc} s_1 & s_2 & s_3 \end{array} \\
\begin{array}{c} A_1 \\ A_2 \\ A_3 \end{array} &
\left[\begin{array}{ccc}
0 & 5 & 9 \\
10 & 0 & 6 \\
5 & 8 & 0
\end{array}\right]
\end{array}
\qquad \text{(payoff in risk units)}
$$

and A_3 *is then optimal* since its maximum risk is a minimum among the maximum risks.

The pessimism–optimism index criterion of Hurwicz

The maximin utility and the minimax risk criteria are each ultra-conservative (or pessimistic) in that, relative to each act, they concentrate upon the state having the worst consequence. Why not look at the best state, or at a weighted combination of the best and worst? This, in essence, is the Hurwicz (1951a) criterion.

For act A_i, let m_i be the minimum and M_i the maximum of the utility numbers u_{i1}, u_{i2}, $\ldots$, u_{in}. Let a fixed number α between 0 and 1, called the pessimism–optimism index, be given. To each A_i associate the index $\alpha m_i + (1 - \alpha)M_i$, which we shall term the α-index of A_i. Of two acts, the one with higher α-index is preferred.

Note that, if $\alpha = 1$, the above procedure is the maximin (utility) criterion, whereas if $\alpha = 0$, it is the maximax (utility) criterion. If neither of these is satisfactory, then how does one decide what α to use? One way is to see what seems reasonable in certain simple classes of d. p. u. u.'s, for example, in the class:

$$
\begin{array}{cc}
 & \begin{array}{cc} s_1 & s_2 \end{array} \\
\begin{array}{c} A_1 \\ A_2 \end{array} & \left[\begin{array}{cc} 0 & 1 \\ x & x \end{array} \right]
\end{array} \quad \text{(utility payoff)}.
$$

The α-indices of A_1 and A_2 are $1 - \alpha$ and x respectively. Consequently, if one can choose an x such that A_1 and A_2 are indifferent, then one can impute an α-level to oneself. For example, if A_1 and A_2 are indifferent for $x = \frac{3}{8}$, then α must be $\frac{5}{8}$. Thus, by resolving a simple decision problem an α-level can be chosen empirically, which, in turn, can be employed in more complicated decisions.

But there are also objections to this criterion; one may be illustrated by the following example:

$$
\begin{array}{cc}
 & \begin{array}{ccc} s_1 & s_2 & s_3 \end{array} \\
\begin{array}{c} A_1 \\ A_2 \\ (\frac{1}{2}A_1, \frac{1}{2}A_2) \end{array} & \left[\begin{array}{ccc} 0 & 1 & 0 \\ 1 & 0 & 0 \\ \frac{1}{2} & \frac{1}{2} & 0 \end{array} \right]
\end{array} \quad \text{(utility payoff)}.
$$

Suppose the α-level of $\frac{1}{4}$ is chosen. The α-indices of A_1 and A_2 are each $\frac{1}{4} \cdot 0 + (1 - \frac{1}{4}) \cdot 1 = \frac{3}{4}$, whereas the index of $(\frac{1}{2}A_1, \frac{1}{2}A_2)$ is $\frac{1}{4} \cdot 0 + (1 - \frac{1}{4}) \cdot \frac{1}{2} = \frac{3}{8}$. Consequently, although A_1 and A_2 are each optimal, the procedure of tossing a fair coin and taking A_1 if heads and A_2 if tails is not optimal. Critics of the Hurwicz criterion claim that any randomization over optimal acts (according to a particular criterion) should itself also be optimal according to that criterion. Remember that a randomization which uses only optimal acts will ultimately cause the decision maker to adopt one of these optimal acts!

A second possible criticism of the Hurwicz criterion is that it resolves the following d.p.u.u. counter to one's best intuitive judgment:

$$
\begin{array}{cc}
 & \begin{array}{ccccccc} s_1 & s_2 & s_3 & \cdots & s_i & \cdots & s_{100} \end{array} \\
\begin{array}{c} A_1 \\ A_2 \end{array} & \left[\begin{array}{ccccccc} 0 & 1 & 1 & \cdots & 1 & \cdots & 1 \\ 1 & 0 & 0 & \cdots & 0 & \cdots & 0 \end{array} \right]
\end{array}.
$$

According to any α-level Hurwicz criterion, both acts A_1 and A_2 have an α-index of $1 - \alpha$, and so they are considered indifferent; however, if one is "completely ignorant" concerning which is the true state, then, the critics argue, A_1 is manifestly better than A_2. But, in defense of Hurwicz, *is* A_1 clearly better than A_2? What seems to be implied here is that the "true" state is "more likely" to be one of the states s_2 to s_{100} than s_1. This, however, is not what Hurwicz intuits about the notion of "complete ignorance," for he would assert that "complete ignorance" implies the above d. p. u. u. is strategically equivalent to

$$
\begin{array}{c}
\quad\;\; s_1' \quad s_2' \\
\begin{array}{c} A_1 \\ A_2 \end{array}
\left[\begin{array}{cc} 0 & 1 \\ 1 & 0 \end{array} \right].
\end{array}
$$

A complete characterization of what he means by the term "complete ignorance" can best be given in axiomatic form (see section 4).

The criterion based on the "principle of insufficient reason"

The criterion of insufficient reason asserts that, if one is "completely ignorant" as to which state among $s_1, s_2, \ldots, s_n$ obtains, then one should behave as if they are equally likely. Thus, one is to treat the problem as one of risk with the uniform *a priori* probability distribution over states, and to each act A_i assign its expected utility index,

$$
\frac{u_{i1} + u_{i2} + \cdots + u_{in}}{n},
$$

and choose the act with the largest index.

At this juncture, it would be apropos to digress into the philosophical foundations of probability and to review the special role of the principle of insufficient reason in relation to these foundations. But we shall resist this temptation, for to do the topic justice would require a sizable digression, and there are already excellent expository accounts of this material. [See, for instance, Arrow (1951b), Nagel (1939), and Savage (1954); each of these references, in turn, gives a relatively complete bibliography.] We will confine ourselves to a few simple remarks.

The principle of insufficient reason, first formulated by Jacob Bernoulli (1654–1705), states in boldest terms that, if there is no evidence leading one to believe that one event from an exhaustive set of mutually exclusive events is more likely to occur than another, then the events should be judged equally probable. This principle is extremely vague, and its indiscriminate use has led to many nonsensical results. Writers since Bernoulli's time have attempted to add qualifications to the principle and to specify limited interpretations so as to avoid some of the more blatant contradictions.

From an empirical point of view, one difficulty with the principle is this: Suppose we are confronted with a real problem in decision making under uncertainty, then our first task is to give a mutually exclusive and exhaustive listing of the possible states of nature. The rub is that many such listings are possible, and in general these different abstractions of the same problem will, when resolved by the principle of insufficient reason, yield different real solutions. For instance, in one listing of the states we might have: s_1, the organism remains fixed; s_2, the organism moves. In another equally good listing we might have: s_1, the organism remains fixed; s_2, the organism moves to the left; s_3, the organism moves to the right. We can further compli-cate our description of the possible states of nature by noting which leg first moves, whether the animal raises its head or not, etc.

There is a counterargument to this objection. Although it may be true that there are various acceptable interpretations as to what constitutes a state in a given real problem, it is not true that we will feel that the states are "equally likely" in each interpretation. In other words, care must be exerted in the choice of states if one wishes to use this principle. As it stands, this defense is weak in that there is a crying need for an empirical clarification of the term "equally likely." Eventually, we shall examine two suggested clarifications. The first, an axiomatic treatment due to Chernoff (1954), characterizes his notion of "complete ignorance" in such a manner as to justify logically the principle of insufficient reason. This will be described in section 4. In the second, the equally likely assignment gains empirical meaning through the "practical" suggestions for probability assignments offered by the personalistic school of probability (see section 5).

Incidentally, the arguments against the principle of insufficient reason become even more cogent when there are an infinite set of pertinent states of nature, for then it is difficult to single out a natural parametrization, or enumeration, of the states for which a suitable generalization of the "equally likely" criterion is appropriate.

Before we turn to the axiomatic studies of decision criteria, what of the poor decision maker who is now totally confused by the pros and cons of the above criteria? Can he, in desperation, compromise by adopting some sort of arbitrary composite of the criteria? Subsequently, we will suggest some plausible composites; however, for the present, the following example must be included as a note of caution, for some *apparently* acceptable compromises may not be so acceptable after all.

Take the case of a decision maker who cannot crystallize his pre-ferences among the maximin criterion, the Hurwicz criterion with $\alpha = \frac{3}{4}$, and the principle of insufficient reason. He thus decides to define one act as preferable to another if and only if a majority of these three criteria register this preference. The following d. p. u. u. establishes that this compromise procedure is not well defined:

$$
\begin{array}{c} & s_1 & s_2 & s_3 \\ \begin{array}{c} A_1 \\ A_2 \\ A_3 \end{array} & \left[\begin{array}{ccc} 2 & 12 & -3 \\ 5 & 5 & -1 \\ 0 & 10 & -2 \end{array}\right] \end{array} \quad \text{(utility payoff).}
$$

Preferences according to:

Maximim criterion	A_2 over A_3 over A_1
Hurwicz criterion[3] ($\alpha = \frac{3}{4}$)	A_3 over A_1 over A_1
Principle of insufficient reason	A_1 over A_2 over A_3

A majority of the criteria select A_1 over A_2, A_2 over A_3, and A_3 over A_1 – an intransitivity. The majority decision principle applied in social welfare contexts (Luce & Raiffa, 1957, Chapter 14) leads to the same embarrassing intransitivities of preference. The reasons are analogous.

3. Axiomatic treatment: the axioms not referring to "complete ignorance"

Instead of applying specific proposed decision criteria to carefully selected decision problems, thereby determining whether or not each criterion complies with our intuitive criteria (which we deem to be reasonable), let us, as so often before, invert the procedure. Let us cull from our intuitions certain reasonable desiderata for decision criteria to fulfill, which we can then investigate both as to compatibility with one another and as to their logical implications. Our axiomatic presentation mainly follows Chernoff (1954), but it is also a curious mixture of the works of Milnor (1954), Hurwicz (1951a), Savage (1954), Arrow (1953), and unpublished comments by Rubin.

There are two distinct types of axiomatic approaches in the literature. In one the criterion must establish for each d.p.u.u. a complete ordering of the available acts. As in the four criteria we have previously mentioned, this is usually effected by attaching a numerical index to each act. In the other approach, a criterion isolates an "optimal" subset of acts, but it does not attempt to rank non-optimal ones. Of course, this can be thought of as a complete ordering of all acts – but into just two categories: optimal and non-optimal! We will follow the latter procedure, for it is closer to the natural demands of the problem area.

Let A' and A'' be two arbitrary but specific acts in a decision problem. We define the following preliminary notions.

1. $A' \sim A''$: means that the acts are *equivalent* in the sense that they yield the same utilities for each state of nature.

[3] The α-indices of A_3, A_1, and A_2 are $\frac{1}{4}(10) + (\frac{3}{4})(-2) = 1$, $\frac{1}{4}(12) + (\frac{3}{4})(-3) = \frac{3}{4}$, and $(\frac{1}{4})(5) + (\frac{3}{4})(-1) = \frac{2}{4}$, respectively.

2. $A' > A''$: means that A' *strongly dominates* A'' in the sense that A' is preferred to A'' for each state of nature.
3. $A' \geqslant A''$: means that A' *weakly dominates* A'' in the sense that A' is preferred to A'' for at least one state and is preferred or indifferent to A'' for all other states.

Since any d. p. u. u. is characterized by a class of acts $\mathscr{A}$, a set of states of nature S, and a utility function u, we may symbolically identify the d. p. u. u. with the triple $(\mathscr{A}, S, u)$. A decision criterion associates to each d. p. u. u., i.e., to each $(\mathscr{A}, S, u)$, a subset $\hat{\mathscr{A}}$ of $\mathscr{A}$; the acts in $\hat{\mathscr{A}}$ are called *optimal* for $(\mathscr{A}, S, u)$ relative to the given criterion. $\hat{\mathscr{A}}$ is called the *choice* or *optimal* set.

Desiderata for criteria

Axiom 1.
For any d. p. u. u. $(\mathscr{A}, S, u)$, the set $\hat{\mathscr{A}}$ is non-empty, i.e., every problem can be resolved.
Axiom 2.
The choice set for d. p. u. u. does not depend upon the choice of origin and unit of the utility scale used to abstract the problem.
Axiom 3.
The choice set is invariant under the labeling of acts, i.e., the real acts singled out as optimal should not depend upon the arbitrary labeling of acts used to abstract the problem.
Axiom 4.
If A' belongs to $\hat{\mathscr{A}}$ and $A'' \geqslant A'$ or $A'' \sim A'$, then A'' belongs to $\hat{\mathscr{A}}$.

Axioms 1 through 4 are quite innocuous in the sense that, if a person takes serious issue with them, then we would contend that he is not really attuned to the problem we have in mind.

An act A' is said to be *admissible* if there is no act A in $\mathscr{A}$ such that $A \geqslant A'$, i.e., A' is admissible if A' is not weakly dominated by any other act.

Axiom 5.
If A' belongs to $\hat{\mathscr{A}}$, then A' is admissible.

Axiom 5 is equivalent to: Given A', if there exists an A such that $A \geqslant A'$ (that is, if A' is not admissible), then A' does not belong to $\hat{\mathscr{A}}$.

It should be noted that as they were originally stated neither the maximin principle nor the Hurwicz α-criteria satisfy axiom 5; however, both can be appropriately modified in a trivial manner. To see the problem, consider the following d. p. u. u.:

$$
\begin{array}{c}
 \quad s_1 \ \ s_2 \ \ s_3 \\
\begin{array}{c} A_1 \\ A_2 \end{array}
\begin{bmatrix} 0 & 1 & \tfrac{3}{4} \\ 0 & 1 & \tfrac{1}{2} \end{bmatrix}.
\end{array}
$$

The strategy A_2 is not admissible, since $A_1 \geqslant A_2$; however, A_1 and A_2, and all randomizations between them, have the same security level, 0, and the same Hurwicz α-index. Consequently, any randomized act is optimal according to these criteria. We can modify them to meet axiom 5 either by deleting all acts which are not admissible, or by deleting from the class of optimal acts those which are not admissible. This point suggests the next axiom.

Axiom 6.
Adding new acts to a d. p. u. u., each of which is weakly dominated by or is equivalent to some old act, has no effect on the optimality or non-optimality of an old act.

Example. A gentleman wandering in a strange city at dinner time chances upon a modest restaurant which he enters uncertainly. The waiter informs him that there is no menu, but that this evening he may have either broiled salmon at $2.50 or steak at $4.00. In a first-rate restaurant his choice would have been steak, but considering his unknown surroundings and the different prices he elects the salmon. Soon after the waiter returns from the kitchen, apologizes profusely, blaming the uncommunicative chef for omitting to tell him that fried snails and frog's legs are also on the bill of fare at $4.50 each. It so happens that our hero detests them both and would always select salmon in preference to either, yet his response is "Splendid, I'll change my order to steak." Clearly, this violates the seemingly plausible axiom 6. Yet can we really argue that he is acting unreasonably? He, like most of us, has concluded from previous experience that only "good" restaurants are likely to serve snails and frog's legs, and so the risk of a bad steak is lessened in his eyes.

This illustrates the important assumption implicit in axiom 6, namely, that adding new acts to a d. p. u. u. *does not alter one's a priori information as to which is the true state of nature.* In what follows, we shall suppose that this proviso is satisfied. In practice this means that, if a problem is first formulated so that the availability of certain acts influences the plausibility of certain states of nature, then it must be reformulated by redefining the states of nature so that the interaction is eliminated.

Axiom 6 can be strengthened to the following form of the principle of the independence of irrelevant alternatives:

Axiom 7.
If an act is non-optimal for a d. p. u. u., it cannot be made optimal by adding new acts to the problem.

A typical violation of axiom 7 is this incongruous exchange.

Doctor: Well, Nurse, that's the evidence. Since I must decide whether or not he is tubercular, I'll diagnose tubercular.

Nurse: But, Doctor, you do not have to decide one way or the other, you can say you are undecided.

Doctor: That's true, isn't it? In that case, mark him not tubercular.

Nurse: Please repeat that!

The example given at the end of the discussion of the minimax risk criterion shows that axiom 7 rules out the minimax risk principle.

Note that axiom 7 does not prevent an optimal act from being changed into a non-optimal one by adding new acts; this is true even if none of the new acts is optimal. Therefore, one might wish to strengthen axiom 7 to:

Axiom 7′.
The addition of new acts does not transform an old, originally non-optimal act into an optimal one, and it can change an old, originally optimal act into a non-optimal one only if at least one of the new acts is optimal.

A further strengthening of axiom 7 is:

Axiom 7″.
The addition of new acts to a d. p. u. u. never changes old, originally non-optimal acts into optimal ones and, in addition, either

1. All the old, originally optimal acts remain optimal,
   or
2. None of the old, originally optimal acts remain optimal.

The all-or-none feature of axiom 7″ may seem a bit too stringent, but one can offer this rationalization for it. Suppose that the merit of each act can be summarized by a single numerical index which is independent of the other acts available. Then the optimal set of the original problem is composed of all the acts with the highest index. Now, among the new acts either there is one with a higher index, which therefore annihilates all the old optimal acts, or there is not and the original optimal set is left intact. A severe criticism of axiom 7″ is that it yields unreasonable results when it is coupled with either of the more palatable axioms 5 and 6. Take, for example, the following d. p. u. u.:

$$
\begin{array}{c@{}c}
 & \begin{array}{cccc} s_1 & s_2 & s_3 & s_4 \end{array} \\
\begin{array}{c} A_1 \\ A_2 \end{array} &
\left[\begin{array}{cccc} 0 & 4 & 2 & 2 \\ 4 & 0 & 0 & 4 \end{array} \right].
\end{array}
$$

It is reasonable that some criterion should allow both A_1 and A_2 in the optimal set. Now add an A_3 whose utilities are

$A_3 \ [4 \quad 0 \quad 0.1 \quad 4]$

Since A_2 is weakly dominated by A_3, axiom 5 implies that act A_2 cannot remain optimal. But one may very well want also to keep A_1 as optimal, in violation of 7''. The rationalization of axiom 7'' (namely, that each act can be fully appraised by a single index) is apparently not suitable. This is suggested by the fact that acts A_2 and A_3 have the same indices according to the maximin (utility), minimax risk or regret, and Hurwicz (for any α-index) criteria. The criterion based on the principle of insufficient reason, however, does satisfy axiom 7''.

▶ There is still another variation on the theme of the independence of irrelevant alternatives, which is especially suited to finding the logical consequences of some combinations of these axioms.

Axiom 7'''.
An act A' is optimal only if it is optimal in the paired comparisons between A' and A, for all A in $\mathscr{A}$.

This axiom enables us to transform the decision problem into a series of paired comparisons between acts and to eliminate those acts which are not optimal in any one of these comparisons. We will not, however, use this condition. ◀

Axiom 7 and its different versions are somewhat controversial. Each of these rules out the minimax risk or regret principle. We are most sympathetic to axioms 7 and 7'''. The others, 7' and 7'', are slightly harder to see through (i.e., they are a little less intuitive), so let us suspend judgment until some of their consequences are stated.

The next axiom is due to Rubin. To suggest it, suppose a decision maker is given two decision problems having the same sets of available acts and states but differing in payoffs. Suppose the second problem is trivial in the sense that the payoff depends only upon the state and not upon the act adopted. In other words, in the array representing problem 2, all entries in the same column are the same. If the decision maker knows only that he is playing problem 1 with probability p and problem 2 with probability $1 - p$ when he has to adopt an act, then he should adopt an act which is optimal for problem 1, since problem 2, which enters with probability $1 - p$, is irrelevant as far as his choice is concerned. It is straightforward to formalize this requirement into an axiom, but we will be content merely with the following suggestive formulation.

Axiom 8.
Consider a probability mixture of two d. p. u. u.'s with the same sets of actions and states. If the second d. p. u. u. has payoffs which do not

depend upon the act chosen, then the optimal set of the mixture problem should be the same as the optimal set of the first d. p. u. u.

Axiom 8 can be shown to imply that *adding a constant to each entry of a column of a d. p. u. u. does not alter the optimal set.* Instead of Rubin's axiom, perhaps it would have been simpler to take the italicized consequence as the axiom; however, we feel, as do Rubin and Chernoff, that this property is not as intuitively compelling as the axiom given.

Axiom 8 goes a long way towards selecting a criterion. For example, it rules out the maximin criterion and all the Hurwicz α-criteria. Therefore, we should be careful before we accept or reject it. First, to argue against the axiom, these points may be raised:

1.   As stated, the axiom is not intuitive enough to be given the status of a basic desideratum.

2.   Consider the following problems:

$$
\begin{array}{ccc}
\text{Problem 1} & \text{Problem 2} & \text{Problem 3} \\[4pt]
\begin{array}{c c c}
 & s_1 & s_2 \\
A_1 & 0 & -9 \\
A_2 & -10 & 0
\end{array}, &
\begin{array}{c c c}
 & s_1 & s_2 \\
A_1 & 1000 & 0 \\
A_2 & 1000 & 0
\end{array}, &
\begin{array}{c c c}
 & s_1 & s_2 \\
A_1 & 500 & -\tfrac{9}{2} \\
A_2 & 495 & 0
\end{array},
\end{array}
$$

where, it will be noted, problem 3 is a mixture of the other two in which each is played with probability ½. Intuitively, a plausible method for analyzing these d. p. u. u.'s is to be somewhat pessimistic and to behave as if the less desirable state is somewhat more likely to arise. The extreme example of this rule is the maximiner who focuses entirely on the undesirable state, but our point holds equally well for one who emphasizes the undesirable state only slightly. In problem 1, s_1 is less desirable, and so one is led to choose A_1. In problem 3, s_2 is less desirable, and so one might be led to choose A_2. But if one subscribes to axiom 8, the same alternative must be chosen in both cases, and so we are led to doubt the axiom.

3. Axiom 8, when added to axiom 3 (i.e., the choice set is invariant under labeling of acts) and to axiom 7 (i.e., the addition of acts cannot make a non-optimal act optimal), both of which are extremely reasonable, yields the following result: *If an optimal act of a given d. p. u. u. is equivalent to a probability mixture of two other acts, then each of these acts is also optimal.*[4] For example, in the d. p. u. u.

$$
\begin{array}{c c c}
 & s_1 & s_2 \\
A_1 & 0 & 2 \\
A_2 & 1 & 0 \\
A_3 & \tfrac{1}{2} & 1
\end{array},
$$

---

[4] This proposition is referred to as the anticonvexity property of the optimal set.

if A_3 is optimal, so are A_1 and A_2, since A_3 is equivalent to ($\frac{1}{2}A_1$, $\frac{1}{2}A_2$). This also implies the result that one need *never resort to randomized acts* in this type of decision problem. Since, it is contended, this consequence is absurd, one should discard the weakest link in the argument leading to it. Therefore, axiom 8 should go.

Now, to argue against these arguments point by point:

1. Rubin's axiom is not only intuitively meaningful but it seems perfectly reasonable. This is a matter of taste!

2. The very compelling *a priori* quality of Rubin's axiom argues against the analysis which led us to choose A_1 in problem 1 and A_2 in problem 3. Certainly, the intuitive analysis cannot be used without restrictions, for it would also lead us to choose A_2 again in

$$\begin{array}{c} & s_1 & s_2 \\ A_1 & \begin{bmatrix} 500 & -0.01 \\ A_2 & 100 & 0 \end{bmatrix}, \end{array}$$

and that seems counterintuitive. We suspect that most people who are unaware of axiom 8 would find it difficult to resolve problem 3 above and that they could easily be persuaded to choose either A_1 or A_2; however, once they become aware of the axiom they will find it acceptable and will use it to decide upon A_1 in that problem.

3. Is the assertion that one need never resort to randomized strategies in a d. p. u. u. so absurd? Maybe not, for one can cite many "reasonable" criteria which lead to an optimal non-randomized act for any d. p. u. u. Furthermore, there are arguments against randomization; for example, part of the discussion found in section 4.10, where we examined the operational interpretation of randomized strategies and cast some doubt upon their applicability, can be taken over almost verbatim. Finally, Chernoff (1954, p. 438) argues as follows "It would seem that the need for randomization depends on the statistician's need to oversimplify the statement of his problem because with limited computational ability he cannot take full advantage of the actual relationships involved. Generally, the simplification has the effect of *combining states of nature which are equivalent when random samples are insisted upon.*" (Italics ours.) This discussion leads naturally to the next axiom.

Axiom 9.
If A' and A'' are both optimal for a d. p. u. u., a probability mixture of A' and A'' is also optimal, i.e., the optimal set is convex.

Remember that a probability mixture using A' or A'' will in fact choose either A' or A'', and, if they are both optimal, certainly any mixture should be. This seems very palatable; however, it rules out all Hurwicz's criteria with $\alpha < 1$. Put in another fashion, if we are committed to using some one of the criteria of the Hurwicz family, and if we impose axiom 9, then we must choose $\alpha = 1$, that is, the maximin (utility) criterion.

Hurwicz would argue, facetiously perhaps, that it does not grieve him too much to be forced into the $\alpha = 1$ camp, for that is where he started from in the first place. He only invented the pessimism–optimism index as a modification of the maximin criterion in order to appease those souls who were unwilling to endorse its pessimistic approach. However, he would continue, axiom 9 is not as innocuous as it seems. If axiom 9 were a consequence of some other more basic axioms, he would not object too much, but it does not seem to him to warrant the status of an axiom. Suppose A_1 and A_2 are both optimal acts. It is true that a mixture such as $(\frac{1}{2}A_1, \frac{1}{2}A_2)$ will, operationally, result in a selection of one of the two optimal acts. Nonetheless, the mixture may evoke a psychological response in its own right, and, before it is known which optimal act is adopted, there is no compelling reason why the anticipation of the mixture must be as good as either A_1 or A_2. For example, an optimist might like both A_1 and A_2 because in each case he can look forward to very desirable returns if certain states obtain; however, with the randomization all expected returns will be mitigated, and so the anticipation is not nearly so pleasant. Of course, the counterargument is that the apparent reasonableness of the axiom simply demonstrates the irrationality of the optimist's wishful thinking. So the battle is joined. The present authors are very partial to the axiom and believe the argument against it is rather weak.

So far we have not tried to characterize the notion of "complete ignorance." Our purpose in postponing this discussion is obvious: Axioms 1 through 9 are pertinent to decision making where one is not "completely ignorant" of the true state. It is interesting that, even without committing ourselves on the notion of "complete ignorance," acceptance of axioms 1 through 9 serves to eliminate the maximin criterion (eliminated by axiom 8), the minimax risk or regret (eliminated by axiom 7 or any of its variations), and the Hurwicz α-criteria (eliminated by axiom 8 and, for $\alpha < 1$, by axiom 9). Nonetheless, axioms 1 through 9 are compatible: the criterion based on the principle of insufficient reason, for example, satisfies all of them.

The following theorem is basic:

To each criterion which resolves all d. p. u. u.'s in such a manner as to satisfy axioms 1, 3, 4, 5, 7', 8, and 9, there is an appropriate a priori distribution over the states of nature which is independent of any new acts which might be added, such that an act is optimal (according to the criterion) only if it is best against this a priori distribution.

Note, this theorem does *not* say that if an act is best against this *a priori* distribution then it is optimal according to the criterion. It only says the converse. The theorem indicates that, if we are committed to axioms 1, 3, 4, 5, 7', 8, and 9, our first step should be to search for a suitable *a priori* distribution. What distribution is chosen will, naturally, depend upon the information we possess concerning the true state of nature.

4. Axiomatic treatment: the axioms referring to "complete ignorance"

Now we turn to the question of "complete ignorance." Consider the following:

Axiom 10.
For any d. p. u. u., the optimal set should not depend upon the labeling of the states of nature.

Obviously, if we have reason to suspect that a given state of nature is quite likely the true state whereas another state is quite likely not the true state, then in any abstraction of the problem we wish to distinguish between these two states. Or, if we number the states of nature in a given problem in such a manner that the lower the number the more likely we feel that it is the "true" state, then certainly we want to keep the labeling of the states in mind and axiom 10 would not be at all appropriate. Loosely speaking, whenever axiom 10 is not appropriate, we are not in the realm of "complete ignorance."

There is a tendency to read too much into this axiom. Some hold that adopting axiom 10 is essentially equivalent to assuming that each state is equally likely. Although this is true when a suitable collection of the other axioms is added to 10 (see below), it is not true for 10 alone, or for 10 and certain of the other axioms. For example, if axiom 7''' is accepted (i.e., A' is optimal only if it is optimal in each paired comparison), then axiom 10 has the following interpretation: If A' is optimal and if the utilities for A'', $[u(A'', s_1), u(A'', s_2), \ldots, u(A'', s_n)]$, are a permutation of those for A', $[u(A', s_1), u(A', s_2), \ldots, u(A', s_n)]$, then A'' is also optimal. This does not require that the states of nature be equally likely, since the maximin criterion, for example, satisfies this requirement.

It is very easy to see the role that axiom 10 plays when appended to axioms 1, 3, 4, 5, 7', 8, and 9. As a consequence of these other axioms, almost everything hinges on an *a priori* probability distribution over the states of nature. Yet, if we must be indifferent to the labeling of the states, it can be shown that the only possible *a priori* distribution must make each state equally likely, i.e., it must be the one which assigns the probability $1/n$ to each state if there are n states in all.

Thus, by coupling axiom 10 with the theorem we stated for these seven axioms, we know that an act is optimal only if it yields the highest average utility (the average being taken over all n utilities associated with the act and where each utility number is given weight $1/n$). But with axiom 10 added it can be shown that the "only if" assertion can be strengthened to "if and only if," i.e., *if an act has the highest average utility, then it is indeed optimal.* To round out the picture, the same result holds if for axiom 7' one substitutes axioms 6 and 7. (Note that 7' implies 7 directly, and when it is bolstered by 4 and 5 it also implies 6.)

In summary, then, axioms 1 through 10 (actually 2 is not needed) characterize the criterion based on the principle of insufficient reason, i.e., it is the unique criterion which satisfies them. This result is due to Chernoff (1954).

The maximiners and minimaxers, however, argue that, although axiom 10 is all right, it does not go far enough in characterizing the notion of "complete ignorance." For example, consider the two d. p. u. u.'s

$$
\begin{array}{cc}
\text{D.P.U.U.1} & \text{D.P.U.U.2} \\[4pt]
\begin{array}{c}
\begin{array}{cccc} s_1 & s_2 & s_3 & s_4 \end{array} \\
\begin{array}{c} A_1 \\ A_2 \end{array}\left[\begin{array}{cccc} 6 & 2 & 2 & 2 \\ 0 & 5 & 5 & 5 \end{array}\right]
\end{array}
&
\begin{array}{c}
\begin{array}{cc} s_1 & s_2 \end{array} \\
\begin{array}{c} A_1 \\ A_2 \end{array}\left[\begin{array}{cc} 6 & 2 \\ 0 & 5 \end{array}\right].
\end{array}
\end{array}
$$

and

According to the criterion based on the principle of insufficient reason, A_2 is optimal for d. p. u. u. 1 and A_1 for d. p. u. u. 2. But *if one is truly completely ignorant* about the true state in each problem aren't these problems identical? In d. p. u. u. 1, s_2, s_3, and s_4 can be strategically lumped into one state – call it s^*. True, s^* is "not less likely" to be true than either s_2, s_3, or s_4, but if we are completely ignorant we cannot say anything about s_1 versus s^*. The principle of insufficient reason interprets complete ignorance as "each state being equally likely," so s^* must be treated as if it were "three times as likely" as s_1, and, therefore, this criterion chooses A_2. But, in considering s^* as more likely than s_1, one admits that he is *not completely ignorant.* According to some, the very essence of complete ignorance is to treat d. p. u. u.'s 1 and 2 as equivalent. They would add that one is almost never in a state of *complete* ignorance, but they would insist that, if one wants to list reasonable desiderata for criteria which purport to handle this case, the following axiom is indispensable.

Axiom 11.
If a d. p. u. u. is modified by deleting a repetitious column (i.e., collapsing two states which yield identical payoffs for all acts into one), then the optimal set is not altered.

Axiom 11 can be strengthened to:

Axiom 11'.

If a d. p. u. u. is modified by deleting a column which is equivalent to a probability mixture of other columns, then the optimal set is not altered.

If one feels strongly about the criterion based on the principle of insufficient reason and also wants to endorse axiom 11, the two can be combined into this criterion: In any d. p. u. u. delete all repetitious columns, and in this modified d. p. u. u. choose those acts having the highest average payoff (equal weights). This criterion fails to satisfy axiom 7 or any of its variations. For example, consider the following d. p. u. u.:

$$
\begin{array}{c}
 \\
A_1 \\
A_2 \\
A_3
\end{array}
\begin{array}{ccc}
s_1 & s_2 & s_3 \\
\left[\begin{array}{ccc}
11 & 0 & 0 \\
0 & 10 & 10 \\
9 & 9 & 0
\end{array}\right].
\end{array}
$$

If the choice is confined to A_1 or A_2, A_1 is optimal (since by axiom 11 s_3 is deleted). If A_3 is added to A_1 and A_2, then s_3 cannot be deleted, and according to this criterion A_2 is changed from non-optimal to optimal whereas A_1 is changed from optimal to non-optimal. Thus, any variant of axiom 7 is contradicted.

Axioms 10 and 11 together are said to characterize "complete ignorance." Although axioms 10 and 11 are compatible, and axioms 1 to 9 are compatible, all eleven obviously are not. Something will have to be deleted, and one possible candidate is Rubin's axiom 8 – which amounts to saying that the addition of a constant to a column has no effect on the optimal set. The Hurwicz α-criteria, modified to the extent of deleting all weakly dominated acts before applying the criteria, satisfy axioms 1 through 6, plus any version of 7, plus 10 and 11. The maximin (utility) criterion, modified in the same way, satisfies these and axiom 9 in addition.

Arrow (1953), modifying a result due to Hurwicz (1951a), has proved the following result: If a criterion satisfies axioms 1, 3, 4, 7", 10, 11, then it takes into account only the minimum and maximum utility associated with each act. However, the particular way these maxima and minima are to be used to select a specific act as best is left unresolved by the group of axioms. For example, all the Hurwicz α-criteria are compatible with this axiom set. Another compatible criterion is: An act is optimal if and only if either its minimum is larger than the minimum of any other act or, when there are ties for the largest minimum, it has the largest maximum among those acts with the largest minimum.

Suppose that we let m denote the minimum utility associated with an

act and M the maximum, then if we accept this axiom set (1, 3, 4, 7",
10, 11) the crux of the problem is to decide upon an ordering between
pairs (m', M') and (m'', M''). If we also demand that axiom 2 be met, the
criterion must yield the same ordering when we change the utilities by
a linear transformation. Thus, if the criterion selects (m', M') over $(m'',
M'')$ then it must also select $(am' + b, aM' + b)$ over $(am'' + b, aM'' + b)$,
where $a > 0$. In this connection, the following can be shown: If, for the
d. p. u. u.

$$
\begin{array}{c}
\begin{array}{cc} s_1 & s_2 \end{array} \\
\begin{array}{c} A_1 \\ A_2 \end{array}
\begin{bmatrix} 0 & 1 \\ x & x \end{bmatrix},
\end{array}
$$

there exists a number α such that we would say A_1 is optimal for all
$x \leq 1 - \alpha$, and A_2 is optimal for all $x \geq 1 - \alpha$, and if we demand that
a criterion yielding this decision also satisfy axioms 1, 2, 3, 4, 7", 10,
and 11, then it must be Hurwicz's with index α.

The approach just used, which will be employed again in the next
section, warrants a comment. We first commit ourselves to a class of
axioms, thereby restricting the class of potential criteria. Second, we
consider a simple class of d. p. u. u.'s for which we feel able to make
subjective commitments as to the optimal sets. If our choice of axioms
and special cases is clever, then by using the axioms we can logically
extend the consistent decisions given for a simple class of d. p. u. u.'s
to a precise formula which resolves all d. p. u. u.'s.

▶ Milnor (1954) states a set of requirements for reasonable decision criteria,
where the criteria do not select an optimal set of acts but yield a complete
(transitive) ordering for all acts. The analysis is much simpler in these terms.
We outline his work here with a minimum of comments. In parenthesis after
each axiom we give the nearest corresponding statement in terms of optimal
sets.

1. **Ordering.** *All acts must be completely ordered.* (1.)
2. **Symmetry.** *The ordering is independent of labeling of rows and columns.* (3 and 10.)
3. **Strong domination.** *Act A' is preferred to A" if A' strongly dominates A".* (4 and 5.)
4. **Continuity.** *If A' is preferred to A" in a sequence of d. p. u. u.'s, then A" is not preferred to A' in the limit d. p. u. u.* [A sequence of d. p. u. u.'s converge to a limiting d. p. u. u. if the utility numbers for each (act, state) pair converge to the utility number of the (act, state) pair of the limit d. p. u. u.] (No correlate.)
5. **Linearity.** *The ordering is not changed by linear utility transformations.* (2.)
6. **Row adjunction.** *The ordering between old rows is not changed by adding a new row.* (7, 7', 7", 7"'.)

7. **Column linearity.** *The ordering is not changed by adding a constant to a column.* (8.)

8. **Column duplication.** *Adding an identical column does not change the ordering.* (11.)

9. **Convexity.** *If A' and A" are indifferent in the ordering, then neither A' nor A" is preferred to (½A', ½A").* (9.)

10. **Special row adjunction.** *Adding a weakly dominated act does not change the ordering of old acts.* (6.)

Milnor summarizes his results in the table.

Axiom	Laplace	Wald	Hurwicz	Savage
1. Ordering	Ⓧ	Ⓧ	Ⓧ	Ⓧ
2. Symmetry	Ⓧ	Ⓧ	Ⓧ	Ⓧ
3. Str. dom.	Ⓧ	Ⓧ	Ⓧ	Ⓧ
4. Continuity	x	Ⓧ	Ⓧ	Ⓧ
5. Linearity	x	x	Ⓧ	x
6. Row adj.	Ⓧ	Ⓧ	Ⓧ	...
7. Col. lin.	Ⓧ	...	...	Ⓧ
8. Col. dup.	...	Ⓧ	Ⓧ	Ⓧ
9. Convexity	x	Ⓧ	...	Ⓧ
10. Sp. row adj.	x	x	x	Ⓧ

In this tabulation Laplace refers to the criterion based on the principle of insufficient reason, Wald to the maximin utility criterion, Hurwicz to the α optimism pessimism criteria, and Savage to the minimax risk or regret criterion. An x means the criterion and the axiom are compatible. Each criterion is characterized by the axioms marked Ⓧ.

Note that, unlike Chernoff's characterization of the Laplace criterion, Milnor's does not require the convexity axiom. This discrepancy seems strange until it is recalled that Milnor's axioms 1 and 6 are stronger then their correlates in Chernoff's system. Milnor demands a complete ordering, not just an optimal set, and his sixth axiom corresponds to axiom 7" [cf. p. 62 this volume] which is stronger than axiom 7 used by Chernoff.

Another point of discrepancy is Milnor's use of strong domination and continuity. All four of the criteria satisfy these conditions, but they would not if weak domination [i.e., axiom 5, p. 60, this volume] were employed instead of strong domination. To see this, consider the d. p. u. u.

$$\begin{array}{c} \\ A_1 \\ A_2 \end{array} \begin{array}{cc} s_1 & s_2 \end{array} \\ \begin{bmatrix} 0 & 4 \\ 1/n & 3 \end{bmatrix}.$$

By the maximum utility criterion, A_2 is preferred to A_1 for all n, but in the limit as n increases we obtain

$$\begin{array}{c} \\ A_1 \\ A_2 \end{array} \begin{array}{cc} s_1 & s_2 \\ \begin{bmatrix} 0 & 4 \\ 0 & 3 \end{bmatrix}, \end{array}$$

so by weak domination A_1 is preferred to A_2. Thus, that criterion cannot satisfy both weak domination and continuity.   ◄

5. The case of "partial ignorance"

A common criticism of such criteria as the maximin utility, minimax regret, Hurwicz α, and that based on the principle of insufficient reason is that they are rationalized on some notion of *complete* ignorance. In practice, however, the decision maker usually has some vague partial information concerning the true state. No matter how vague it is, he may not wish to endorse any characterization of complete ignorance (e.g., axiom 10 or 11), and so the heart is cut out of criteria based on this notion. The present section is devoted to suggestions for coping with this hiatus between complete ignorance and risk.

As background for this discussion, consider a contestant on the famous \$64,000 quiz show who has just answered the \$32,000 question correctly. His problem is whether to choose act A_1, to try for \$64,000, or to choose act A_2, to stop at \$32,000. His d. p. u. u. takes the form:

The \$64,000 question is one that the contestant

	s_1 = could answer	s_2 = could not answer
A_1 = try for \$64,000	Obtain \$64,000 (taxable plus prestige, publicity, etc.	Obtain a consolation prize of a Cadillac, plus knowledge that \$32,000 (taxable) was lost
A_2 = stop	Obtain \$32,000 (taxable), get less prestige and publicity than for the (A_1, s_1) pair	Same as (A_2, s_1) pair

We assume that in utility terms the problem reduces to the form:

$$\begin{array}{c} \\ A_1 \\ A_2 \end{array} \begin{array}{cc} s_1 & s_2 \\ \begin{bmatrix} 1 & 0 \\ x & x \end{bmatrix}. \end{array}$$

Let us suppose, further that no other contestant has ever tried for the \$64,000 question. For all our contestant knows, the difficulty of the question can run the gamut from the impossible to "What was the color of Washington's white horse?" Everything hinges on his appraisal of the relative possibilities of s_1 and s_2. He might take the point of view that he is completely ignorant of the true state, but it is much more likely that he would take into consideration such intangibles as: (a) the public reaction against the sponsor if the question were too difficult; (b) the bad precedent that would be set if the question were too easy; and (c) the trend in question difficulty in going from \$4000 to \$8000, from \$8000 to \$16,000, and from \$16,000 to \$32,000. Although the problem surely is not in the realm of complete ignorance, it is not obvious how this vague information can be systematically processed.

Suppose, after due deliberation, the contestant chooses A_1. We can then assert that he behaved as if it were meaningful to assign an *a priori* probability to s_1 of x or greater.[5] Conversely, one is tempted to say that, if the "subjective probability" of s_1 is x or greater, then A_1 should be chosen. It is this net of ideas which will be partially formulated now.

We shall first report on the school led by Savage (1954), which holds the view that by processing one's partial information (as evidenced by one's responses to a series of simple hypothetical questions of the Yes-No variety) one can generate an *a priori* probability distribution over the states of nature which is appropriate for making decisions. This reduces the decision problem from one of uncertainty to one of risk. The *a priori* distribution obtained in this manner is called a *subjective* probability distribution.

Savage, in his *The Foundations of Statistics*, "develops, explains, and defends a certain abstract theory of behavior of a highly idealized person faced with uncertainty." The theory is based on a synthesis of the works of Bruno de Finetti on a personalistic view of probability and of the modern theory of utility due to von Neumann and Morgenstern. Since Savage expounds his position with vigor and clarity, we shall merely attempt to capture what, to our minds, is the most salient contribution of his school. Furthermore, we shall not follow Savage's development of the subject; rather we shall graft the new concepts onto the development given in the two previous sections.

Let $s_1, s_2, \ldots, s_n$ be a labeling of the possible states of nature for some concrete decision problem. Each of these labels refers to specific real world phenomena and we (in the role of a decision maker) might feel that some states are more plausible than others. Suppose, furthermore, that after reflection we are convinced that we want to be consistent when facing problems of this type – consistent in the sense that

---

[5] An equally valid interpretation of this single choice is that the subject applied a Hurwicz criterion with index $\alpha \leq 1 - x$.

our adopted decision criterion should satisfy axioms 1, 3, 4, 5, 7', 8, and 9. Since these axioms do not in any way refer to our state of ignorance concerning the true state of nature, we are free to commit ourselves to them independent of any information we possess or subjective feelings we have as to the relative plausibility of the different states. Now, as we previously noted, any criterion which satisfies these axioms must select as optimal a subset of the acts which are best against some specific *a priori* distribution. Furthermore, this *a priori* distribution is independent of the particular acts available in a given problem (as long as the states $s_1, s_2, \ldots, s_n$ are involved) since adding new acts does not change non-optimal acts into optimal ones. Thus, it is reasonable to assert that if there exists an "appropriate" *a priori* probability distribution over the states, then this distribution depends solely upon our state of information concerning $s_1, s_2, \ldots, s_n$. The strategy now is to consider a series of simple hypothetical d. p. u. u.'s with these states of nature, to resolve them according to our best intuitive judgement, and then to use these commitments to infer a plausible *a priori* distribution.

Let us illustrate the procedure by a case which involves three specific states s_1, s_2 and s_3. In order to generate an "appropriate" *a priori* distribution over these states let us introduce two hypothetical acts, A_1 and A_2, such that their consequences for the various states have the following monetary equivalences:

$$\begin{array}{c} & s_1 & s_2 & s_3 \\ A_1 & \$0 & \$0 & \$100 \\ A_2 & \$y & \$y & \$y \end{array}.$$

Adjust act A_2, i.e., y, until we are indifferent between A_1 and A_2. Suppose the point of indifference (which is assumed to exist) is at \$65. Suppose, further, that we are indifferent between obtaining \$65 for certain and getting \$100 with an objective probability of 0.8 and \$0 with an objective probability of 0.2. Hence the utilities of \$0, \$65, and \$100 can be taken as 0, 0.8, and 1. In utility payoffs we have

$$\begin{array}{c} & s_1 & s_2 & s_3 \\ A_1 & 0 & 0 & 1 \\ A_2 & 0.8 & 0.8 & 0.8 \end{array}.$$

Now, indifference between A_1 and A_2 is compatible with an *a priori* distribution only if the *a priori* probability of s_3 is 0.8. If we have no preferences about the states themselves then, as a check and possible short cut, we could ask ourselves: "If we were given the alternative (*a*) of obtaining a prize of x dollars if s_3 turns out to be true and nothing if s_1 or s_2 were true, versus the alternative (*b*) of obtaining a prize of x dollars with objective probability p and nothing with objective probability $1 - p$, for what p would we be indifferent?" To check, we would require that indifference come at $p = 0.8$ independent of the value of x, so long as it is positive! In a similar manner, we could force ourselves to accept a probability assignment for s_2 and for s_1. In practice, however, one's choices for a series of problems – no matter how simple – usually are not consistent. For example, the *a priori* probability assignments for s_1, s_2, s_3 may not add up to 1. Once confronted with such inconsistencies, one should, so the argument goes, modify one's initial decisions in such a manner as to be consistent. Let us assume that this jockeying – making snap judgments, checking on their consistency, modifying them, again checking on consistency, etc. – leads ultimately to a bona fide *a priori* distribution. Now, if we wish our decision criterion both to satisfy the axioms stated above and to yield results that agree with our by now consistent set of preferences for simple hypothetical problems, then we are committed to a criterion which selects as optimal only acts which are best against this *a priori* distribution.

To describe precisely what Savage means by a consistent set of preferences, we must outline briefly his postulates for a personalistic theory of decision. The assumed ingredients of the decision problem are:

1. The set of *states* of the world – a set S with (an infinite number of) elements $s, s', \ldots$ and with subsets $E, E', \ldots$ called *events*.
2. The set of *consequences* – a set C with elements $c, c', \ldots$.
3. The set of *acts* – a set $\mathscr{A}$ with elements $A, A', \ldots$.
4. An assignment to each act-state pair (A, s) of a consequence from C which is denoted by $A(s)$.
5. A binary relation $\succeq$ between pairs of acts which is interpreted to mean "is preferred or indifferent to."

Savage then postulates and defines the following:

Postulate 1. The relation $\succeq$ is a weak ordering of the acts, i.e., every pair of acts is comparable and the relation is transitive.

Definition. The expression "$A \succeq A'$ given E" means that, if acts A and A' are modified so that their consequences are the same for every state not included in the event E, but if they are not changed for the states in E, then the modification of A is preferred or indifferent to the modification of A'.

This definition is not well defined unless the preference relation between modified acts is required not to depend upon the particular agreement selected for states not in E. The next postulate makes this assumption indirectly.

Postulate 2. Conditional preference, as defined above, is well defined.

Definition. If $A(s) = c$ and $A'(s) = c'$ for every s in S, then we define $c \gtrsim c'$ if and only if $A \gtrsim A'$.

The given A and A' of this definition are called "constant" acts since their consequences are independent of which state holds. The relation $\gtrsim$ is extended to the set of consequences by identifying each consequence with the constant act which yields it for each state.

Definition. An event ϕ is called *null* if every pair of acts are indifferent given ϕ, i.e., for every A and A', $A \gtrsim A'$ given ϕ and $A' \gtrsim A$ given ϕ.

Postulate 3. If E is a non-null event and $A(s) = c$ and $A'(s) = c'$ for all s in E, then $A \gtrsim A'$ given E if and only if $c \gtrsim c'$.

This asserts that conditional preferences do not affect consequence preferences.

Definition. The event E is said to be *not more probable than* the event E' if, whenever

   1. c and c' are any two consequences such that $c > c'$,
   2. $A(s) = c$ for s in E and $A(s) = c'$ for s not in E,
and
   3. $A'(s) = c$ for s in E' and $A'(s) = c'$ for s not in E', then $A' \gtrsim A$.

Postulate 4. Probabilitywise, any two events are comparable.

Postulate 5. There is at least one pair of acts which are not indifferent.

Postulate 6. Suppose $A > A'$. For each consequence c, no matter how desirable or undesirable it may be, there exists a sufficiently fine partitioning of S into a finite number of events such that if either A or A' is modified to yield c for any single event of the partition the preference for A over A' is not changed.

Postulate 7. Let A' be an act and let A_s' be the constant act which agrees with A' for the state s. Then,

   1. $A \gtrsim A_s'$ given E for all s in E implies $A \gtrsim A'$ given E,
and
   2. $A_s' \gtrsim A$ given E for all s in E implies $A' \gtrsim A$ given E.

From these seven postulates Savage is able to show (among other things) the following two theorems.

Theorem. There exists a unique real-valued function P defined for the set of events (subsets of S) such that

1. $P(E) \geqslant 0$ for all E,
2. $P(S) = 1$,
3. If E and E' are disjoint, then $P(E \cup E') = P(E) + P(E')$, and
4. E is not more probable than E' if and only if $P(E) \leqslant P(E')$.

P is called the *personalistic probability* measure reflecting the individual's reported feelings as to which of a pair of events is more likely to occur.

Theorem.   There exists a real-valued function u defined over the set of consequences having the following property: If E_i, where $i = 1, 2, \ldots,$ n, is a partition of S and A is an act with consequence c_i on E_i, and if E_i', where $i = 1, 2, \ldots, m$, is another partition of S and A' is an act with consequence c_i' on E_i', then $A \gtrsim A'$ if and only if

$$\sum_{i=1}^{n} u(c_i)P(E_i) \geqslant \sum_{i=1}^{m} u(c_i')P(E_i')$$

The function u is called a *utility* function. As in the von Neumann-Morgenstern theory, it is unique up to a positive linear transformation.

A primary, and elegant, feature of Savage's theory is that no concept of objective probability is assumed; rather a subjective probability measure arises as a consequence of his axioms. This in turn is used to calibrate utilities, and it is established that it can be done in such a way that expected utilities correctly reflect preferences. Thus, Savage's contribution – a major one in the foundations of decision making – is a synthesis of the von Neumann-Morgenstern utility approach to decision making and de Finetti's calculus of subjective probability.

To transform vague information concerning the states of nature into an explicit *a priori* probability distribution, the decision maker has had to register consistent choices in a series of simple hypothetical problems involving these states. No one claims that this is an easy task, but some go so far as to assert that in some contexts even these preliminary choices are too difficult to make with any confidence. They hold, further, that, if consistent responses are forced, the results are not very reliable and to build upon them is a mistake. They feel, introspectively, that, if one could instantaneously wipe out the memory of one's past choices and if the process for obtaining a subjective *a priori* distribution were immediately repeated, the new *a priori* distribution could easily be quite different from the old one.

There are two suggestions in the literature, Hurwicz (1951*b*) and Hodges and Lehman (1952), designed to cope partially with this problem. Let A be a generic act in $\mathscr{A}$ (the decision maker's strategy set);

let **x** denote the generic randomized act in X (the set of all randomized acts); let s be a generic state of nature in S (nature's state set); let **y** denote an *a priori* probability distribution over S; and let Y be the set of all *a priori* probability distributions. As we have seen, Savage suggests that partial knowledge can be utilized to find a unique *a priori* distribution $\mathbf{y}^{(0)}$, and the decision maker is to choose an A which is best against $\mathbf{y}^{(0)}$. Hurwicz goes in the other direction: he suggests that *partial ignorance* over S can be effectively processed to yield *complete ignorance* over some subset $Y^{(0)}$ of Y. That is, although our knowledge may be insufficient to choose a specific *a priori* distribution in Y, it may be adequate to eliminate certain *a priori* distributions – let the remaining class be $Y^{(0)}$. Hurwicz proposes that the *a priori* distribution in $Y^{(0)}$ should be treated as new states of nature about which one is totally ignorant, and that a criterion based on complete ignorance over these states should be utilized. For example, let $M(\mathbf{x}, \mathbf{y})$ be the utility payoff when the decision maker chooses the randomized act **x** and when **y** is the *a priori* distribution. To apply the Hurwicz α-criterion, associate to each act **x** the α-index,

$$\alpha m_{\mathbf{x}} + (1 - \alpha)M_{\mathbf{x}},$$

where $m_{\mathbf{x}}$ and $M_{\mathbf{x}}$ are, respectively, the minimum and maximum payoffs[6] which result from **x** as the *a priori* distribution **y** runs over its domain $Y^{(0)}$. Choose an act which yields the highest α-index.

The spirit of Hurwicz's proposal is quite clear, and there are contexts[7] where we feel his specific proposal can be employed. In general, however, we feel that his suggestions are too vague to resolve the problem. Operationally, how does one characterize the elements of $Y^{(0)}$? Even if all "reasonable" **y** are included in $Y^{(0)}$, can't some **y**'s be "more reasonable" than others? Maybe one could capture this differential plausibility for **y**'s in $Y^{(0)}$ by an *a priori* distribution on $Y^{(0)}$. But why stop there? There is a next level, and a next, etc. Of course, expedient compromises can be made, and Hurwicz's original hope still has merit: that from a lot of special decisions about $Y^{(0)}$, one will come closer to extracting faithfully one's partial information about the states than by a forced choice of an *a priori* distribution.

Independently of Hurwicz, Good (1950) has offered much the same

---

[6] There is some question here of the existence of the minimum and maximum; however, from a mathematical point of view, this can be taken care of easily.

[7] Let the two states of nature be whether a subject does or does not have tuberculosis, and suppose that from medical statistics the proportion of people having T. B. is known to be π. Because the subject is self-selected, we may be unwilling to say that the *a priori* probability of T.B. is π; but we may find it acceptable to say that it is anything greater than or equal to π, and, conceivably, we might behave as if we were completely ignorant as to which value it has in this interval.

suggestion for processing information; however, he subsequently used the maximin criterion rather the α-criteria.

Hodges and Lehmann (1952) also take the position that, in practice, information about states of nature often lies somewhere between complete ignorance and a precise specification of an *a priori* distribution. For example, an *a priori* distribution $\mathbf{y}^{(0)}$ might seem likely and yet not be sufficiently reliable to base decisions on. An act which is best against $\mathbf{y}^{(0)}$ might involve a large risk if some state actually turns out to be true. (Note that Hodges and Lehmann, like most statisticians, phrase their results in terms of risk payoffs rather than in utility payoffs.) So they propose that: (*a*) An act (maybe randomized) be found which minimizes the maximum risk; let its maximum risk be C. (*b*) On the basis of the quantity C and the context of the problem, choose a quantity C_0, greater than C, to serve as the maximum tolerable risk. (*c*) Choose an act $\mathbf{x}$ which is best against $\mathbf{y}^{(0)}$ *subject to the condition that the act has a maximum risk not greater than* C_0.[8] Naturally, the choice of C_0 will depend upon how much confidence we have in $\mathbf{y}^{(0)}$.

---

[8] Once $\mathbf{y}^{(0)}$ is chosen, the payoff for $\mathbf{x}$ is a linear function. Mathematically, then, the problem is one of minimizing a linear function subject to linear inequalities – i.e., a linear-programing problem. Because of the equivalence between linear programing and two-person zero-sum game theory, it is reasonable that game theory should be pertinent in proving theorems in this area. It is!

4.   The sure-thing principle

Leonard J. Savage

A businessman contemplates buying a certain piece of property. He considers the outcome of the next presidential election relevant to the attractiveness of the purchase. So, to clarify the matter for himself, he asks whether he would buy if he knew that the Republican candidate were going to win, and decides that he would do so. Similarly, he considers whether he would buy if he knew that the Democratic candidate were going to win, and again finds that he would do so. Seeing that he would buy in either event, he decides that he should buy, even though he does not know which event obtains, or will obtain, as we would ordinarily say. It is all too seldom that a decision can be arrived at on the basis of the principle used by this business-man, but, except possibly for the assumption of simple ordering, I know of no other extralogical principle governing decisions that finds such ready acceptance.

Having suggested what I shall tentatively call the *sure-thing principle*, let me give it relatively formal statement thus: If the person would not prefer **f** to **g**, either knowing that the event B obtained, or knowing that the event $\sim B$ obtained, then he does not prefer **f** to **g**. Moreover (provided he does not regard B as virtually impossible) if he would definitely prefer **g** to **f**, knowing that B obtained, and, if he would not prefer **f** to **g**, knowing that B did not obtain, then he definitely prefers **g** to **f**.

The sure-thing principle cannot appropriately be accepted as a pos-tulate because it would introduce new undefined technical terms referring to knowledge and possibility that would render it mathemati-

Reprinted from *The Foundations of Statistics*, Dover Publications, Inc., New York, 1972, pp. 21–26, by kind permission of the publisher.

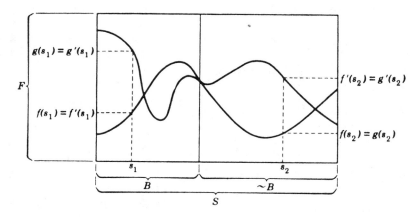

Figure 4.1

cally useless without still more postulates governing these terms. It will be preferable to regard the principle as a loose one that suggests certain formal postulates.

What technical interpretation can be attached to the idea that **f** would be preferred to **g**, if B were known to obtain? Under any reasonable interpretation, the matter would seem not to depend on the values **f** and **g** assume at states outside of B. There is, then, no loss of generality in supposing that **f** and **g** *agree* with each other except in B, that is, that $f(s) = g(s)$ for all $s \; \varepsilon \; \sim B$. Under this unrestrictive assumption, **f** and **g** are surely to be regarded as equivalent given $\sim B$; that is, they would be considered equivalent, if it were known that B did not obtain. The first part of the sure-thing principle can now be interpreted thus: If, after being modified so as to agree with one another outside of B, **f** is not preferred to **g**; then **f** would not be preferred to **g**, if B were known. The notion will be expressed formally by saying that $\mathbf{f} \leq \mathbf{g}$ *given* B.

It is implicit in the argument that has just led to the definition of $\mathbf{f} \leq \mathbf{g}$ given B that, if two acts **f** and **g** are so modified in $\sim B$ as to agree with each other, then the order of preference obtaining between the modified acts will not depend on which of the permitted modifications was actually carried out. Equivalently, if **f** and **g** are two acts that do agree with each other in $\sim B$, and $\mathbf{f} \leq \mathbf{g}$; then, if **f** and **g** are modified in $\sim B$ in any way such that the modified acts **f'** and **g'** continue to agree with each other in $\sim B$, it will also be so that $\mathbf{f'} \leq \mathbf{g'}$. This assumption is made formally in the postulate P2 below and illustrated schematically in Figure 4.1, a kind of diagram I find suggestive in many such contexts.

In Figure 4.1, the set S of all states s and the set F of all consequences f are represented by horizontal and vertical intervals respectively. In any such diagram an act **f**, being a function attaching a value $f(s) \; \varepsilon \; F$ to each $s \; \varepsilon \; S$ is represented by a graph. This particular diagram graphs two acts **f** and **g** that agree with each other in $\sim B$, and two other acts **f'** and **g'** that also agree with each other in $\sim B$ and arise by modifying **f** and **g** respectively only in $\sim B$, that is, acts agreeing with **f** and **g** respectively in B.

P2 If **f, g,** and **f', g'** are such that:

1. in $\sim B$, **f** agrees with **g**, and **f'** agrees with **g'**,
2. in B, **f** agrees with **f'**, and **g** agrees with **g'**,
3. $\mathbf{f} \le \mathbf{g}$; then $\mathbf{f'} \le \mathbf{g'}$.

Each of the relations "$\le$ given B" is now easily seen to be a simple ordering, and the relations "$\ge, <, >, \doteq$ given B" are to be defined mutatis mutandis. It is noteworthy though obvious that, if $f(s) = g(s)$ for all $s \; \varepsilon \; B$, then $\mathbf{f} \doteq \mathbf{g}$ given B.

It is now possible and instructive to give an atemporal analysis of the following temporally described decision situation: The person must decide between **f** and **g** after he finds out, that is, observes, whether B obtains; what will his decision be if he finds out that B does in fact obtain?

Atemporally, the person can submit himself to the consequences of **f** or else of **g** for all $s \; \varepsilon \; B$, and, independently, he can submit himself to the consequences of **f** or else of **g** for all $s \; \varepsilon \sim B$; which alternative will he decide upon for the s's in B?

Finally, describing the situation not only atemporally but also quite formally, the person must decide among four acts defined thus:

> h_{00} agrees with **f** on B and with **f** on $\sim B$,
> h_{01} agrees with **f** on B and with **g** on $\sim B$,
> h_{10} agrees with **g** on B and with **f** on $\sim B$,
> h_{11} agrees with **g** on B and with **g** on $\sim B$.

The question at issue now takes this form. Supposing that none of the four functions is preferred to the particular one $\mathbf{h}_{ij}$, is $i = 0$, or is $i = 1$; that is, does $\mathbf{h}_{ij}$ agree with **f** on B or with **g** on B?

It is not hard to see that i can be 1, if and only if $\mathbf{f} \le \mathbf{g}$ given B. Indeed, if $i = 1$, $\mathbf{h}_{0j} \le \mathbf{h}_{ij}$, which means that $\mathbf{f} \le \mathbf{g}$ given B. Arguing in the opposite direction, if $\mathbf{f} \le \mathbf{g}$ given B; then $\mathbf{h}_{00} \le \mathbf{h}_{10}$, and $\mathbf{h}_{01} \le \mathbf{h}_{11}$. Suppose now, for definiteness, $\mathbf{h}_{10} \le \mathbf{h}_{11}$, then none of the four possibilities is preferred to $\mathbf{h}_{11}$; this proves the point in question.

It may fairly be said that the person considers B *virtually impossible*, or that B is *null*; if and only if, for all **f** and **g**, $\mathbf{f} \le \mathbf{g}$ given B. Indeed, if B is null in this sense, the values acts take on elements of B are irrelevant to all decisions.

Several trivial conclusions about null events are listed as a compound theorem, all components but the last of which have immediate intuitive interpretations.

Theorem 1.

1. The vacuous event, 0, is null.
2. B is null, if and only if, for every $\mathbf{f}$ and $\mathbf{g}$, $\mathbf{f} \doteq \mathbf{g}$ given B.
3. If B is null, and $B \supset C$; then C is null.
4. If $\sim B$ is null; $\mathbf{f} \leq \mathbf{g}$ given B, if and only if $\mathbf{f} \leq \mathbf{g}$.
5. $\mathbf{f} \leq \mathbf{g}$ given S, if and only if $\mathbf{f} \leq \mathbf{g}$.
6. If S is null, $\mathbf{f} \doteq \mathbf{g}$ for every $\mathbf{f}$ and $\mathbf{g}$.

Component 6 of Theorem 1 requires comment, because it corresponds to a pathological situation. In case S is null, it is not really intuitive to say that S (and therefore every event) is virtually impossible. The interpretation is rather that the person simply doesn't care what happens to him. This is imaginable, especially under a suitably restricted interpretation of F, but it is uninteresting.

A finite set of events B_i is a *partition* of B; if $B_i \cap B_j = 0$, for $i \neq j$, and $\cup_i B_i = B$. With this definition, it is easily proved by arithmetic induction that

Theorem 2. If B_i is a partition of B, and $\mathbf{f} \leq \mathbf{g}$ given B_i for each i, then $\mathbf{f} \leq \mathbf{g}$ given B. If, in addition, $\mathbf{f} < \mathbf{g}$ given B_j for at least one j, then $\mathbf{f} < \mathbf{g}$ given B.

Corollary 1. The union of any finite number of null events is null.

There are still other interesting consequences of Theorem 2, which may be most conveniently mentioned informally. If, in Theorem 2, $B = S$ (or, more generally, if $\sim B$ is null), it is superfluous to say "given B" in the conclusions of the theorem. If $\mathbf{f} \doteq \mathbf{g}$ given B_i for each i, then $\mathbf{f} \doteq \mathbf{g}$ given B. So much for the consequences of P2.

Acts that are *constant*, that is, acts whose consequences are independent of the state of the world, are of special interest. In particular, they lead to a natural definition of preference among consequences in terms of preference among acts. Following ordinary mathematical usage, $\mathbf{f} \equiv g$ will mean that $\mathbf{f}$ is identically g, that is, for every s, $f(s) = g$. A formal definition of *preference among consequences* can now conveniently be expressed thus. For any consequences g and g', $g \leq g'$; if and only if, when $\mathbf{f} \equiv g$ and $\mathbf{f}' \equiv g'$, $\mathbf{f} \leq \mathbf{f}'$.

In the same spirit, meaning can be assigned to such expressions as $f \leq g$, $g \leq f$ given B, etc., and I will freely use such expressions without defining them explicitly. In particular, $f \leq g$ given B has a natural meaning, but one that is rendered superfluous by the next postulate.

Incidentally, it is now evident how awkward for us it would be to use $f(s)$ for $\mathbf{f}$; because $f(s) \leq g(s)$ is a statement about the consequences $f(s)$ and $g(s)$, whereas $\mathbf{f} \leq \mathbf{g}$ is a statement about acts, and we will have frequent need for both sorts of statements.

Suppose that $\mathbf{f} \equiv g$, and $\mathbf{f}' \equiv g'$, and that $g \leq g'$, is it reasonable to admit that, for some B, $\mathbf{f} > \mathbf{f}'$ given B? That depends largely on the interpretation we choose to make of our technical terms, as an example helps to bring out.[1]

Before going on a picnic with friends, a person decides to buy a bathing suit or a tennis racket, not having at the moment enough money for both. If we call possession of the tennis racket and possession of the bathing suit consequences, then we must say that the consequences of his decision will be independent of where the picnic is actually held. If the person prefers the bathing suit, this decision would presumably be reversed, if he learned that the picnic were not going to be held near water. Thus the question whether it can happen that $\mathbf{f} > \mathbf{f}'$ given B would be answered in the affirmative. But, under the interpretation of "act" and "consequence" I am trying to formulate, this is not the correct analysis of the situation. The possession of the tennis racket and the possession of the bathing suit are to be regarded as acts, not consequences. (It would be equivalent and more in accordance with ordinary discourse to say that the coming into possession, or the buying, of them are acts.) The consequences relevant to the decision are such as these: a refreshing swim with friends, sitting on a shadeless beach twiddling a brand-new tennis racket while one's friends swim, etc. It seems clear that, if this analysis is carried to its limit, the question at issue must be answered in the negative; and I therefore propose to assume the negative answer as a postulate. The postulate is so couched as not only to assert that knowledge of an event cannot establish a new preference among consequences or reverse an old one, but also to assert that, if the event is not null, no preference among consequences can be reduced to indifference by knowledge of an event.

P3   If $\mathbf{f} \equiv g$, $\mathbf{f}' \equiv g'$, and B is not null; then $\mathbf{f} \leq \mathbf{f}'$ given B, if and only if $g \leq g'$.

Applying Theorem 2, it is obvious that

Theorem 3.   If B_i is a partition of B; and if (for all i and s) $f_i \leq g_i$, $f(s) = f_i$, and $g(s) = g_i$ when $s \in B_i$; then $\mathbf{f} \leq \mathbf{g}$ given B. If, in addition, $f_j < g_j$ for some j for which B_j is not null, then $\mathbf{f} < \mathbf{g}$ given B.

Theorem 3 has been widely accepted by the British-American School of statisticians, special emphasis having been given to it, in connection

---

[1] The role of such freedom throughout science is brilliantly discussed by Quine (1951).

with his notion of admissibility, by the late Abraham Wald. I believe, as will be more fully explained later, that much of its particular significance for that school stems from the implication that, if several different people agree in their preferences among consequences, then they must also agree in their preferences among certain acts.

This brings the present chapter to a natural conclusion, since the further postulates to be proposed can be more conveniently introduced in connection with the uses to which they are put in later chapters.

5.   Probable knowledge

Richard C. Jeffrey

The central problem of epistemology is often taken to be that of explaining how we can know what we do, but the content of this problem changes from age to age with the scope of what we take ourselves to know; and philosophers who are impressed with this flux sometimes set themselves the problem of explaining how we can get along, knowing as little as we do. For knowledge is sure, and there seems to be little we can be sure of outside logic and mathematics and truths related immediately to experience. It is as if there were some propositions – that this paper is white, that two and two are four – on which we have a firm grip, while the rest, including most of the theses of science, are slippery or insubstantial or somehow inaccessible to us. Outside the realm of what we are sure of lies the puzzling region of probable knowledge – puzzling in part because the sense of the noun seems to be cancelled by that of the adjective.

The obvious move is to deny that the notion of knowledge has the importance generally attributed to it, and to try to make the concept of belief do the work that philosophers have generally assigned the grander concept. I shall argue that this is the right move.

1. A pragmatic analysis of belief

To begin, we must get clear about the relevant sense of "belief". Here I follow Ramsey: "the kind of measurement of belief with which probability is concerned is ... a measurement of belief *qua* basis of action" (Ramsey, 1931, p. 171 [p. 28, this volume]).

Ramsey's basic idea was that the desirability of a gamble G is a weighted average of the desirabilities of winning and of losing in

Reprinted from *The Problem of Inductive Logic*, ed. by I. Lakatos, North-Holland Publishing Company, Amsterdam, 1968, pp. 166–180, by kind permission of the author and the publisher.

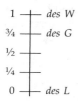

Figure 5.1

which the weights are the probabilities of winning and of losing. If the proposition gambled upon is A, if the prize for winning is the truth of a proposition W, and if the penalty for losing is the truth of a proposition L, we then have

$$prob\ A = \frac{des\ G - des\ L}{des\ W - des\ L}. \tag{1}$$

Thus, if the desirabilities of losing and of winning happen to be 0 and 1, we have prob $A = des\ G$, as illustrated in Figure 5.1 for the case in which the probability of winning is thought to be $\frac{3}{4}$.

On this basis, Ramsey (1931) is able to give rules for deriving the gambler's subjective probability *and* desirability functions from his preference ranking of gambles, provided the preference ranking satisfies certain conditions of consistency. The probability function obtained in this way is a probability measure in the technical sense that, given any finite set of pairwise incompatible propositions which together exhaust all possibilities, their probabilities are non-negative real numbers that add up to 1. And in an obvious sense, probability so construed is a measure of the subject's willingness to act on his beliefs in propositions: it is a measure of degree of belief.

I propose to use what I take to be an improvement of Ramsey's scheme, in which the work that Ramsey does with the operation of forming gambles is done with the usual truth-functional operations on propositions (Jeffrey, 1965, the mathematical basis for which can be found in Bolker, 1965, 1966). The basic move is to restrict attention to certain 'natural' gambles, in which the prize for winning is the truth of the proposition gambled upon, and the penalty for losing is the falsity of that proposition. In general, the situation in which the gambler takes himself to be gambling on A with prize W and loss L is one in which he believes the proposition

$$G = AW \vee \bar{A}L.$$

If G is a natural gamble we have $W = A$ and $L = \bar{A}$, so that G is the necessary proposition, $T = A \vee \bar{A}$:

$$G = AA \vee \bar{A}\bar{A} = T.$$

Now if A is a proposition which the subject thinks good (or bad) in the sense that he places it above T (or below T) in his preference ranking, we have

$$prob\ A = \frac{des\ T - des\ \bar{A}}{des\ A - des\ \bar{A}}, \tag{2}$$

corresponding to Ramsey's formula (1).

Here the basic idea is that if $A_1, A_2, \ldots, A_n$ are an exhaustive set of incompatible ways in which the proposition A can come true, the desirability of A must be a weighted average of the desirabilities of the ways in which it can come true:

$$des\ A = w_1\ des\ A_1 + w_2\ des\ A_2 + \cdots + w_n\ des\ A_n, \tag{3}$$

where the weights are the conditional probabilities,

$$w_i = prob\ A_i/prob\ A. \tag{4}$$

Let us call a function *des* which attributes real numbers to propositions a *Bayesian desirability function* if there is a probability measure *prob* relative to which (3) holds for all suitable $A, A_1, A_2, \ldots, A_n$. And let us call a preference ranking of propositions *coherent* if there is a Bayesian desirability function which ranks those propositions in order of magnitude exactly as they are ranked in order of preference. One can show (Jeffrey, 1965, chs. 6, 8) that if certain weak conditions are met by a coherent preference ranking, the underlying desirability function is determined up to a fractional linear transformation, i.e., if *des* and *DES* both rank propositions in order of magnitude exactly as they are ranked in order of preference, there must be real numbers a, b, c, d such that for any proposition A in the ranking we have

$$DES\ A = \frac{a\ des\ A + b}{c\ des\ A + d}. \tag{5}$$

The probability measure *prob* is then determined by (2) up to a certain quantization. In particular, if *des* is Bayesian relative to *prob*, then *DES* will be Bayesian relative to *PROB*, where

$$PROB\ A = prob\ A\ (c\ des\ A + d). \tag{6}$$

Under further plausible conditions, (5) and (6) are given either exactly (as in Ramsey's theory) or approximately by

$$DES\ A = a\ des\ A + b, \tag{7}$$

$$PROB\ A = prob\ A. \tag{8}$$

I take the principal advantage of the present theory over Ramsey's to be that here we work with the subject's actual beliefs, whereas Ramsey needs to know what the subject's preference ranking of relevant propositions would be if his views of what the world is were to be changed by virtue of his having come to believe that various arbitrary and sometimes bizarre causal relationships had been established via gambles (Jeffrey, 1965, pp. 145–150).

To see more directly how preferences may reflect beliefs in the present system, observe that by (2) we must have *prob A* > *prob B* if the relevant portion of the preference ranking is

$$
\begin{array}{cc}
A, & B \\
T \\
\bar{B} \\
\bar{A}
\end{array}
$$

In particular, suppose that A and B are the propositions that the subject will get job 1 and that he will get job 2, respectively. Pay, working conditions, etc., are the same, so that he ranks A and B together. Now if he thinks himself more likely to get job 1 than job 2, he will prefer a guarantee of ($\bar{B}$) not getting job 2 to a guarantee of ($\bar{A}$) not getting job 1; for he thinks that an assurance of not getting job 2 leaves him more likely to get one or the other of the equally liked jobs than would an assurance of not getting job 1.

2. Probabilistic acts and observations

We might call a proposition *observational* for a certain person at a certain time if at that time he can make an observation of which the *direct* effect will be that his degree of belief in the proposition will change to 0 or to 1. Similarly, we might call a proposition *actual* for a certain person at a certain time if at that time he can perform an act of which the *direct* effect will be that his degree of belief in the proposition will change to 0 or to 1. Under ordinary circumstances, the proposition that the sun is shining is observational and the proposition that the agent blows his nose is actual. Performance of an act may give the agent what Anscombe calls "knowledge without observation" (1957, §8) of the truth of an appropriate actual proposition. Apparently, a proposition can be actual or observational without the agent's knowing that it is; and the agent can be mistaken in thinking a proposition actual or observational.

The point and meaning of the requirement that the effect be "direct", in the definitions of "actual" and "observational", can be illustrated by considering the case of a sleeper who awakens and sees that the sun is shining. Then one might take the observation to have shown him, directly, that the sun is shining, and to have shown him indirectly that it is daytime. In general, an observation will cause numerous changes in the observer's belief function, but many of these can be construed as consequences of others. If there is a proposition E such that the *direct* effect of the observation is to change the observer's degree of belief in E to 1, then for any proposition A in the observer's preference ranking, his degree of belief in A after the observation will be the conditional probability

$$prob_E\ A = prob\ (A/E) = prob\ AE/prob\ E, \tag{9}$$

where *prob* is the observer's belief function before the observation. And conversely, if the observer's belief function after the observation is $prob_E$ and $prob_E$ is not identical with *prob*, then the *direct* effect of the observation will be to change the observer's degree of belief in E to 1. This completes a definition of *direct*.

But from a certain strict point of view, it is rarely or never that there is a proposition for which the direct effect of an observation is to change the observer's degree of belief in that proposition to 1; and from that point of view, the classes of propositions that count as observational or actual in the senses defined above are either empty or as good as empty for practical purposes. For if we care seriously to distinguish between 0.999,999 and 1.000,000 as degrees of belief, we may find that, after looking out the window, the observer's degree of belief in the proposition that the sun is shining is not quite 1, perhaps because he thinks there is one chance in a million that he is deluded or deceived in some way; and similarly for acts where we can generally take ourselves to be as best *trying* (perhaps with very high probability of success) to make a certain proposition true.

One way in which philosophers have tried to resolve this difficulty is to postulate a phenomenalistic language in which an appropriate proposition E can always be expressed, as a report on the immediate content of experience; but for excellent reasons, this move is now in low repute (see, e.g., Austin, 1962). The crucial point is not that 0.999,999 is so close to 1.000,000 as to make no odds, practically speaking, for situations abound in which the gap is more like one half than one millionth. Thus, in examining a piece of cloth by candlelight one might come to attribute probabilities 0.6 and 0.4 to the propositions G that the cloth is green and B that it is blue, without there being any proposition E for which the direct effect of the observation is anything near changing the observer's degree of belief in E to 1. One might think of some such proposition as that (E) *the cloth looks green or possibly blue*, but this is far too vague to yield $prob\ (G/E) = 0.6$ and $prob\ (B/E) = 0.4$. Certainly, there is *something* about what the observer sees that leads him to have the indicated degrees of belief in G and in B, but there is no reason to think the observer can express this something by a statement in his language. And physicalistically, there is some perfectly definite pattern of stimulation of the rods and cones of the observer's retina which prompts his belief, but there is no reason to expect him to be able to describe that pattern or to recognize a true description of it, should it be suggested.

As Austin (1962, ch. 10) points out, the crucial mistake is to speak seriously of the *evidence* of the senses. Indeed the relevant experiences have perfectly definite characteristics by virtue of which the observer comes to believe as he does, and by virtue of which in our example he comes to have degree of belief 0.6 in G. But it does not follow that there is a proposition E of which the observer is certain after the observation and for which we have *prob* (G/E) = 0.6, *prob* (B/E) = 0.4, etc.

In part, the quest for such phenomenological certainty seems to have been prompted by an inability to see how uncertain evidence can be used. Thus C. I. Lewis:

If anything is to be probable, then something must be certain. The data which themselves support a genuine probability, must themselves be certainties. We do have such absolute certainties, in the sense data initiating belief and in those passages of experience which later may confirm it. But neither such initial data nor such later verifying passages of experience can be phrased in the language of objective statement – because what can be so phrased is never more than probable. Our sense certainties can only be formulated by the expressive use of language, in which what is signified is a content of experience and what is asserted is the givenness of this content (Lewis, 1946, p. 186).

But this motive for the quest is easily disposed of (Jeffrey, 1965, ch. 11). Thus, in the example of observation by candlelight, we may take the direct result of the observation (in a modified sense of "direct") to be, that the observer's degrees of belief in G and B change to 0.6 and 0.4. Then his degree of belief in any proposition A in his preference ranking will change from *prob A* to

$$PROB\ A = 0.6\ prob\ (A/G) + 0.4\ prob\ (A/B).$$

In general, suppose that there are propositions $E_1, E_2, \ldots, E_n$, in which the observer's degrees of belief after the observation are $p_1, p_2, \ldots, p_n$; where the Es are pairwise incompatible and collectively exhaustive; where for each i, *prob* E_i is neither 0 nor 1; and where for each proposition A in the preference ranking and for each i the conditional probability of A on E_i is unaffected by the observation:

$$PROB\ (A/E_i) = prob\ (A/E_i). \tag{10}$$

Then the belief function after the observation may be taken to be PROB, where

$$PROB\ A = p_1\ prob\ (A/E_1) + p_2\ prob\ (A/E_2) + \cdots + p_n\ prob\ (A/E_n), \tag{11}$$

if the observer's preference rankings before and after the observation are both coherent. Where these conditions are met, the propositions $E_1, E_2, \ldots, E_n$, may be said to form a *basis* for the observation; and the notion of a basis will play the role vacated by the notion of *directness*.

The situation is similar in the case of acts. A marksman may have a fairly definite idea of his chances of hitting a distant target, e.g. he may have degree of belief 0.3 in the proposition H that he will hit it. The basis for this belief may be his impressions of wind conditions, quality of the rifle, etc.; but there need be no reason to suppose that the marksman can express the relevant data; nor need there be any proposition E in his preference ranking in which the marksman's degree of belief changes to 1 upon deciding to fire at the target, and for which we have $prob\ (H/E) = 0.3$. But the pair H, $\bar{H}$ may constitute a *basis* for the act, in the sense that for any proposition A in the marksman's preference ranking, his degree of belief after his decision is

$$PROB\ A = 0.3\ prob\ (A/H) + 0.7\ prob\ (A/\bar{H}).$$

It is correct to describe the marksman as *trying* to hit the target; but the proposition that he is trying to hit the target can not play the role of E above. Similarly, it was correct to describe the cloth as *looking* green or possibly blue; but the proposition that the cloth looks green or possibly blue does not satisfy the conditions for directness.

The notion of directness is useful as well for the resolution of un-philosophical posers about probabilities, in which the puzzling element sometimes consists in failure to think of an appropriate proposition E such that the direct effect of an observation is to change degree of belief in E to 1, e.g. in the following problem reported by Mosteller (1965, Problem 13).

Three prisoners, a, b, and c, with apparently equally good records have applied for parole. The parole board has decided to release two of the three, and the prisoners know this but not which two. A warder friend of prisoner a knows who are to be released. Prisoner a realizes that it would be unethical to ask the warder if he, a, is to be released, but thinks of asking for the name of *one* prisoner *other than himself* who is to be released. He thinks that before he asks, his chances of release are ⅔. He thinks that if the warder says "b will be released," his own chances have now gone down to ½, because either a and b or b and c are to be released. And so a decides not to reduce his chances by asking. However, a is mistaken in his calculations. Explain.

Here indeed the possible cases (in a self-explanatory notation) are

AB, AC, BC,

and these are viewed by a as equiprobable. Then *prob* A is ⅔ but *prob* $(A/B) = prob\ (A/C) = $ ½, and, since the warder must answer either "b" or "c" to a's question, it looks as if the direct result of the "observation" will be that a comes to attribute probability 1 either to the proposition B that b will be released, or to the proposition C that c will be released. But this is incorrect. The relevant evidence-proposition would be more like the proposition *that the warder says, "b"*, or *that the warder says, "c"* even though neither of these will quite do. For it is only in cases *AB* and *AC* that the warder's reply is dictated by the facts: in case *BC*, where *b* and *c* are both to be released, the warder must somehow choose *one* of the two true answers. If *a* expects the warder to make the choice by some such random device as tossing a coin, then we have *prob* (*A*/the warder says, "*b*") = *prob* (*A*/the warder says, "*c*") = *prob A* = ⅔; while if *a* is sure that the warder will say "*b*" if he can, we have *prob* (*A*/the warder says "*b*") = ½ but *prob* (*A*/the warder says "*c*") = 1.

3. Belief: reasons vs. causes

Indeed it is desirable, where possible, to incorporate the results of observation into the structure of one's beliefs via a basis of form E, $\bar{E}$ where the probability of E after the observation is nearly 1. For practical purposes, E then satisfies the conditions of directness, and the 'direct' effect of the observation can be described as informing the observer of the truth of E. Where this is possible, the relevant passage of sense experience *causes* the observer to believe E; and if *prob* (*A*/*E*) is high, his belief in E may be a *reason* for his believing A, and E may be spoken of as (inconclusive) *evidence* for A. But the sense experience is evidence neither for E nor for A. Nor does the situation change when we speak physicalistically in terms of patterns of irritation of our sensory surfaces, instead of in terms of sense experience: such patterns of irritation *cause* us to believe various propositions to various degrees; and sometimes the situation can be helpfully analyzed into one in which we are caused to believe E_1, E_2, ..., E_n, to degrees p_1, p_2, ..., p_n, whereupon those beliefs provide *reasons* for believing other propositions to other degrees. But patterns of irritation of our sensory surfaces are not reasons or evidence for any of our beliefs, any more than irritation of the mucous membrane of the nose is a *reason* for sneezing.

When I stand blinking in bright sunlight, I can no more believe that the hour is midnight than I can fly. My degree of belief in the pro-position that the sun is shining has two distinct characteristics. (a) It is 1, as close as makes no odds. (b) It is compulsory. Here I want to emphasize the second characteristic, which is most often found in conjunction with the first, but not always. Thus, if I examine a normal coin at great length, and experiment with it at length, my degree of belief in the proposition that the next toss will yield a head will have two characteristics. (a) It is ½. (b) It is compulsory. In the case of the coin as in the case of the sun, I cannot decide to have a different degree of belief in the proposition, any more than I can decide to walk on air.

In my scientific and practical undertakings I must make use of such compulsory beliefs. In attempting to understand or to affect the world, I cannot escape the fact that I am part of it: I must rather make use of that fact as best I can. Now where epistemologists have spoken of observation as a source of *knowledge*, I want to speak of observation as a source of compulsory *belief* to one or another degree. I do not propose to identify a very high degree of belief with knowledge, any more than I propose to identify the property of being near 1 with the property of being compulsory.

Nor do I postulate any *general* positive or negative connection between the characteristic of being compulsory and the characteristic of being sound or appropriate in the light of the believer's experience. Nor, finally, do I take a compulsory belief to be necessarily a permanent one: new experience or new reflection (perhaps, prompted by the arguments of others) may loosen the bonds of compulsion, and may then establish new bonds; and the effect may be that the new state of belief is sounder than the old, or less sound.

Then why should we trust our beliefs? According to K. R. Popper,

> ... the decision to accept a basic statement, and to be satisfied with it, is causally connected with our experiences – especially with our *perceptual experiences*. But we do not attempt to *justify* basic statements by these experiences. Experiences can *motivate a decision*, and hence an acceptance or a rejection of a statement, but a basic statement cannot be *justified* by them – no more than by thumping the table. (Popper, 1959, p. 105)

I take this objection to be defective, principally in attempting to deal with basic statements (observation reports) in terms of *decisions* to *accept* or to *reject* them. Here acceptance parallels belief, rejection parallels disbelief (belief in the denial), and tentativeness or reversibility of the decision parallels *degree* of belief. Because logical relations hold between statements, but not between events and statements, the relationship between a perceptual experience (an event of a certain sort) and a basic statement cannot be a logical one, and therefore, Popper believes, cannot be of a sort that would justify the statement:

> Basic statements are accepted as the result of a decision or agreement; and to that extent they are conventions. (Popper, 1959, p. 106)

But in the absence of a positive account of the nature of acceptance and rejection, parallel to the account of partial belief given in section 1, it is impossible to evaluate this view. Acceptance and rejection are apparently acts undertaken as results of decisions; but somehow the decisions are conventional – perhaps only in the sense that they may be *motivated* by experience, but not *adequately* motivated, if adequacy entails justification.

To return to the question, "Why should we trust our beliefs?" one must ask what would be involved in *not* trusting one's beliefs, if belief is analyzed as in section 1 in terms of one's preference structure. One way of mistrusting a belief is declining to act on it, but this appears to consist merely in lowering the degree of that belief: to mistrust a partial belief is then to alter its degree to a new, more suitable value.

A more hopeful analysis of such mistrust might introduce the notion of sensitivity to further evidence or experience. Thus, agents 1 and 2 might have the same degree of belief – ½ – in the proposition H_1 that the first toss of a certain coin will yield a head, but agent 1 might have this degree of belief because he is convinced that the coin is normal, while agent 2 is convinced that it is either two-headed or two-tailed, he knows not which[1]. There is no question here of agent 2's expressing his mistrust of the figure $\frac{1}{2}$ by lowering or raising it, but he can express that mistrust quite handily by aspects of his belief function. Thus, if H_i is the proposition that the coin lands head up the ith time it is tossed, agent 2's beliefs about the coin are accurately expressed by the function $prob_2$ where

$$prob_2\, H_i = \tfrac{1}{2},\ prob_2\, (H_i/H_j) = 1.$$

while agent 1's beliefs are equally accurately expressed by the function $prob_1$ where

$$prob_1\, (H_{i_1}, H_{i_2}, \ldots, H_{i_n}) = 2^{-n},$$

if $i_1 < i_2 < \cdots < i_n$. In an obvious sense, agent 1's beliefs are *firm* in the sense that he will not change them in the light of further evidence, since we have

$$prob_1\, (H_{n+1}/H_1, H_2, \ldots, H_n) = prob_1\, H_{n+1} = \tfrac{1}{2},$$

while agent 2's beliefs are quite tentative and in that sense, mistrusted by their holder. Still, $prob_1\, H_i = prob_2\, H_i = \tfrac{1}{2}$.

After these defensive remarks, let me say how and why I take compulsive belief to be sound, under appropriate circumstances. Bemused with syntax, the early logical positivists were chary of the notion of truth; and then, bemused with Tarski's account of truth, analytic philosophers neglected to inquire how we come to believe or disbelieve simple propositions. Quite simply put, the point is: coming to have suitable degrees of belief in response to experience is a matter of training – a *skill* which we begin acquiring in early childhood, and are never quite done polishing. The skill consists not only in coming to have appropriate degrees of belief in appropriate propositions under paradigmatically good conditions of observation, but also in coming to have appropriate degrees of belief between zero and one when conditions are less than ideal.

---

[1] This is a simplified version of 'the paradox of ideal evidence', Popper (1959, pp. 407–409).

Thus, in learning to use English color words correctly, a child not only learns to acquire degree of belief 1 in the proposition that the cloth is blue, when in bright sunlight he observes a piece of cloth of uniform hue, the hue being squarely in the middle of the blue interval of the color spectrum: he also learns to acquire appropriate degrees of belief between 0 and 1 in response to observation under bad lighting conditions, and when the hue is near one or the other end of the blue region. Furthermore, his understanding of the English color words will not be complete until he understands, in effect, that blue is between green and violet in the color spectrum: his understanding of this point or his lack of it will be evinced in the sorts of mistakes he does and does not make, e.g. in mistaking green for violet he may be evincing confusion between the meanings of "blue" and of "violet", in the sense that his mistake is linguistic, not perceptual.

Clearly, the borderline between factual and linguistic error becomes cloudy, here: but cloudy in a perfectly realistic way, corresponding to the intimate connection between the ways in which we experience the world and the ways in which we speak. It is for this sort of reason that having the right language can be as important as (and can be in part identical with) having the right theory.

Then learning to use a language properly is in large part like learning such skills as riding bicycles and flying aeroplanes. One must train oneself to have the right sorts of responses to various sorts of experiences, where the responses are degrees of belief in propositions. This may, but need not, show itself in willingness to utter or assent to corresponding sentences. Need not, because e.g. my cat is quite capable of showing that it thinks it is about to be fed, just as it is capable of showing what its preference ranking is, for hamburger, tuna fish, and oat meal, without saying or understanding a word. With people as with cats, evidence for belief and preference is behavioral; and speech is far from exhausting behavior (Jeffrey, 1965, pp. 57–59).

Our degrees of beliefs in various propositions are determined jointly by our training and our experience, in complicated ways that I cannot hope to describe. And similarly for conditional subjective probabilities, which are certain ratios of degrees of belief: to some extent, these are what they are because of our training – because we speak the languages we speak. And to this extent, conditional subjective probabilities reflect *meanings*. And in this sense, there can be a theory of degree of confirmation which is based on analysis of meanings of sentences. Confirmation theory is therefore semantical and, if you like, logical.

Part II

Conceptualization of probability and utility

This part focuses on some conceptual problems concerning the notions of probability and utility assumed by Bayesian decision theory. One of the founders of the subjective theory of probability, Bruno de Finetti (1979, p. x), proclaimed: *Probability does not exist*! He obviously meant that *objective* probability does not exist. But what about subjective probability and utility? In what way do they exist? How are they conceptualized? The three papers in this part all contain critical examinations of the basic notions and assumptions of the Bayesian program.

To introduce one of the problems, let us assume that you are offered two options and your task is to determine which is the more preferable:

Option 1: In front of you on the table there is an opaque urn containing 30 black and 70 white balls. You are allowed to inspect the balls and the urn. In doing so, you find that the balls are identical except for their color. If, after mixing the balls carefully, you draw a black ball from the urn, you receive $100; if the drawing results in a white ball, you get nothing.

Having read Part I of this book, you argue that your subjective probability that you will draw a black ball is 0.30. If, for simplicity, we assume that, for small sums of money, dollars and utilities are exchangeable, the subjective expected utility of this option is thus $30.

Option 2: Consider the possibility that there will be a bus strike in Verona, Italy, during the next month. If there is a strike, you receive $100; if not, you get nothing.

For the sake of the argument, let us assume that, after considering it carefully, you believe that the probability that there will be a bus strike in Verona next month is 0.30. Thus the expected utility of the second option equals that of the first. In both cases $30 would be a fair price to pay for the options.

However, although the two games have the same expected utility for you, we are pretty sure that you will prefer Option 1 to Option 2. But this preference conflicts with the recommendations of the MEU principle since both options have the same expected utility.

Why do you prefer Option 1? It is not because you have assigned too high a value to the probability of the bus strike. You believe that 0.30 is, under the circumstances, the best assessment you can make. The crux of the matter is that you feel you know more about the urn than about Italian wages, working conditions, and other factors that may provoke a bus strike in Verona. There is an important difference in degree between your knowledge concerning these two random events. The obvious *unreliability* of the probability assessment in Option 2 cannot be ignored when it comes to choosing between the options.

As Henry Kyburg points out in his paper, one problem for the Bayesian or subjectivistic view of probability is that it cannot account for this obvious difference in information which prevails in the two options. Savage is certainly clear about the problem:

... there seem to be some probability relations about which we feel relatively "sure" as compared with others. When our opinions, as reflected in real or envisaged action, are inconsistent, we sacrifice the unsure opinions to the sure ones. The notion of "sure" and "unsure" introduced here is vague, and my complaint is precisely that neither the theory of personal probability, as it is developed in this book, nor any other device known to me renders the notion less vague. (1954/1972, pp. 57–58)

The strict Bayesian approach to probability also faces another problem, which is related to the one above. It is mentioned by Savage in the following way:

A second difficulty, perhaps closely associated with the first one, stems from the vagueness associated with judgments of the magnitude of personal probability. The postulates of personal probability imply that I can determine, to any degree of accuracy whatsoever, the probability (for me) that the next president will be a Democrat. (1954/1972, p. 59)

It is a central tenet of the traditional subjectivistic theory that the probability of any statement can be determined to any degree of accuracy we like. But isn't this demanding too much? It may work well for drawings from urns or card decks, but are there any events in real life for which you can assess the probability with the required accuracy?

Kyburg argues that examples like the ones above show that the Bayesian theory of probability is defective. Kyburg's discussion of these problems and his hint at a solution have very much influenced the new types of theories presented in Part IV of this volume.

The problems – whether the Bayesian theory demands too much of its users, and whether the axioms of the theory are defective because no one can reasonably be assumed to obey them – are also discussed in Ian Hacking's paper. Another problematic consequence of the Bayesian theory is that if p logically entails q, one should be at least as confident of q as of p. This implies that one should be able to see all the logical consequences of one's beliefs. Hacking's solution to these problems is "slightly more realistic personal probabilities." At the end of his article, however, Hacking emphasizes that his alternative should be viewed as a solution to a special problem, not as an alternative to the traditional theory.

As regards the conceptualization of utility, a fundamental question is what kinds of *entities* can be assigned utilities. Or in other words, what are the carriers of utility? In the axiomatizations presented in Part I, Savage assigns utilities to outcomes, Ramsey assigns utilities to "worlds" [which he defines as "different possible totalities of events between which our subject chooses – the ultimate organic unities" (1931), p. 32 this volume], and Jeffrey, finally, attributes desirabilities to propositions. Among economists, utilities are sometimes assigned to specific commodity bundles and sometimes to total wealth (or some other economically comprehensive totality) and sometimes these two interpretations are not kept apart.

In response, it may be argued that it does not matter much what is chosen as the carriers of utility. But it does. If utilities are not assigned to total descriptions like Savage's outcomes or Ramsey's worlds, the resulting theory will become vulnerable to counterexamples directed against the transitivity and additivity of values. On the other hand, if it is accepted that utilities can only be assigned to total descriptions of the world, then it becomes very difficult to employ the theory for practical purposes. Bengt Hansson's contribution spells out some of these difficulties although he discusses the problem in more detail in section 2 of his 1975 presentation. It is therefore tempting to connive at the conceptual problems of utility in order to have an applicable theory.

Hansson's contribution also deals with the question, touched upon in the introduction, whether utility should be interpreted "realistically" as a measure of subjective usefulness or "formalistically" as a convenient way to represent the decision maker's preferences. It is quite clear that the *explanatory power* of decision theory increases if we opt for the realistic interpretation of utility.

Another related problem discussed in Hansson's paper [see also Hansson (1975), (1981)] is whether a decision maker's utility measure can be used to identify his attitude toward *risk*. The answer to this problem seems to depend on how the utility measure is obtained. If it is obtained by an operational method using gambles and lotteries, it can be expected that the (formalistically) elicited utility function is affected, up to a point, by the decision maker's attitude toward risk. On the other hand, if the utility measure is obtained by a (realistically oriented) method not involving gambles or other types of probabilistic prospects, it may result in a different utility measure not in any way reflecting the decision maker's risk attitudes. As mentioned in the introduction, this kind of distinction cannot be made within the traditional Bayesian theory.

6. Bets and beliefs

Henry E. Kyburg

The subjectivistic or personalistic interpretation of probability is playing a larger and larger role both in statistics and in philosophy these days. In statistics it is associated with the recent resurgence[1] of Bayesian techniques which began with the work of Bruno de Finetti (1937/1964), and which has been enthusiastically championed in English-speaking statistical circles by L. J. Savage (1954). In philosophy the importance of this interpretation of probability has been felt in certain problems concerning induction, rationality, and decision-making, into which it has injected new spirit. It also has important connections with the (now) traditional logical approach of Carnap (1950) and his followers. Most of those who hold a logical view of probability – that is, those who think that there is at least one sense in which probability statements reflect logical relationships – now find in the personalistic arguments first stated by Ramsey (1931), and developed and refined by Savage, Lehman (1955), Kemeny (1955), and Shimony (1955) in recent years, a justification for the principles or axioms that any conventional logical probability theorist takes his logical probability relations to satisfy.

I think this subjectivistic interpretation of probability is mistaken, but I also think it is an important interpretation, and I think it is very important indeed to be very clear about why it is mistaken. A good part of this paper will be concerned therefore with a sympathetic presentation of the subjectivistic point of view, and a rebuttal of some

---

[1] H. Jeffreys, long an enthusiastic champion of Bayesian statistical techniques, wrote from a logical rather than a subjectivistic point of view.

Reprinted from *American Philosophical Quarterly*, no. 5 (1968), pp. 63–78, by kind permission of the author and the publisher.

common objections to it. The remainder will be concerned with what I take to be its serious drawbacks, which are also, in view of the fact that the subjectivistic theory may be regarded as merely a weakened form of the confirmation theories, drawbacks to those theories.

Just as we consider ordinary logic to be a logic of full belief, so let us consider an extension of it, probability logic, to be a logic of partial belief. Even a frequentist can agree to this, if he is willing to regard partial beliefs as *estimates* of relative frequencies. (Indeed this very approach is adopted by Reichenbach in his *Theory of Probability*, 1949.) If we believe that a is a crow, then we must (by ordinary logic) believe that a is a crow or a is black, and we must refrain from having the belief that a is not a crow. Ordinary logic is a logic of consistency in a very strong sense; in the sense that if my beliefs violate the laws of ordinary logic, some of those beliefs are flatly wrong.

Now if I am about to toss a coin, I can consider the statement that it will land heads. Do I believe this? It seems reasonable to want to avoid answering this question with either "Yes" or "No," and instead to say, "I believe it to *a certain degree*," or "I *partially* believe it." We can lay down criteria for partial beliefs on intuitive grounds – and this is very much what Keynes (1921) and Koopman (1940) have done – without specifying how these beliefs are to be measured or detected. For both Keynes and Koopman these partial beliefs represent probability relations that can be detected introspectively, and can even be compared with one another in certain (but not in all) cases. But there is no objective, interpersonal, way of measuring them. There are those, like Ramsey, who say that they simply cannot detect probability relations, however hard they introspect. I rather feel that I can, in certain special cases, and I gather that such intuitions were what provided the original basis for Carnap's confirmation functions. Yet even if Keynes is right and Ramsey wrong about being able to uncover certain simple probability relations introspectively, Ramsey is surely right in feeling that such introspective measures will not be sufficient as a basis for making the probability calculus into an instrument for achieving interpersonal agreement and for making decisions under circumstances of uncertainty. We need some way of determining when the beliefs of an individual conform to the conditions that we intend to lay down, and when they don't.

The time-honored way of finding out how seriously someone believes what he says he believes is to invite him to put his money where his mouth is. Ramsey's proposal for measuring partial beliefs is simply a refinement of this technique. Because there is a considerable amount of misunderstanding concerning just what is presupposed by Ramsey's proposal to define degrees of belief behavioristically, it is worthwhile going into some detail.

Let us first define an ethically neutral proposition as one such that its truth or falsity is a matter of indifference – i.e., all other things being equal (everything that affects us), we do not care whether this proposition is true or false. (For example, the statement "The number of hairs on my head is odd" is surely an ethically neutral proposition.) We define belief of degree ½ in an ethically neutral position P as follows: Suppose a person is not indifferent between two states of affairs, say α and β; i.e., that he prefers the state of affairs represented by α to the state of affairs represented by β, or he prefers the state of affairs represented by β to the state of affairs represented by α. Suppose that he has no preference between (1) α-if-proposition-P-and-β-if-proposition-not-P, and (2) β-if-proposition-P-and-α-if-proposition-not-P. Then he believes P to the degree ½.

Given, now, an ethically neutral proposition, believed to degree ½, we need only an arbitrary starting point: say the value of the total situation, α-if-P, β-if-not-P. Take the value of this gamble to be ½. Now we need merely find a state of affairs γ which is just as desirable as this gamble: its value, too, must be ½. Consider the gamble γ-if-P-and-β-if-not-P. (where β is preferred to α); it has the value of ¾. And the gamble α-if-P and γ-if-not-P, will have the value ¼. If there are other states of affairs preferable to β or more detestable than α, the scale may be extended to include them as well. What matters is simply that the existence of this single ethically neutral proposition P, which is believed to the degree ½, suffices to determine the whole scale of values for possible states of affairs.[2]

We can now use these values of total states of affairs to determine degrees of belief in *any* proposition: including those that are *not* ethically neutral: if the option of α-for-sure has the same value for a person as the option β-if-S-and-γ-if-not-S then his degree of belief in S is the ratio of the difference in value between α and γ, to the difference in

value between β and γ: $\dfrac{\gamma - \alpha}{\beta - \gamma}$ (we suppose $\beta > \alpha$). $\Rightarrow \gamma < \alpha \leq \beta \Rightarrow$ this ratio is $-ve\,!!$

On this basis it is possible to *prove* that the usual axioms of probability theory hold for degrees of belief. Ramsey (p. 181, pp. 35–36, this volume) proves as "fundamental laws of probable belief":

1. The degree of belief in P, plus the degree of belief in not-P equals 1.
2. The degree of belief in P given Q, plus the degree of belief in not-P given Q equals 1.
3. The degree of belief in P and Q equals the degree of belief in P multiplied by the degree of belief in Q given P.

[2] Ramsey actually employs seven other axioms; these concern only requirements of consistency in assigning values to total states of affairs.

$$\alpha = \beta\, p(s) + \gamma\, p(\bar{s}) = \beta\, p(s) + \gamma - \gamma p(s)$$
$$p(s)\,(\beta - \gamma) = \alpha - \gamma$$
$$p(s) = \frac{\alpha - \gamma}{\beta - \alpha} \quad , \text{ if } \beta - \alpha > 0.$$

4. The degree of belief in P and Q plus the degree of belief in P and not-Q equals the degree of belief in P.

Ramsey writes:

These are the laws of probability, which we have proved to be necessarily true of any consistent set of degrees of belief. Any definite set of degrees of belief which broke them would be inconsistent in the sense that it violated the laws of preference between options, such as that preferability is a transitive asymmetrical relation, and that if α is preferable to β, β for certain cannot be preferable to α if p, β if not-p. If anyone's mental condition violated these laws, his choice would depend on the precise form in which the options were offered him, which is absurd. He could have a book made against him by a cunning bettor, and would then stand to lose in any event. (p. 182) [p. 36, this volume]

This has been christened the Dutch Book Theorem, by Isaac Levi.[3]

This is as far as the formal logic of subjective probability takes us. From the point of view of the "logic of consistency" (which for Ramsey includes the probability calculus), no set of beliefs is more rational than any other, so long as they both satisfy the quantitative relationships expressed by the fundamental laws of probability. Thus I am free to assign the number $\frac{1}{3}$ to the probability that the sun will rise tomorrow; or, more cheerfully, to take the probability to be $\frac{9}{10}$ that I have a rich uncle in Australia who will send me a telegram tomorrow informing me that he has made me his sole heir. Neither Ramsey, nor Savage, nor de Finetti, to name three leading figures in the personalistic movement, can find it in his heart to detect any logical shortcomings in anyone, or to find anyone logically culpable, whose degrees of belief in various propositions satisfy the laws of the probability calculus, however odd those degrees of belief may otherwise be. Reasonableness, in which Ramsey was also much interested, he considered quite another matter.[4] The connection between rationality (in the sense of conformity to the rules of the probability calculus) and reasonableness (in the ordinary inductive sense) is much closer for Savage and de Finetti than it was for Ramsey, but it is still not a strict connection; one can still be wildly unreasonable without sinning against either logic or probability.

Now this seems patently absurd. It is to suppose that even the most simple statistical inferences have no logical weight where my beliefs are concerned. It is perfectly compatible with these laws that I should have a degree of belief equal to $\frac{1}{4}$ that this coin will land heads when next I toss it; and that I should then perform a long series of tosses

---

[3] Isaac Levi. Paper read at the 1965 meeting of the American Philosophical Association, Eastern Division.

[4] Indeed, he suggests that sometimes it may be *reasonable* to sin against the probability calculus, or even against the standards of deductive logic (p. 191 [p. 42, in this volume]).

(say, 1000), of which ¾ should result in heads; and then that on the 1001st toss, my belief in heads should be unchanged at ¼. It *could* increase to correspond to the relative frequency in the observed sample, or it could even, by the agency of some curious maturity-of-odds belief of mine, decrease to ⅛. I think we would all, or almost all, agree that anyone who altered his beliefs in the last-mentioned way should be regarded as irrational. The same is true, though perhaps not so seriously, of anyone who stuck to his beliefs in the face of what we would ordinarily call contrary evidence. It is surely a matter of simple *rationality* (and not merely a matter of instinct or convention) that we modify our beliefs, in some sense, some of the time, to conform to the observed frequencies of the corresponding events.

This sort of objection, which may have been responsible for the lack of attention Ramsey's theory received for over a decade, was partially met by the first statistician to promote the subjectivist cause, Bruno de Finetti. De Finetti's theorems concern sequences of events that he calls "equivalent" or "exchangeable." A sequence of exchangeable events is a sequence in which the probability of any particular distribution of a given property in a finite set of events of the sequence depends only on the *number* of events in that distribution that have that property. An easy way of characterizing sequences of exchangeable events is merely to say that they are sequences in which the indices are held to be irrelevant. "Held to be," because whether or not a particular sequence is regarded as being composed of exchangeable events is a subjective matter – one which might be determined, for a given person, by proposing bets *à la* Ramsey or Savage.

An illustration will make clear what I mean: consider an irregular die. One of the ways it can land is with a one-spot uppermost. Let us call this type of event a success. Now consider any throw of the die. The absolute, unconditional, prior (*a priori*, if you will) probability that this particular throw will yield a success may be anything you want, according to the judgment you form from the particular shape and feel of the die; but if you regard the throws of the sequence to be exchangeable, as most would, you must attribute the same prior probability to success on the first throw, on the 7th throw, on the 300th throw, and in general on the i-th throw. This does not mean that you are prohibited from learning from experience; indeed it is precisely to allow for and to explain learning from experience that de Finetti introduces the concept of exchangeability, to take the place of the traditional independence and equiprobability. The *conditional* probability of success on the n-th throw, given knowledge of the results of the preceding $n - 1$ throws, will *not* in general be the same as the prior probability of a success on the n-th throw. This conditional probability will generally reflect the frequency of past occurrences.

The "problem of induction" is solved for the subjectivist, with respect to sequences of exchangeable events, by various laws of large numbers which de Finetti has proved for these sequences. One pointed consequence of these theorems is the following: If two people agree that events of a certain sequence are exchangeable, then, with whatever opinions they start, given a long enough segment of the sequence as evidence, they will come arbitrarily close to agreement concerning the probability of *future* successes, provided that neither of them is so pigheaded that his probability evaluations are totally insensitive to evidence provided by frequencies.[5] This is to say, after they have both observed a sequence of n of the events under consideration, they will attribute probabilities that differ by at most ε to success on the $(n + 1)$-st event. And indeed, however small ε may be made, there is an n which will make this statement true.

There are two perverse ways of avoiding this natural result. One is to deny that the events are exchangeable; the other is to deny the influence of past observations on present probability assignments. Neither of these gambits is ruled out by the axioms developed by Ramsey and his successors. The present holders of subjectivistic views seem to feel that this is simply a fact of logic. This is all that logic can do for us, and that's that. (But as Professor Nagel has pointed out in discussion, a judgment of exchangeability amounts to a good old-fashioned *a priori* judgment of equiprobability: we judge every combination of n successes and m failures to be equiprobable.) Logic cannot determine our views, even relative to a body of evidence; it can only stipulate that certain combinations of statements are inconsistent (deductive logic) or that certain combinations of beliefs are inconsistent (the Dutch Book Theorem).

The typical response of the logical probability theorist, such as Carnap, is to argue that there are more standards of logical adequacy than those expressed by the standard axioms of the probability calculus. These logical theorists welcome the subjectivist approach as providing a *proof* of the soundness of the basic postulates of probability theory, but they do not think that these postulates go far enough. Carnap, for example, wishes to incorporate in his system postulates reflecting not only the symmetries corresponding to exchangeability, but also symmetries concerning predicates. But these refinements and developments would certainly be irrelevant if I could show that even the few axioms accepted by the subjectivists were inappropriate as guides to rational belief. This is precisely what I intend to do in the remainder of this paper. I shall try to point up those respects in which even the basic axioms accepted by the subjectivists seem to be too

---

[5] Strictly speaking, there are other pathological distributions of belief that must also be ruled out.

strong; in doing so, I shall at the same time be attacking both subjectivistic and logical points of view, and also attacking those relative frequency views in which *estimates* of relative frequencies play the same role as logical probabilities, and also satisfy the same axioms.

There are three respects in which I take these axioms to be defective: (1) in taking it to be possible to evaluate a degree of belief to any desired degree of accuracy; (2) in supposing it sufficient to characterize a degree of belief by a single number, thus disregarding considerations such as the weight of the evidence; and (3) in leading to very strong beliefs about matters of fact on purely *a priori* grounds. The first two of these points are well known and have been discussed by subjectivists.[6]

In the formalized systems of subjective probability – Savage's, for example – it is a theorem that the degree of a person's belief in any statement can be evaluated to any degree of accuracy whatever. A simple consequence of this theorem which will illustrate the point at issue is that my degree of belief in the statement, "The President of the United States in 1984 will be a Republican," is fixed with just the same precision as my degrees of belief in the various possible outcomes on tosses of a coin that I am fully convinced is fair. Thus the probability of the political statement in question may be just the same as the probability that between 456 and 489 tosses out of 1027 tosses of this coin will result in heads; or it may be a bit less, say, between the probability of this coin-tossing statement and the statement: Between 455 and 488 tosses out of 1027 tosses of this coin will result in heads. Now the subjectivist would argue that this is not altogether unrealistic; after all, if I were presented with the choice of a gain (to be paid in 1984) if there is a Republican president in 1984, or the same gain if between 456 and 489 of 1027 tosses result in heads, I might prefer to stake the gain on the coin tossing statement; and if I were presented with the choice between staking the gain on the political statement, or on the slightly less probable coin tossing statement that there will be between 455 and 488 heads on 1027 tries, I might prefer to stake the gain on the political statement. The point is that if I am *forced* to choose what event I shall stake a gain on, I shall make such a choice.

But it seems unreasonable to demand that such choices be *consistent* in the sense in which the subjectivist uses the term. Thus I see no reason why I should not stake my possible gain on the political statement in the first case, and then turn right around and stake my possible gain on the coin statement in the second case, despite the fact that I know perfectly well that the former coin statement is more probable than the second. It seems reasonable, because my opinions as

---

[6] See, e.g., Savage (1954, p. 57 [see the quotations on p. 98, this volume]); the third I take to be novel.

to the probability of the President being Republican in 1984 simply aren't that definite.

The subjectivist can argue that my opinions have changed between the first choice and the second; and there is nothing wrong with that. But I don't *feel* as if my opinions have changed; I have acquired no new *evidence* that bears on the likelihood that the President in 1984 will be a Republican. *One* way of looking at the matter would be that my degree of belief has changed; but an equally good way of looking at it, I should think, would be that my degree of belief was simply *not* altogether definite. It might be that the best we could say was that it was on the same order of magnitude as the probability of either of the two coin-tossing statements.

How about what the subjectivist calls consistency? If my belief that the President in 1984 will be a Republican is characterized merely by a vague interval, say as falling between .3 and .4, and if I can appropriately accept odds corresponding to any real number within these limits, then why shouldn't I accept a bet that there will be a Republican president in 1984 at odds of 7:3 and simultaneously a bet that there will be a non-Republican president at odds of 4:6? If the bets are in dollars, and I make both of them, I can have a book made against me, and I shall lose a dollar in either event. It is only if beliefs correspond simply to real numbers (rather than to intervals, say) that we can formulate a condition that will preclude my having a book made against me; so the subjectivist argues.

But this argument for precision of beliefs is irrelevant. The fact that it is irrational to make both bets simultaneously (which fact depends on *no* probabilistic considerations at all) doesn't imply in the least that it is unreasonable to make either bet alone. The mere fact that I can be forced into making a choice upon which of two eventualities I would prefer to stake a gain does not really imply that my belief is best represented by a single real number.[7]

There is another argument against both subjectivistic and logical theories that depends on the fact that probabilities are represented by real numbers. This has been discussed here and there in the literature as the problem of total evidence. The point can be brought out by considering an old fashioned urn containing black and white balls. Suppose that we are in an appropriate state of ignorance, so that, on the logical view, as well as on the subjectivistic view, the probability that the first ball drawn will be black, is a half. Let us also assume that the draws (with replacement) are regarded as exchangeable events, so that the same will be true of the i-th ball drawn. Now suppose that we

---

[7] Savage considers this problem (p. 59), and is clearly made uncomfortable by it, but does not regard it as fatal to the theory.

draw a thousand balls from this urn, and that half of them are black. Relative to this information both the subjectivistic and the logical theories would lead to the assignment of a conditional probability of ½ to the statement that a black ball will be drawn on the 1001st draw. Nearly the same analysis would apply to those frequency theories which would attempt to deal with this problem at all (Salmon, 1957). Although there would be no posit appropriate before any balls had been drawn, we may suppose that of the first two balls one is black and one is white; thus the "posited" limiting relative frequency after two balls have been drawn is ½, and this is the weight to be assigned to the statement that the next ball will be black. Similarly, after 1000 balls have been drawn, of which half have been black, the appropriate posit is that the limiting frequency is one half, and the weight to be assigned to the statement that the next ball will be black is ½.

Although it does seem perfectly plausible that our bets concerning black balls and white balls should be offered at the same odds before and after the extensive sample, it surely does not seem plausible to characterize our beliefs in precisely the same way in the two cases.

To put the matter another way: If we say that we are going to take probability as legislative for rational belief (or even as legislative for relations among degrees of beliefs in related statements), then to say that the probability of statement S is ½ and that the same is true of statement T is simply to say that our epistemological attitude toward the two statements should be the same. But for the two statements in question, this is clearly wrong. The person who offers odds of two to one on the first ball is not at all out of his mind in the same sense as the person who offers odds of two to one on the 1001st ball. What is a permissible belief in the first case is not the same as what is a permissible belief in the second case.

This is a strong argument, I think, for considering the measure of rational belief to be two dimensional; and some writers on probability have come to the verge of this conclusion. Keynes, for example, considers an undefined quantity he calls "weight" to reflect the distinction between probability relations reflecting much relevant evidence, and those which reflect little evidence. Carnap considers a similar quantity when he suggests that we regard probabilities as estimates of relative frequencies, for we can distinguish between more and less *reliable* estimates. But it is still odd to think that "$c (h, e) = p$" is *logically true*, considered as a degree of confirmation statement, and at the same time only *more or less reliable* considered as an estimation statement.

Such curious gambits are not even open to the subjectivist. Though Savage distinguishes between these probabilities of which he is sure and those of which he is not so sure, there is no way for him to make this distinction within the theory; there is no internal way for him to reflect the distinction between probabilities which are based on many instances and those which are based on only a few instances, or none at all. The fact that corresponds to increase in weight for the logical theorist and the frequentist is the increasing uniformity of probability judgments of different individuals with an increase in the evidential basis for them, and the decreasing impact of new evidence. But this yields no reason, for example, for not gambling with people who have better information than you have. Nor does this give any reason for preferring to bet for large stakes when our beliefs are based on a large amount of information, than when we have only a little information. Or consider a case in which our degree of belief is ½, and the odds we are offered are 10:9. If our degree of belief is based on a good deal of information it makes good sense to take advantage of this positive mathematical expectation; but if our belief is based only on a small amount of information, it would surely be regarded as foolish to risk much, even though our expectation is quite positive. But on either subjectivistic or logical theories, strictly considered, it would be rational to bet any amount (subject to the limitations imposed by the fact that money is not the same as utility), simply because the mathematical expectation is positive.

I come now to the third sort of objection. This is the objection that *a priori* subjective or logical probabilities implicitly seem to contain information or predictions about future frequencies. It is a difficult objection to state, in its general form, for subjectivists and logical theorists are quick to point out that a prior probability does not constitute any sort of a prediction about long-range frequencies. It is true that if my (or the) prior probability of E is ¼, I shall act in many ways just as I would if I knew that the long-run frequency of E's was ¼. But this is not to say that the statements are equivalent in meaning, because there are also crucial differences. The one statement is logical or psychological; the other is an empirical statement about a certain sequence of events. Furthermore, the statement about the long-run frequency would not be changed as we observed members of the sequence: If we take the long-run frequency of heads on tosses of a particular coin to be ½, the observation of a dozen tosses, eight of which resulted in heads, will not make us assign a different probability to heads on the 13th toss. But the conditional probability that interests us when we have observed a number of elements of the sequence is generally quite different from the prior probability. There is no ground on which we can say that a prior probability gives us any *knowledge* about future frequencies. Put another way: a subjectivistic or logical probability statement does not *entail* any (empirical) statement about relative frequencies.

On the other hand, we may have a lingering feeling that the division between logical probabilities and relative frequencies is not quite as clean and sharp as Carnap would have us think. A clue to this fact lies before our eyes in de Finetti's laws of large numbers: if we can draw conclusions about subjective probabilities on the basis of relative frequencies, can't we perhaps draw some conclusions about relative frequencies on the basis of subjective (or logical) probabilities?

Although there is no relation of entailment between statements about subjective probabilities and statements about frequencies, we may consider a weaker relation. Thus if it were to follow from a statement about subjective or logical probabilities that a certain statement about frequencies was overwhelmingly probable, I would regard this as constituting a relationship between the subjectivistic statement and the empirical statement about frequencies only slightly less strong than the entailment relation. (After all, this is just what subjective and logical probabilities are supposed to reflect: a relation between evidence and conclusion that is weaker than the entailment relation.) Furthermore, if, from an *a priori* logical or subjectivistic probability statement based on (essentially) no evidence whatsoever, it followed that a certain empirical statement was overwhelmingly probable, and worthy of a very high degree of rational belief, I would take this as evidence of *a priorism* of the worst sort, and evidence that there was something seriously wrong with the theory. Surely we no longer expect, or even desire, to be able to draw conclusions about the world from statements that are not about the world (in the same sense). From statements that are logically true, and do not even *refer* to our evidence, we do not expect to be able to draw (even probabilistic) inferences about the relative frequency with which black balls are drawn from urns; and the same is surely true of psychological statements.

What I shall next do is show that in fact there is a broad class of cases in which just this occurs within the subjectivistic theory, and *a fortiori* within the logical theory. To do this, I must state and prove a theorem. It isn't very long or very complicated; it is quite closely connected to theorems of de Finetti.

Theorem: Let E_1, E_2, ... be a sequence of exchangeable events with respect to some characteristic C, represented by a random quantity X_i which takes the value 1 if E_i has the characteristic C and the value 0 if E_i lacks the characteristic C. Define $y_h = \frac{1}{h} \sum_{i=1}^{h} X_i$; Y_h is thus just the relative frequency of the characteristic in question among the first h events. Let m_1 be the prior probability of C: $P(X_i = 1) = m_1$, and let m_2 be the conditional probability $P(X_i = 1 \mid X_j = 1)$ for $i \neq j$. Then we have:

$$P\left(|Y_h - m_1| > k \sqrt{\frac{m_1}{h} + \frac{h-1}{h} m_1 m_2 - m_1^2}\right) < \frac{1}{k^2}$$

Proof:

1. It is easy to show that the mean of Y_h is just the prior probability of the characteristic in question,

$$m_1 = P(X_i = 1)$$

2. We calculate the variance of Y_h:

$$E(Y_h - m_1)^2 = E(Y_h^2 - 2m_1 Y_h + m_1^2) = \frac{1}{h^2} (hE(X_i^2) + h(h-1)E(X_i X_j)) - 2m_1^2 + m_1^2$$

3. Since the random quantity X_i either has the value 0 or the value 1, $X_i^2 = X_i$ and $E(X_i^2) = E(X_i) = m_1$.

4. Since the events are regarded as exchangeable, $P(X_i = 1 \cdot X_j = 1)$ always has the same value when $i \neq j$; let us write it in the form

$$P(X_i = 1) \cdot P(X_j = 1 \mid X_i = 1), = m_1 m_2$$

5. Thus we have, finally,

$$E(Y_h - m_1)^2 = \frac{m_1}{h} + \left(\frac{h-1}{h}\right)m_1 m_2 - m_1^2$$

6. Tchebycheff's inequality gives us:

$$P\left(|Y_h - m_1| > k \sqrt{\frac{m_1}{h} + \frac{h-1}{h} m_1 m_2 - m_2^2}\right) < \frac{1}{k^2} \text{ Q.E.D.}$$

We now have an expression which gives us an upper limit for the probability of there existing more than a certain difference between a prior probability, m_1 and a relative frequency, Y_h. The difference itself involves both the prior probability m_1, and the conditional probability m_2. To fix our ideas, let us suppose that the prior probability of C is $\frac{1}{100}$, and that the conditional probability of C_i given C_j is double that. For example, we might consider an urn filled with balls, and we might very well attribute a probability of $\frac{1}{100}$ to the statement that the first ball to be drawn is purple. Then, given that one ball is purple, we might attribute the probability $\frac{2}{100}$ to the statement that another ball is purple.

Now if h is very large – and there is no reason for not letting h approach infinity – the expression for the variance of Y_h boils down to:

$$m_1(m_2 - m_1)$$

Taking advantage of the fact that in this particular case $m_1 = m_2 - m_1$, we have the further simplification,

$$E(Y_h - m_1)^2 = m_1^2$$

Using Tchebycheff's inequality, but reversing the inequality signs, we have:

$$P(|Y_h - m_1| < km_2) > 1 - \frac{1}{k^2}$$

Taking $k = 10$, for example, what we have just proved, in plain words, amounts to:

If the prior probability of getting a purple ball from this urn is $\frac{1}{100}$, and if the conditional probability of getting a purple ball, given that one other ball is purple, is $\frac{2}{100}$, and if we regard the sequence of draws as exchangeable, then the prior probability is at least .99 that in an *arbitrarily long run*, or in the limit, not more than 11 percent of the balls will turn out to be purple. The probability is at least .9996 that no more than half will be purple.

Here is a clear case in which we have arrived at a practical certainty – surely a probability of .99 may amount to practical certainty at times – that a certain relative frequency will hold (not more than 11 percent), for an indefinitely long sequence of events. It is a clear case, it seems to me, of claiming to obtain knowledge about the world from *a priori* premises. It is true that the claim is not a categorical one; we do not arrive at the conclusion that the relative frequency of purple balls *will* be less than 11 percent but only at the conclusion that it is *overwhelmingly probable* that this relative frequency will be less than 11 percent. But this does not save the plausibility of the theory, for almost all of our empirical knowledge is of this sort, particularly from the logical and subjectivistic point of view. Our scientific knowledge is not *certain* but is *overwhelmingly probable*. It is no more plausible to claim to be able to deduce on *a priori* grounds (from psychological or logically true premises) that the probability is .99 that no more than 11 percent of the balls drawn from an urn will be purple, than it is to claim to be able to deduce on similar *a priori* grounds that the probability is .99 that the number of planets is seven.

It may be wondered whether there is any difference *in principle* between being able to be 99 percent sure on *a priori* grounds that less than 11 percent of the balls will be purple, and being able to be 1 percent sure on *a priori* grounds that the first ball to be drawn will be purple. To this there are two answers: First, that there is no difference in principle, but a very great intuitive difference between being practically sure about a very general hypothesis, and just having a certain small degree of belief about a particular event. The subjectivist will be quick to point out that all the theorem says – all any subjectivistic theorem says – is that *if* you have such and such beliefs about individual draws and pairs of draws, *then* you ought to have such and such other beliefs about long sequences of draws. But there seems to me all the difference in the world, intuitively, between being pretty sure that a given ball in the urn is not purple, and being equally sure that in the next thousand, or ten thousand, or million(!) draws from the urn, not more than 11 percent will result in a purple ball. Both the "given ball" and the "million draws" are, strictly speaking, particulars; but the latter is just the sort of particular knowledge which is properly termed general scientific knowledge. Indeed knowledge about – i.e., in subjectivistic terms, very strong beliefs about – the statistical character of the next million draws is precisely the sort of thing we should expect to get only by painstaking research, sampling, etc., and not at all the sort of thing we should expect to be able to deduce from tautologies concerning logical ranges or from the data of introspection. The subjectivist will tell me that my belief about the simple event, together with the axioms of the probability calculus, simply *entails* my having such and such a belief about the complex event; to which I reply, so much the worse for the axioms.

The subjectivist may say that he proposes rather to rearrange his initial beliefs than to reject any axioms. As even a subjectivist must admit, this is a perilous practice. It also greatly reduces the regulative force of personalistic probability, for the whole point of de Finetti's laws of great numbers is to show how the theory is not subjective in a *bad* sense, because ordinary people whose opinions are not very extreme will eventually be forced into agreement with each other and with the empirical relative frequency. But if we can withdraw our judgments of exchangeability whenever we begin to feel uncomfortable – even in such straightforward cases as drawing balls out of urns or rolling dice – then our pet opinions need have nothing to fear from the evidence even if the opinions are not extreme.

The answer that the original opinion of the person will change as he makes draws from the urn is irrelevant: he has not made any draws from the urn, and if all he wants is to be sure (let us say) that less than half of the draws will result in a purple ball, he doesn't even have any motivation to make draws from the urn; he can be extremely sure *a priori*. The answer that prior probability assignments and beliefs about long-run relative frequencies go hand in hand is also not quite to the point. In a sense, that is just what has been shown above. But in another sense, one may find beliefs about long-run frequencies to be relatively inaccessible to behavioristic assessment, while such simple beliefs as those we started with are very easy to discover. It is the fact that those simple, accessible, natural prior beliefs entail, through the axioms of the probability calculus, such strangely strong beliefs about long-run frequencies, that seems counter-intuitive.

The second answer is more interesting and also more controversial.

It is that if we have a rule of acceptance in inductive logic, it is very difficult to see how the case in question could be distinguished in point of generality or high probability from the most conventional acceptable statistical hypotheses. The distinction between a statistical hypothesis, for the subjectivist, and a statistical hypothesis as understood by an objectivist, is that the subjectivist (at least, de Finetti) wants to reject the whole notion of an unbounded or indefinite or completely general statistical hypothesis. According to this view, we have only to deal with specific propositions. But some of these propositions are statistical in form (e.g., about 51 percent of next year's births will be births of males), and do refer to rather large groups of instances (e.g., in the case in question, to millions). Let us leave entirely aside the question of the general statistical hypothesis (the *indefinitely* long run of births will produce just 51 percent males); whether such completely general hypotheses are ever accepted or not, there are many who would agree that hypotheses like the one about next year's births can, given a sufficiently cogent body of evidence, come to be reasonably accepted. The ground on which such a statement might come to be accepted is generally taken to be its probability, relative to a body of evidence. And in precisely the same way, a statement (such as, of the next million balls drawn from the urn, no more than 11 percent will be purple) may be highly probable relative to the evidence for it (even though that evidence be no more than tautologies or the results of introspection). This material statistical statement, then, making a specific assertion about the number of purple balls that will be encountered in a rather long future, may perfectly well come to be accepted on the basis of no empirical evidence at all. (Surely introspection does not provide *empirical evidence* for the statement.) We would then have to say that the statement was accepted on *a priori* grounds, in the sense that it was accepted independently of any empirical evidence for it. If we adopt both subjectivistic probabilities and a rule of detachment, we are committed to *a priori* synthetic knowledge.

One way out is to eschew a rule of acceptance for inductive logic. There are many who think that no such rule (rules) can be found. (But it is an interesting fact, I think, that the largest concentration of such people is to be found in the ranks of subjectivists!) Much has been made of the fact that no consistent (and otherwise adequate) rule has been proposed. There are plenty of consistent acceptance rules, such as "Accept anything you want, provided it doesn't conflict with what you've already accepted." "Accept anything entailed by observation statements that you can vouch for." "Accept the strongest hypotheses that you can think of that are consistent with your present body of beliefs, and accept them only so long as they are not refuted by observation." Indeed, the only subjectivist I know who allows no form of acceptance rule is Richard Jeffrey (1965). The problem of an inductive acceptance rule is the problem of spelling one out in practical detail; a purely probabilistic rule faces very serious difficulties, and these difficulties may either be solved by employing a probability concept for which they do not arise, or circumvented by proposing other criteria than probability as conditions of acceptance.

But whether or not one believes that a rule of acceptance can be formulated that is satisfactory, the consequences of the theorem are not satisfactory. If one has a probabilistic rule of acceptance, then one can be led to *accept* very far-reaching empirical statements, despite the fact that one has not a shred of empirical evidence for them; if one supposes there is no acceptance rule of this sort, one is still led to have very high degrees of belief in these far-reaching empirical statements on the basis of no empirical evidence whatsoever. It is one thing to have a degree of belief based on no evidence about the color of a certain ball; it is something else again to have a high degree of belief based on no evidence concerning a relative frequency in an enormous empirically given population.

The argument I have presented does depend on certain peculiarities of the case: we require that the events considered be exchangeable, in de Finetti's sense. But most of the sequences of events that we deal with are reasonably regarded as sequences of exchangeable events; it is only in virtue of this fact that de Finetti's laws of large numbers are interesting. We also require that the *a priori* probability of the characteristic in question be small; but surely we shall have no trouble finding sequences of events and characteristics for which the prior probability of the characteristic is small. Finally I made use of the supposition that the conditional probability would be conveniently double the prior probability. We need merely suppose that it be *no more than* double the prior probability. And surely we can find sequences for which this is true, too. The assumptions that I have made are not very special, or very farfetched. Furthermore, we shall of course get very similar results in circumstances which are merely *quite* similar to those that I have described.

As providing grounds for rejecting subjectivistic theories of probability, however, we do not have to suppose that situations like the one that I have described crop up all the time; the mere fact that they *can* crop up seems to me to provide good enough reason for abandoning the theory which allows them.

Now observe that we must also reject logical theories like Carnap's, for Carnap's systems are essentially built on the subjectivistic theory by the addition of further conditions. These conditions not only do not interfere with the above argument, but would make it even easier to carry through. Symmetry conditions would allow us to *prove* that certain sets of individual events are exchangeable, for example.

Furthermore, frequency theories that are embellished with a super-structure of posits, like those of Salmon and Reichenbach, must also go by the board. If the posits are to serve as basis for determining betting quotients, they must satisfy the same axioms as the subjectivistic pro-babilities; and therefore the above proof will go through for them too.

Where do we go from here? At this place I can only hint at an answer. A probability concept which took "degrees" of rational belief to be measured by intervals, rather than by points, would avoid all of these difficulties. Although on this sort of theory, a belief, as reflected behavioristically, is still either rational or irrational (the betting ratio either corresponds to a point inside the allowable interval, or it doesn't), there is nothing in the theory itself which demands that degrees of belief in a behavioristic sense be sharply defined. Secondly, the interval conception provides a very natural way of reflecting the weight of the evidence. The prior probability that a ball drawn from an urn of unknown composition will be black could be the whole interval from 0 to 1. In a rational corpus (of a given level) based on a certain amount of evidence consisting of a sequence of draws from the urn, the probability may be narrowed down to the interval between ⅓ and ⅔; relative to a rational corpus (of the same level) based on a large amount of evidence of the same sort, the probability may be the interval between .45 and .55. Finally, if we take probabilities to be intervals, the theorem proved above simply will not work. Perhaps if one wishes to apply probabilities in the conduct of everyday life and in the evaluation of beliefs, one may simply have to put up with the inconvenience of an interval conception of probability.

7. Slightly more realistic personal probability

Ian Hacking

A person required to risk money on a remote digit of π would, in order to comply fully with the theory [of personal probability] have to compute that digit, though this would really be wasteful if the cost of computation were more than the prize involved. For the postulates of the theory imply that you should behave in accordance with the logical implications of all that you know. Is it possible to improve the theory in this respect, making allowance within it for the cost of thinking, or would that entail paradox?[1]

Like each of Professor Savage's difficulties in the theory of personal probability, his problem about the remote digit of π is entirely general. It concerns logical consequence as much as logical truth: his theory implies that if e entails h you should be as confident of h as of e. His own example is one of three distinct cases which militate against this part of his theory. In his example there is a known algorithm for working out of the relevant logical implications, but it is too costly for sensible use. A second case arises when there is no known algorithm

This is a substantial revision of my symposium contribution at the meeting of the Western Division of the A.P.A., Chicago, May 4–6, 1967. James Cargile, John Vickers and Bruno de Finetti are among those whose letters provoked some of the changes. The present version owes a special debt to L. J. Savage's meticulous line-by-line criticism of the earlier draft.

Reprinted from *Philosophy of Science*, **34** (1967), pp. 311–325, by kind permission of the author and the publisher.

[1] L. J. Savage (1967). Unless otherwise specified all references to Savage's work are to this article.

for finding out whether the hypothesis h follows from the evidence e. Perhaps there are two subcases: in the first, the algorithm is not known to anyone; in the second, it is not accessible to the person who is making decisions. In either case the person who is as confident of h as of e is, though lucky, not reasonable, but prejudiced; a man who is less confident may be the sensible man who tailors his beliefs to the available evidence.

Intuitionist mathematicians offer ready examples for the first form of the second case. Does 777 occur in the decimal expansion of π? According to classical logic, any analytical definition either entails that 777 occurs, or entails that it does not, but we know no procedure sure to settle which it is. Complete confidence in either outcome is absurd. Yet complete confidence is demanded by personalism. If it is hard to imagine real life betting on such a question, recall the 15th century algorithm competitions. When Tartaglia knew the algorithm for solving cubic equations and Cardano did not, Cardano had to "risk money," or at least his reputation, on problems that could be solved only by an algorithm he did not know (Cox, 1961, ch. 5).

A third case arises from undecidability. Suppose a man is to have a set of betting rates over a whole class of problems for which there exists no algorithm. It must be an infinite class because algorithms exists for all finite classes of problems. Such a man is prevented from systematically satisfying the demands of personal probability. For a concrete example, let our man have to bet about assertions of the form "F is a theorem of the predicate calculus," where F ranges over all well formed formulae of the calculus.

These three cases make distinct versions of the difficulty suggested by Savage. The third one, though it will appeal to logicians, might be discounted by a practical personalist on the grounds that we never do have to risk money over the whole range of an infinite undecideable class. Hence I shall attend mainly to the first two cases, although the third will also be kept in mind. The first and second cases do arise in serious practical matters. Many questions in probability theory are answered by Monte Carlo methods that yield only probable solutions with a range of uncertainty. Yet a computer technologist will often decide to use Monte Carlo methods: both when expensive exact solutions are theoretically available, and also when no algorithm for the exact solution *is* known. In either case he is rationally deciding to act against the axioms of personal probability. A slightly more realistic theory must show that his decision is reasonable.

Savage fears that any theory which is, in this respect, more realistic, will "entail paradox." This is especially plausible in the first two cases, for although we expect the precise analysis of recursive functions to help with the third, no analysis is already tailored for the other two.

The difficulty seems to arise from some feature of what Savage calls "logical implication." Philosophers know, to their cost, the difficulty of getting any intuitively adequate analysis of relations among logical truths. The best known analysis of logical implication, namely C. I. Lewis' theory of strict implication, says that a self-contradictory proposition entails everything (Lewis and Langford, 1932, p. 250). Many philosophers balk at that result, but none has circulated an alternative which is, at present, widely accepted. It is plausible to guess that attempts to patch up personalism will sink into the same quagmires that have, in my opinion, swallowed up students of entailment.

1. A priori and a posteriori reasoning. Plausible though such defeatism is, I shall argue against it. The argument goes near many philosophical quagmires, but we can skirt most of them in the way which, as Savage reminds us, so many other philosophical difficulties are evaded by personalism. A main strand in the argument can be sent out at once. Personalism is, says Savage, a theory for policing one's own potential decisions and systems of belief. Hence we distinguish between the theory and what it is about. In logician's parlance personalism is a metatheory. It is about, in part, various beliefs that are represented by propositions. Some aspects of Savage's problem may stem from over-willing acceptance of philosophical dogmas about propositions and our knowledge of them.

In particular I do not believe that the theory should acknowledge any distinction between facts found out by *a priori* reasoning and those discovered *a posteriori*. I am not referring to the current controversy as to whether there is a sharp distinction between analytic and synthetic truths. I insist only that actions based ultimately upon knowledge need not distinguish ways in which the knowledge is acquired.

Consider the problem of finding the surface of least area bounded by a closed curve in space. It is hard to establish even that there is a least area. Yet in the early 19th century the Belgian physicist Plateau could often answer by determining the film a soap bubble forms on a closed loop of wire; he knew enough about soap bubbles to be sure the film was of least area. The complete mathematical solutions had to wait for over a century (Courant, 1941, p. 386). Yet the empirically obtained results should provide as much confidence for practical decisions as the later mathematical proofs – maybe more, considering several debacles that from time to time occurred in the calculus of variations! What matters to the decision maker is what he knows or can find out; philosophical distinctions among the means of discovery are of no moment.

Take a pair of examples directly related to Savage's problem about the remote digit of π. Imagine a man taught binary notation, but not even told that it is a system of numbering. He is taught only the natural ordering of the binary numerals. He is also taught how to add and multiply in this notation, although he is not told what the operation means. He is asked to speculate on the relative magnitude of products of pairs of five-digit binary numbers. It does not matter to him much; say he risks no money at all, but can make a little every time he is right. His beliefs can be represented by betting odds in the way that Savage has taught us. Suppose that considering any pair of products of two five-digit binary numbers, his betting rate is 0.2 on the two products being equal, and 0.4 on each of the other two alternatives.

This man, whom we shall recall from time to time in what follows, is to be compared with another: someone who is first introduced to the mysteries of underground city transport, say that of the city of London. He is asked questions like, "Are there more stops travelling between Gloucester Rd. and King's Cross on the Piccadilly or on the Circle line?" His odds parallel those of the first man, the binary computer. The two have much in common. Their betting rates hardly fit the facts as we know them. In each case, an elementary algorithm answers each question which can be put to them; each declines it as too expensive considering the trifling gains. In each case some "insight" short of working out complete answers would lead to more profitable betting odds.

Despite the parallel, personalism treats one man as sensible and the other as incoherent. We need a theory that puts both on a par. It should also explain why each man should find out more before wagering, if investigation is cheap enough. In trying to lessen the formal distinction between the two men, a remark of Savage's may suggest a fallacy we should avoid. He says that "the example about π does not adequately express the utter impracticality of knowing our own minds in the sense implied by the theory." I believe the example about π does not express the impracticality of knowing our own minds at all: it has nothing to do with knowing our own minds; it is a matter of knowing π. And our speculator on binary products may know his own mind full well; what he does not know is binary arithmetic.

2. *Classical personalism.* Personalists attribute probabilities to events, but Savage's problem arises out of logical implication, which is a relation between propositions. So it is natural to work in one of the formalisms that attribute probability to propositions rather than to events.

Classical personalism offers a theory of rational belief and reasonable decision. At any moment in his life a man will know a body of facts f. He is interested in some set of propositions. Associated with this set is a Boolean algebra A. $Prob_f(h)$ is to be a number representing the person's personal probability for h, when he knows f; for short, his probability given *f*. In one behavioural analysis, confidence is measured by the least favourable rate at which the person will bet about *h*. This leads to a well known argument for what I shall call the *static assumption* of personalism: For any *h*ε*A*, and at least any $f \subset A$, $Prob_f(h)$ is defined and satisfies the probability axioms. As de Finetti proved, the probability axioms give necessary and sufficient conditions that a person's odds not be open to a Dutch book, i.e. not open to a book against him which is guaranteed a net gain (de Finetti, 1964). Perhaps other arguments for the static assumption are more profound. Many readers will prefer those of F. P. Ramsey's (1931) or Savage's (1954, ch. 3). But de Finetti's argument is so familiar, so simple, and by comparison so brief that it serves as a convenient reference point for the rest of this paper. I believe each point made in connection with the Dutch book argument can be transferred to the other famous arguments for the static assumption.

Probability given facts is not to be confused with conditional probability, which is defined in the usual way:

$$Prob_f (h/e) = \frac{Prob_f (he)}{Prob_f (e)}$$

for positive denominators. Conditional probabilities indicate how confident a person knowing only *f* judges that he would be if he knew *e* as well. The distinction between probability given facts and conditional probabilities is not found in the usual personalist writings. The terminology is copied from an objectivist paper of J. S. Williams (1966, p. 276). Formally the distinction is clear. The probability of *h* given *f* is a primitive to be circumscribed by the axioms of Kolmogorov. Conditional probability is defined as above. The latter is extraneous to the system, and introduced solely for convenience, the former is basic.

I say the distinction is fundamental to personalism yet personalists never use it explicitly. They never write "*f*" as a subscript to probabilities, nor express the idea in other ways. Why then introduce it? Because it will be crucial to our treatment of Savage's problem, and also because it makes explicit something fundamental to that part of Savage's theory which leads one to call his work Bayesian. Let me explain this after stating an implicit assumption of personalists which connects conditional probability with probability given facts. I call it the *dynamic assumption*:

$$Prob_{f \cup [e]} (h) = Prob_f (h/e).$$

The meaning is as follows. Suppose I know only *f*. I judge that if I knew *e* as well, I would be confident of *h* to degree *p*; behaviourally this judgement is shown by the conditional bets I would place. Now I find out that *e* is the case. The dynamic assumption asserts that now my confidence in h is p, as behaviourally shown in a readiness to place unconditional bets.

This assumption is not a tautology for personalism. It is a tautology for theories like Harold Jeffreys' (1939), where a unique probability is associated with any pair h, e. Those theories do not need our distinction between probability given evidence and conditional probability. But personalists do need the distinction, and do need the dynamic assumption.

Since the assumption seems never to be stated explicitly in the classic personalist studies, how dare I say it is needed? Because it is essential to that "model of how opinion is modified in the light of experience" to which Savage refers above. This requires a digression, but it is so important to understanding personalism, and my modification of it, that the point deserves a section of its own.

3. Conditional and given. Savage's model of modifying opinion employs Bayes' theorem; that is why we speak of Bayesians today. Savage has stated the theorem "somewhat informally" in the following way (I use an innocent paraphrase of Savage, et al., 1962, p. 15).

Prob (h/e) $\propto$ Prob (e/h) $\cdot$ Prob (h)

In words, the probability of h given the datum e is proportional to the product of the probability of observing e given h multiplied by the initial probability of h.

Well known properties of this theorem lead us to a model of learning from experience. My own catalogue of the properties, guilty of exactly the same confusion as I shall attribute to Savage's presentation, is given in (Hacking, 1965, ch. XIII). The idea of the model of learning is that *Prob (h/e)* represents one's personal probability after one learns e. But formally the conditional probability represents no such thing. If, as in all of Savage's work, conditional probability is a defined notion, then *Prob (h/e)* stands merely for the quotient of two probabilities. It in no way represents what I have learned after I take e as a new datum point. It is only when we make the dynamic assumption that we can conclude anything about learning from experience. To state the dynamic assumption we use probability given data, as opposed to conditional probability.

The conflation of two distinct concepts may explain why people favourable to personalism can say both that conditional probability is an "extraneous" defined notion, and also that, as D. V. Lindley puts it in discussing an address of Savage's "All probabilities are conditional" (Savage, et al., 1962, p. 83).

It may seem as if Lindley's position could let us avoid the distinction I have been urging. I said Jeffreys' interpersonal theory could get along with conditional probabilities taken as primitive. Why cannot the personalist do the same, as Lindley does in his own recent book (1965)? Unfortunately we find the equivocation in a new guise. Lindley gives a betting rate justification of his axioms along personalist lines (1965, vol. I, pp. 32–36). It relies on reading *Prob* (*h*/*ef*) as the rate, all conditional on *f*, at which I would bet on *h* conditional on *e*. Later in his Bayesian statistics, the same conditional probability symbol represents my confidence or betting rate for *h* when I know both *e* and *f*; when *e* is a sample, *Prob* (*h*/*ef*) shows how beliefs are "changed by the sample according to Bayes' theorem" (1965, vol. II, p. 2).

The equivocation can be explained but not excused by the fact that a man knowing *e* would be incoherent if the rates offered on *h* unconditionally differed from his rates on *h* conditional on *e*. But no incoherence obtains when we shift from the point before *e* is known to the point after it is known. Thus, suppose to begin with both *h* and *e* are uncertain. A man offers odds of *p*,*q*,*r*, and $1-p-q-r$ on *he*, ~*he*, *h*~*e* and ~*h*~*e* respectively. His conditional rates fit in with this. Then *e* is found out to be true. The man revises his rates, betting 1 on *e*, 0 on ~*e*, and $p+s/p+q+s$ on *h*, and $q/p+q+s$ on ~*h* for some positive *s*. These new rates show how much the man has "learned" from *e*. His learning violates the dynamic assumption. It is non-Bayesian. But since the man announces his post-*e* rates only after *e* is discovered, and simultaneously cancels his pre-*e* rates, there is no system for betting with him which is guaranteed success in the sense of a Dutch book. It is of no avail to express all rates as conditional: then the man's *Prob* (*h*/*ef*) before learning *e* differs from his *Prob* (*h*/*ef*) after learning *e*. Why not, he says: the change represents how I have learned from *e*!

I am not here quarrelling with the dynamic assumption, although I know of no personalist defence of it. Probability dynamics is too little studied, although Richard Jeffrey's (1965, ch. 11) is a good start at clarifying another aspect of the problem which I am here ignoring. Patrick Suppes' (1966b, sec. 4) is well aware of the matter I have just described, although the axiom he proposes does not seem sufficient to guarantee the dynamic assumption. One non-personalist defence of the dynamic assumption can, I believe, be derived from the continuity and differentiability argument of R. T. Cox (1961, ch. 1) to which Shimony alludes in his essay in the present issue of *Philosophy of Science*. But that argument has never been favoured by personalists. And neither the Dutch book argument, nor any other in the personalist arsenal of proofs of the probability axioms, entails the dynamic assumption. Not one entails Bayesianism. So the personalist requires the dynamic assumption in order to be Bayesian. It is true that in consistency a personalist could abandon the Bayesian model of learning from experience. Salt could lose its savour.

4. *The betting rate interpretation.* Our digression into the concept of probability given facts was needed for our overall view of Savage's problem and its solution. For we propose a trivially Bayesian treatment of mathematical learning, in agreement with our view that learning mathematical facts, and learning empirical facts, are both learning facts. The model of how learning facts modifies opinion will be the same in each case, namely Bayesian. We can achieve this only by weakening the axioms for personal probability, but in such a way that no practical application of the classical theory is impeded. For a hint of how to proceed, re-examine the betting rate interpretation, where $Prob_f(h) = p$ if and only if p is the largest number such that for any relatively small S I would exchange pS for the right to collect S if h is true, and nothing if h is false.

Under the usual interpretation of a betting rate, de Finetti's theorem is valid: betting rates must satisfy the probability axioms or else be open to a Dutch book. But the usual interpretation involves a trifling idealization. In real life betting I will not collect on h merely if it is true. It must be seen to be true. The bettors (or their heirs) must find out that h is true, or at worst abide by the decision of a trusted arbiter who claims to know about h. This idea has, I think, been implicit in de Finetti's insistence that we can only bet on hypotheses of the sort that can be settled in finite time. But that insistence is not enough, for only a few of the hypotheses that can be settled in theory are ever in fact settled. Even if something can be in principle settled but in fact never is, there will be no pay-offs.

More realistically my personal probability for h must be measured by p when p is the largest number such that I will contract with another party as follows. I agree to pay him pS if we find out that h is false. He agrees to pay me S in exchange for pS if we find out that h is true. No money changes hands until we settle the truth value of h. Of course like any other contract the "we" is less than literal: contracts can be inherited, bought, or adjudicated. But we discard the custom of leaving the stake in the hands of a bookmaker until the issue is settled: that custom is due to human dishonesty and has nothing essential to do with betting.

With this reinterpretation in mind, examine two of the probability axioms, say in a form adapted from Shimony's (1955).

1. If some elements of f logically imply h, then $Prob_f(h) = 1$.
2. If some elements of f logically imply that h and i are incompatible, then $Prob_f(h \lor i) = Prob_f(h) + Prob_f(i)$.

The only other axiom for probabilities in finite algebras says that probabilities lie between 0 and 1. The axioms are sensible for the usual betting rate interpretation, for if my rates fail to satisfy either (1) or (2), then, in the usual interpretation, a Dutch book can be made against me. This does not hold for the more realistic interpretation. In the extreme case suppose there is no available way to find out if elements of f logically imply h; f could even be the null class, and h a proposition of logic. Then, on the basis of knowledge of f there is no absurdity in having a betting rate on h less than 1, nor is there any known way to make a book against me with guaranteed profit. Though sufficient, the probability axioms are not necessary for avoiding a real life Dutch book.

John Vickers noticed this and in (1965) suggested weakening (2). He proposed additivity only if there is a proof that the incompatibility of h and i follows from f. He rightly said that even this is too strong for strictly personal probability. To extend Vickers' line of thought we need to analyse more closely the possible states of affairs contemplated by a decision maker.

5. *Possibilities.* Axioms (1) and (2) both use the concept of logical implication. As Shimony's (1955) takes for granted in presenting probability, strict implication is the appropriate formal analysis of logical implication in this context. C. I. Lewis explained strict implication in terms of possibility: $e \rightarrow h$ if it is not logically possible for e to be true while h is false (Lewis and Langford, 1932, p. 124). This implicit falling back on possibility should make us prick up our ears. Aristotle had a scale of modes: impossible, possible, probable, necessary. It is a tradition, which I do not admire, always to consider this as a scale of logical possibility, logical probability, etc. Savage snapped tradition by going to an opposite extreme: personal probability. Perhaps he gets into trouble because he is not completely radical. Just as logical probability is related to logical possibility, so personal probability demands a concept of personal possibility.

There is nothing sacred about logical possibility. We know how Quine has mocked it (1953, chs. 1, 2). A recent attempt to define what we commonly mean by possibility argues that though the concept is "objective" it falls short of logical possibility and is an epistemic concept (Hacking, 1967). That work was a by-product of trying to define "objective" probability short of logical probability. Likewise some concept of personal possibility should be a by-product of personal probability.

6. *Personal possibility.* The personalist wants to choose among acts, given a partition into possible states of the world. As Savage says, a possible state of the world is a "possible list of all answers to questions that might be pertinent to the decision situation at hand." But the partition need not consist of distinct logical possibilities. It should consist of states of affairs each of which is "possible to the agent." Of course in English we don't say "possible to him" (and

"possible for him" is something different; what Hacking, 1967, calls an M-occurrence of the word). But personal probability requires the odd "probable to him" or "probable for him" and personal possibility will need new locutions too.

For me, when is a proposition possible? When I do not know it to be false. Hence p may be possible for me although, to use the rubric of Jaakko Hintikka's (1962, p. 3), it is not possible for all that I know that p (i.e., p may be possible for me when it is incompatible with facts I do know, so long as I do not know the incompatibility).

What are the objects of personal probability? If, as in Carnap's (1947, p. 27), logically equivalent propositions are identical, then propositions cannot be the objects. For h and i may be logically equivalent, and I may know h, yet, because I am ignorant of the equivalence, I may not know i; hence $\sim h$ would not be personally possible while $\sim i$ is. This is absurd if personal possibility applies to propositions. No tighter criterion of propositional identity has ever succeeded. Hence we must cast about for other objects for personal probability. Sentences are the obvious choice. When p is an unambiguous sentence that a person understands, I shall speak of p being possible for him, and of his knowing p. This is not our normal way of speaking, but in the present context the meaning will be quite clear. We pretend that, as in a formal language, all sentences are unambiguous.

To attach knowledge to sentences is a blow against sound epistemology but is fine for personal probability, the theory of a person's choice. One can deliberate among only those possibilities expressed in sentences he can understand. Hence we abandon the idea of choosing within a Boolean algebra of propositions, and think of choosing among sentences in a language or "personal language" closed under what, in that language, correspond to the forming of conjunctions, negations, conditionals and alternations.

For an artificial example, recall the person comparing products of five-digit binary numbers. He need never employ any number over 961. Hence he need use only the following language. The terms are the first 961 binary numbers and the recursive result of writing "+" or "×" between two bracketted terms. The atomic sentences result from writing "=" or ">" between two terms. The closure of this under the Boolean sentential operations would be what I have called a language within which the person forms his beliefs about the problem at hand. It is not Boolean since the equivalence classes of sentential logic are not admitted.

It is not realistic to permit unending iteration of sentential operations, for there is an upper bound to the length of the sentences one can understand. A more realistic "personal language" would be the intersection of the closure under sentential operations, with the class of sentences a person understands. This can be characterized artificially, e.g. by limiting the sentences to 10,000 or fewer symbols. But I know of no difficulty in personal probability caused by ceaseless iteration, and I know no formal characterization of intelligibility which is not hopelessly artificial. Hence I shall not strive for realism in this matter.

We may notice, without elaboration, that tying personal probability to a personal language of sentences, or of intelligible sentences, makes one defect of personalism more transparent. Much scientific learning consists in devising new hypotheses or forming new concepts. The personalist difficulty over the unexpected hypothesis is explained in Hacking (1965, p. 221); Patrick Suppes examines concept formation and personalism in (1966a). Since new hypotheses and new concepts typically lead to newly intelligible sentences, they lead to a new personal language. So we should restrict Bayesian learning to that learning which occurs when the personal language is unchanged; when experience or thought prompts a change in one's language, quite another analysis is called for.

7. *Knowledge.* It is fine to relate personal probability to sentences, but it is not inviting to explain "p is personally possible for me" as "I do not know that p is false." For philosophers have never agreed on what knowledge is. They have agreed, at least since the *Gorgias*, that only what is true can be known. No other necessary condition is universally accepted. There is a long tradition of analysing knowledge as justified belief: for a man to know p, it is said, he must have good reasons for believing that p, must see these reasons to be good reasons, and must believe or even be certain that p. But this tradition is in a bad way, and like many other people, I suspect it is on the wrong track entirely.

The problem of what is knowledge is already a problem for personal probability, as noted by Savage above. Hence we will have achieved our aim of reducing Savage's list of difficulties by one, even if our treatment of the problem about π takes for granted the meaning of "knowledge." But one question we cannot evade. What are the closure conditions of knowledge? Despite the enduring argument of the *Meno*, knowledge is not closed under logical consequence. It is a tribute to Socrates' rhetoric that even today a good many philosophers agree with him, but at most they can be proposing a new, "divine," sense of knowledge. Using the verb "to know" in anything like its customary sense, it is at best a bad joke to say that once a student learns Peano's axioms he knows all their consequences.

Yet knowledge must surely have some closure conditions? If a man knows both p and $p \supset q$, does he not thereby know q as well? For him not to know q would be for him to betray misunderstanding of the conditional, and hence to show that he does not know $p \supset q$ after all.

So much is a natural conclusion to draw from Lewis Carroll's riddle about Achilles and the tortoise (1895) when taken together with work like Gilbert Ryle's (1950). Yet closure under *modus ponens* leads disastrously near to the divine sense of knowledge. I think the solution to this dilemma is to say that indeed a man must know how to use *modus ponens* (the cash value being that when presented with p and $p \supset q$ he can unswervingly infer q). If not, he does not understand the conditional. It in no way follows that knowledge is closed under *modus ponens*. Thinking otherwise must stem from confusing knowing how to get something (when certain conditions are met) and knowing that one gets it (when the conditions are met).

Hence in what follows I adopt the very harsh view that a man can know how to use *modus ponens*, can know that the rule is valid, can know p, and can know $p \supset q$, and yet not know q, simply because he has not thought of putting them together. We should call this an examiner's view of knowledge.

8. Slightly more realistic personal probability. To sum up: A *personal language* based on a set of sentences which a person understands is the closure of the set under the sentential operations which, in his language, correspond to the formation of conjunctions, negations, conditionals and alterations. An element of a personal language is personally possible to the person if he does not know it to be false, in an examiner's sense of knowledge. Paralleling the Lewis definition of strict implication, we could say that an element e of the personal language *personally implies* an element h if $e(\sim h)$ or $(\sim h)e$ or e or $\sim h$ is not personally possible to the person. Then slightly more realistic personal probability satisfies the dynamic assumption and also the static assumption restricted to the case in which personal implication replaces logical implication in the first two axioms.

Since personal implication is a degenerate concept with no closure conditions, it is more natural to express the axioms in terms of the fundamental concept of possibility. Then axioms (1) and (2) take the form:

1. If given facts f, $\sim h$ is not possible, $Prob_f (h) = 1$.
2. If given facts f, hi or ih is not possible, then $Prob_f (h \lor i) = Prob_f (h) + Prob_f (i)$.

In the theory of slightly more realistic personal probability, "possible" is construed as personally possible; in the classical theory it is construed as logically possible; other points are noted in a list below. First let us see how our theory works for the trifling example of a man comparing products of pairs of five-digit binary numbers.

We have settled on a personal language sufficient for his problem. What does he know? Nothing but the initial rules of calculation. For convenience of the example, we describe this knowledge as a set of facts about binary arithmetic, plus knowledge of how to infer by *modus ponens*. Let us represent the facts he knows as follows. I. All substitution instances of axioms for "=" which are sentences of the personal language. II. The ordering of the binary digits within the personal language, *viz.*, up to 961, he knows every true instance of m > n. III. The relation between "=" and ">", *viz.*, every true instance of $t > u \vee t = u \vee u > t$ for all terms of the personal language. IV. Simple addition, *viz.*, every true instance of $m + n = k$ up to 961. V. Recursive multiplication, *viz.* every true instances of $m \times n = k + (m \times j)$ up to 961, where, if n has $r + 1$ digits, k is the result of writing r zeros to the right of m. VI. All substitution instances in the language of some set of axioms for the propositional calculus.

Evidently in this model we have amply idealized this man's knowledge, but even so, not up to the point of classical personalism. Specifically, we were concerned with this stupid man's betting rates on the 16^4 elements of the form $m \times n > j \times k$, and on the 16^4 elements of the form $m \times n = j \times k$, where m,n,j, and k are five-digit binary numbers. A man can be consistent with the axioms of slightly more realistic personal probability if he assigns a betting rate of 0.4 on each inequality, and 0.2 on each equality, except, (to give him minimum good sense) if m and n are the same as j and k, when he plumps for equality, and if $m > j$ while $n > k$ (and the like), when he bets solidly on the appropriate inequality. Although he cannot assign arbitrary odds to remaining elements of his language, there is a wide range of assignments that leaves him consistent with the axioms of slightly more realistic personal probability. Such a man is stupid, but speaking personally not much stupider than I. Personally, I would give lower odds for equality, larger for the inequalities, but otherwise my behaviour would not differ much. I know hardly any binary arithmetic.

9. *A hierarchy.* According to how we construe "possible" in the axioms stated above we get a lattice of theories which includes the points in this list.

1. Realistic personalism. Possible = personally possible = not known to be false.
2. The theory of Vickers' (1965): possible = not proven to be inconsistent with given facts.
3. Hacking's theory: possible = possible (as analysed in Hacking, 1967).
4. An algorithmic theory: possible = not provably inconsistent with the given facts according to any available algorithm.
5. Classical personalism: possible = logically possible.
6. God's theory: possible = not known by God to be false = true.

Note that theory 4 would avoid the second and third versions of Savage's difficulty – the cases where an algorithm is unknown or is impossible. But 4 remains open to Savage's objection, as do 2 and 3.

There are many more ways to fill in this epistemological list. People who are concerned with Savage's problem, and annoyed by my harsh examiner's sense of knowledge, will want to find more plausible points between 1 and 3. I hope they succeed. I must first show that even 1 evades some criticisms that might arise from devotion to 5. If you have a better theory than 1, which falls short of 5, there is every reason to expect that it too will avoid these criticisms. I have three criticisms in mind: the objection that anything short of 5 is too weak for personalism, a Dutch book objection, and the objection that anything less than 5 permits logical sloth. Each objection is unsound.

10. Is slightly more realistic personal probability mathematically weak? Not for the purposes for which Savage recommends classical personalism. His theory is for policing one's own potential decisions and degrees of confidence. Might not the weaker theory be less good at detecting blunders? No. In the course of his personal police work a person proves theorems from the classical axioms and adapts his degree of belief accordingly. But any correction deemed necessary by the classical personalist will be available to the realist. Suppose the classicist who knows f settles on a coherent betting rate of p on h because he works out that, for him, in consistency, $Prob_f(h) = p$. Then the realist knowing f' (f plus the known logical truths which the classicist never bothers to mention) will settle on p as well, proving that in consistency $Prob_{f'}(h) = p$.

In detail take the first time the classicist reasons, "I know f, which includes e. I prove e logically implies h. By axiom (1), $Prob_f(h) = 1$; for me, in possession of f and no more, the betting rate on h is 1." The realistic *alter ego*, who includes logic among his store of facts, begins with some facts; like the classicist he proves $e \supset h$ and infers h from his known e. By now he possesses f', namely f plus some logic and logical consequences of f', and concludes, by the realistic axiom (1), that $Prob_{f'}(h) = 1$. His metatheory differs from that of the classicist but he ends up with the same betting rates. Similarly for uses of axiom (2). Note that we are using a degenerate case of the dynamic assumption; the realist's reasoning can be represented as an application of Bayes' theorem. As an exercise one can apply this story to the model of the binary bettor when he takes the trouble to work out some binary products.

11. What about the Dutch Book argument? It is said that necessary and sufficient conditions for a set of betting rates to escape a Dutch book is that they satisfy the classical axioms. We remarked earlier how this theorem fails for a more realistic betting rate interpretation. But even more skepticism needs to be expressed. To quote de Finetti's original formulation:

> Once an individual has evaluated the probabilities of certain events, two cases can present themselves: either it is possible to bet with him in such a way as to be assured of gaining, or else this possibility does not exist. In the first case one clearly should say that the evaluation of the probabilities given by this individual contains an incoherence, an intrinsic contradiction. (1964, p. 103)

Taken literally, the words are not quite right. For in order to bet with a person so as to be assured of winning, all that is required is that I know more than he does. If you bet on the outcome of a coin, but I know it is double-headed while you do not, and you offer odds on both heads and tails, I shall bet against tails and be assured of winning. But you are not incoherent or intrinsically inconsistent; you had the bad luck to bet with a crook.

It will be protested that I quibble: of course de Finetti meant "logically assured." Exactly such an interpretation is guaranteed for example by Shimony's (1955) definitions, although few other writers have been quite as careful as he. But I do not quibble. I urge that de Finetti's actual words are closer to an appropriate definition of coherence than the logician's gloss on them. Obviously I am not incoherent merely if someone knowing more than I can bet with me so as to be assured of winning. But, I contend, a man is incoherent if a person *knowing no more than that man does* is assured of winning.

If this is correct, it follows that a definition of incoherence must be tied to a definition of knowledge. Since no precise sense of knowledge is stronger than the examiner's sense, we want the following theorem. Suppose X knows no more (in the examiner's sense) than Y; then if Y's betting rates satisfy the slightly more realistic axioms, X cannot bet with Y in such a way that X knows (in the examiner's sense) that he will win from Y. This theorem holds.

Every stronger sense of "knowledge" will determine both a stronger definition of incoherence and a stronger set of probability axioms; thus whatever analysis you give to knowledge which takes you up the list from theory 1, you will discover a corresponding Dutch book theorem. The theorem does not discriminate among points on an epistemological list.

12. What about logic?   We can surely insist that we do some logic: does not the slightly more realistic theory excuse a man from any cogent reasoning whatsoever? No. In the classical theory, the Dutch book argument is used to club a man into reasoning. There may be a better club to hand.

Notice that even for classical personalism, we need more than the Dutch book argument to make a man open his eyes and collect the information around him. Since realistic personalism makes no distinction between finding out logical and empirical facts, we will require the same reason for harvesting logical information as for collecting empirical information. There are not two distinct kinds of decision, shall I do logic, and, shall I experiment. The question is, shall I find out what I can? The question is answered by a single maxim already accepted by personalists. I. J. Good (1967) calls it the principle of rationality. It says one should act so as to maximize expected subjective utility. Good shows that if information is essentially free, acts based on more information cannot have less but can have more expected subjective utility. This is, incidentally, the first formal reason in the literature for Carnap's requirement of total evidence (1947, p. 211), although the idea is anticipated at several places in Savage's (1954, e.g., p. 114, ex. 15). It follows that slightly more realistic axioms for personal probability, plus Good's principle, give a reason for getting facts. One is stupid if one declines to reason, not on account of the realistic version of the Dutch book argument, but because one is cutting down on expected utility. But the very judge which calls you stupid here does not call you stupid if you choose not to find out everything. It does not call you irrational if you fail to find out all the logical consequences of what you know. If the cost of information exceeds the gain in expected utility, you should decline the information.

13. How to allow for the cost of thinking. Good's theorem shows why to think when thinking is free, but thinking takes time and time is money. How should our model of the binary bettor allow for the cost of thinking? To answer we must import costs and prizes into the model. For each pair $m \times n$ and $j \times k$ in question, let our man be offered \$4 if he rightly calls them equal, \$2 if he rightly calls the first greater, and \$2 if he rightly calls the second greater. Recall that his personal odds where .4 on each inequality and .2 on equality, except for a few cases where I supposed that he found that right answer evident. The three simple strategies – bet on "equal" or "$m \times n$ greater" or "$j \times k$ greater" – each have subjective expectation of 80c. But our man may also undertake a calculating strategy: calculate which product is greater and bet accordingly. How does a calculation cost? Every calculation is a sequence of detachments, at least as we have constructed our model bettor. Now applying *modus ponens* is not simply a matter of detaching q from p and $p \supset q$; in the course of a significant calculation you must select the right p's and q's and that is not so easy. Indeed for me it is so time consuming that I personally set a price of 25c on every appropriate application of *modus ponens* needed by the binary computer. Now let u_n be the number of occurrences of the digit one in n, and u_k the same for k. As we have set up the model of our bettor, then, assuming he has efficient axioms for equality, he requires $u_n - 1$ detachments to evaluate $m \times n$, and $u_k - 1$ for $j \times k$; once he has evaluated each product, he requires two more detachments to be able to infer their relative magnitude. Thus it requires $u_n + u_k$ detachments in all.

The subjective expectation of any simple strategy, or mixture thereof, is 80c. The expected gross profit of the calculating strategy is $2.40. Hence it is sensible to calculate when the cost of doing so is less than $1.60; that is to say, when there are six or fewer occurrences of *one* in n and k together. When there are seven, it is better not to calculate.

14. The cost of police work. We called personalism a metatheory whose objects are beliefs and potential decisions. Our last calculation allowed for the cost of object level thinking. None of the costed detachments involved probability theory. We who look down on the binary bettor can say what his best strategy is. But personalism is for policing one's own decisions. Policing the bettor is not the same as the bettor policing himself. For among his costs will be what, in this special case, is the high cost of police work. He has to think harder to discover his best strategy than he does to work out binary products. He has to allow for the high cost of metatheoretic thinking?

This is not Savage's question. His example concerned π. The cost of working out π can be analysed as in my simple model. But in real life there is a curious problem. The decision maker is faced with an initial meta-option. Should he invest in finding out his best object strategy, or would it be cheaper to gamble blindly? This question induces a formal regress. So it may be as Savage feared: accounting for the cost of thinking leads to "paradox," if regress be accounted paradox.

I do not find the regress paradoxical. You can allow for the cost of as much thinking as you like, up a long string of meta-metas. But you have to disregard the cost of thinking out the ultimate meta-decision. It is true that in our model of the binary bettor, first level meta-thinking costs more than object level thinking, and only a fool would disregard it. It is quite otherwise for the computer programmer who arranges a Monte Carlo solution rather than an exact one. His meta-thinking may take ten minutes of pencil time, while the object level thinking may take hours of computer time. It makes good sense to forget about the pencil time. All practical Bayesian business decision has to round off estimates of costs to one or two per cent. The cost of meta-thinking gets rounded off.

15. Disclaimer. Slightly more realistic personal probability is intended as a solution to Professor Savage's problem about the remote digit of π. It is not proposed that personalists should change the opening chapters of their books. The classical axioms plus the dynamic assumption provide a highly instructive model of scientific inference, especially of statistical inference. Like all models, this is both idealization and approximation. It is characteristic of the theory of personal probability that even when philosophical scruples invite one to replace the axioms by slightly more realistic assumptions, the entire substance of the theory remains. Now anyone who, like Professor Savage, thinks of personalism as a normative theory, may find this attitude complacent. For he has two difficulties which, though they seem separate, are closely related. "In what sense is this theory normative?" he asks. Later he questions the idea of the theory being "approximately valid." Let me, in closing, question whether there are *any* normative theories. I think there are only descriptive models of reasonable behaviour. If any were normative, I do not see how they could be approximately valid. But I believe there are not any. There are models of reasonable behaviour, and all models only approximate the truth. For all its defects, personalism is a good proxy.

8. Risk aversion as a problem of conjoint measurement

Bengt Hansson

1. A dialogue

Gambler to decision analyst: I would sure like to have a copy of that new book "Speculator's Almanack" by Lit L. Nohn. It received such a favourable review in *Portugaliae Oeconomicae*.

Decision analyst: You are a lucky man – I incidentally happen to have a few copies with me. In fact, I think I will offer you the following gamble: you have the option of either having one copy for sure or an even chance of getting either no copy or two copies. Which option do you choose?

G: What a silly question! Of course I choose the one copy for sure!

DA: I see. What about if I change the second option to an even chance of getting zero or three copies?

G: Same answer, of course. I can't understand why you ask! I have no more use for a third copy than for a second one.

DA: So you are pretty risk averse then. The expected value of the last option is 1½ copies, as you know, but yet you choose something with an expected value of only one copy. This must be accounted to your aversion to taking risks.

G: Come on, don't be silly! You know me, you know that I am a gambler and that I have a temperament for taking risks. But that has nothing to do with *this* gamble. It is simply that a second or a third copy of a book is practically useless. Of course you can burn it in the fire to save fuel, or you can steady a rickety table with it, or you can give it to a friend, so it has *some* value, but very little as compared with the first copy. And that's why I choose as I do, and it has nothing to do with my attitude towards risk. It would be a dumb thing to endanger something good (a first copy) in order to get something useless (an extra copy), no matter what your gambling temperament is.

DA: Eh, maybe, but try to look at it this way: we can plot your utility for books, so that $f(n)$ is the amount of utility you get from having n copies, and it will look like this:

This curve is concave, and that means that your expected utility of a gamble will always be smaller than the utility of the expected number of copies, and, if you try to maximise your utility, you would rather not play. So you *are* risk averse!

In fact, enonomic theory defines a risk aversion coefficient which can be computed once you know the utility curve for a given commodity. Essentially, it is the second derivative over the first.

G: Now what, is my risk aversion a property of my utility curve? Think a little yourself and don't quote so much economic theory! My utility curve is *one* thing. It has to do with how useful different quantities of a commodity are to me and depends on what commodity it is and on my tastes and values and it can, as an abstract concept at least, be determined completely without reference to gambling situations. And my gambling temperament, that's my attitude towards risk, is *another* thing. It doesn't enter the picture until that commodity is somehow made part of a gambling situation. So the two are different – and even independent.

DA: You are arguing on general principles, but in the particular example I have given, you must admit that you did show a risk averse behaviour.

G: No sir! You are looking at cases where I prefer a prospect which is certain, but has lower actuarial value, to one which is risky and has a higher value, and you then try to infer risk aversion from that. That's a mistake, at least in my case. Of course, risk aversion could be a *possible* reason for such a preference, but it is certainly not the only possible reason – lacking relevance of the actuarial value for actual usefulness, i.e., practical uselessness of excess copies is another. What you have

are two sorts of considerations, both of which produce the same preference. All you can observe is their combined effect and there is no
way for you to tell how much it depends on risk aversion and how
much on utility concavity. And in my case it is only the latter.

DA: But those economists . . . ?

G: I don't doubt that they are able to twist words in some way and
come out right at the end, in a purely technical sense. But their notion
of risk aversion is an artificial one. They have forgotten to ask the
important question: aversion to risking what?, and they tacitly assume
that the relevant answer is loss of actuarial value. But that is not
what risk means. Risk, I would say, is the exposure to the possibility of
losing something desirable (or of having something undesirable
imposed upon one). And if the utility curve is very flat in one region,
this means that the possibility of a loss of some amount of the
commodity is not a real risk, because what is lost is not very desirable.
Likewise, there is no risk involved if I am offered an even chance of A
and B if A and B are of equal value to me (though there *is* uncertainty).
And if my utility curve for something, like Nohn's book, is extremely
concave in one region, it means that there is almost no difference in
value for copies after the first one, and I am not exposed to a possible
loss of something very desirable if I go from two copies to only one,
nor to a gain if I go the other way. And that is what was important in
my decision: 50 percent probability of going down from one to no copy
is a 50 percent probability of a real loss, but 50 probability of going up
from one to two copies is not a real gain. This makes the choice so clear
that my gambling temperament never enters the picture.

DA: I think I am beginning to see the point. Your decision *could* have
been a result of risk aversion, for, if we had considered a commodity
for which your utility were linear, say bottles of beer, a strong aversion
to the riskiness of the gamble *per se* (rather than a fear of losing some
amount of beer) could have made you decide in the same way. And
if I observe only your decision, I am in no position to say whether this
or that factor were the cause of it, or maybe a combination. The
economist's "risk aversion coefficient" measures propensity to avoid
actuarial risk. Such a propensity could be a consequence of either a
cautious gambling temperament, or a lack of linear connection between
physical amount and usefulness. But is this important? Maybe propensity to avoid actuarial risk is the interesting thing to measure, rather
than gambling temperament?

G: Only in rather special cases, as when you want to make a purely
descriptive characterisation of a decision pattern, without any explanatory pretensions. But if you want to *explain* a behaviour, i.e., to put it

into relationship with other things that you know independently, then the unanalysed propensity won't do, because it is just a restatement of the decision behaviour that has already been observed. And, above all, when you want to use decision theory as a tool for decision-*making*, i.e., to apply it to real life problems in order to help you see the structure of the problem and to arrive at a rational decision, then decision theory has to be a bridge from actually perceived values to the decision problem. And propensity to avoid actuarial risk is no such actually perceived thing and it cannot function as a starting-point for decision-makers. Rather, it usually is the endpoint of an analysis of decisions already made. Can you see any application of the economic theory of risk aversion to the *making* of a decision, rather than the analysis of one which has already been made?

DA: All right, all right. But what can you do about it? Can you define a numerical measure, can you give an axiomatic treatment that compares in exactness to that of expected utility theory?

G: Maybe not quite, but I think I can make a reasonably good start. The way I'll go about it is like this: First I'll make clear the mathematical basis for expected utility theory and how "risk aversion" is now defined by economists, then discuss the formalistic and realistic ways of interpreting axiomatic expected utility theory and the general problems of conjoint measurement. Then I'll try to find the reasons why people's utility curves are concave, as a matter of decreasing marginal usefulness and not as a technical construction from observed behaviour. Having understood this factor, and knowing the total effect, I ascribe the difference to risk aversion.

DA: I would like to see that done before I give in completely.

G: You will.

2. The mathematical basis for expected utility theory

If we assume that a decision-maker assigns a numerical value to each prize in a lottery with known probabilities, and then chooses among lotteries according to the mathematical expectation of that value, then one can formulate a number of general principles of behaviour for this decision-maker. Conversely, from a suitable set of such principles, one can infer the existence of a numerical value for each prize. Such studies were initiated in Ramsey (1931) and the first full-fledged axiomatisation is that of von Neumann and Morgenstern (1944).

Many minor variations in the set of axioms are possible, but there have to be representatives of each category below. The following set is rather standard, except for the formulation of the sure-thing principle.

First of all, there must be some assumption about *the structure of decision situations*. Usually, they are thought of as lotteries, with a given set of basic prizes and known probabilities of winning these. Then there are second-order lotteries with first-order lotteries as prizes, and so on. It is thus assumed that the set of lotteries is algebraically closed under the operation of "probability mixture".

The decision-maker is also assumed to be able to *order* all lotteries from better to worse in a consistent way (i.e., the ordering has to be a preorder).

Furthermore, ordinary rules for *reducing compound probabilities* are assumed to be respected, e.g., a p-chance of winning a lottery which consists of a q-chance of winning A and a $(1 - q)$-chance of B, and a $(1 - p)$-chance of a lottery, consisting of an r-chance of C and a $(1 - r)$-chance of D is supposed to be indifferent to the lottery consisting of a $p \cdot q$-chance of A, a $p(1 - q)$-chance of B, a $(1 - p)r$-chance of C and a $(1 - p)(1 - r)$-chance of D. This axiom can be viewed as requiring the decision-maker to abstract from the technical arrangement for producing given probability combinations.

A *continuity axiom* could be formulated thus: if A, B and C are three prospects (basic prizes or lotteries) which are preferred in that order, then there exists a real number p, such that the decision-maker is indifferent between a p-chance of A and a $(1 - p)$-chance of C on the one hand, and B for certain on the other.

The *general sure-thing principle* is the final axiom. It says that if L is a lottery with prizes A_1, A_2, $\ldots$, A_n (basic or other lotteries), each having a positive probability, and if L' is exactly like L, except that A_k' is substituted for A_k, where A_k' is preferred to A_k, then L' is preferred to L.

This principle is a frequent source of misunderstandings. In an axiomatic framework it can be replaced by two other axioms: the *restricted sure-thing principle*, which is exactly like the general one, except that it only considers lotteries with basic prizes, and a *principle of context independence*, which says that if two prospects are substitutable for each other in the context of one lottery, they are so in the context of every other lottery too. In particular, if they are substitutable in the null context, i.e. indifferent to each other, then they are substitutable in all contexts.

The mathematical theorem is that if these axioms are satisfied, then there exists a numerical value for each basic prize, such that the given preference ordering coincides with the ordering according to decreasing mathematical expectation of this value. There is no question about the validity or mathematical correctness of this result, though, as we shall see in subsequent sections, there is considerable disagreement about the correct interpretation.

In particular, if the basic prizes are different sums of money, one can construct a *utility curve for money*, showing the connection between a sum of money and its corresponding numerical value, conveniently called its utility. In the same fashion a utility curve for different amounts of any other commodity can be construed.

The misunderstanding that I mentioned above is that the *restricted* sure-thing principle would suffice in a set of axioms. This is not true, and this fact is of some importance, since the restricted principle has a much greater support than the general one in a prescriptive interpretation of utility theory. The difference is roughly the principle of context independence, which can be said to require the decision-maker to disregard any holistic property of the gambling situation, and in particular, not to consider any dispersion measure. Obviously, this is a potentially crucial point for an analysis of the concept of risk aversion.

The distinction between the restricted and the general sure-thing principle is not always made sufficiently clear. The word "prize" usually refers to basic prizes, but could refer to other lotteries too in the case of a higher-order lottery, and a formulation of the principle which uses this word alone is therefore ambiguous. However, the arguments being used for supporting the principle and the use made of it can sometimes help to determine the precise meaning. A case in point is Howard Raiffa (1968), where the principle is presented as a generalisation of an example with basic prizes. (Raiffa actually substitutes an indifferent prize, rather than a better one, but the distinction between a general and a restricted form applies equally to his version.) This fact, together with a consistent use of different symbols for basic prizes and lotteries in the rest of the book, makes it reasonable to say that the principle is presented in a deceptively simplified way; if interpreted restrictedly, it is insufficient for its purpose, but well supported by a relevant argument; if interpreted generally, it will do its job, but lacks relevant justification.

Another case in point is Harsanyi (1977). The formulation of the sure-thing principle given here is almost *verbatim* copied from that paper, except for the clarifying phrase in parentheses. Later, Harsanyi identifies his sure-thing principle with a dominance principle: "if one strategy yields a better outcome than another does under *some* conditions, and never yields a worse outcome under *any* conditions, then always choose the first strategy, in preference over the second". It is thus fair to assume that Harsanyi is thinking of the restricted principle, in which case his argument contains a mathematical mistake, since the point of his paper is to spell out sufficient conditions for the existence of a utility function. What is missing is the principle of context independence, which is precisely the point attacked by John Watkins in his (1977), to which Harsanyi's paper is addressed. This point is closely connected to the concept of risk aversion and that is the reason why I chose to mention Harsanyi (1977) here, and also why I wish to return to the discussion between Watkins and Harsanyi in the next few sections.

3. The formalistic interpretation of expected utility theory

It has frequently been noticed that the above theorem admits of both a descriptive and a prescriptive interpretation. This is not what I mean when I make a distinction between a formalistic and a realistic interpretation of expected utility theory. In fact, I think that each of these interpretations can be further split up in a descriptive or prescriptive direction. Rather, my distinction has to do with the degree to which the interpretation provides a bridge between the formal apparatus and some outside, empirical phenomenon.

I will present two views on this, and in order to help sort out the implications of these views, I will compare them from three aspects in particular: their explanatory force, their prescriptive content and the degree to which they can serve as guides to practical action.

The formalistic view is that the preferences among the lotteries in a given set form a self-contained whole, which does not derive its legitimacy from anything outside, but is only subject to internal rules of coherence and consistency.

Perhaps the most eloquent defender of this viewpoint is John Harsanyi. In his (1977) he stresses that maximisation of expected utility is not a *reason* why someone makes a choice, but rather a construed way to *describe* a given choice:

In other words, a Bayesian need not have any special desire to maximize his expected utility *per se*. Rather, he simply wants to act in accordance with a few very important rationality axioms; and he knows that this fact has the inevitable mathematical *implication* of making his behavior equivalent to expected-utility maximization. As long as he obeys these rationality axioms, he simply cannot help acting *as if* he assigned numerical *utilities*, at least implicitly, to alternative possible outcomes of his behavior, and assigned numerical *probabilities*, at least implicitly, to alternative contingencies that may arise, and *as if* he then tried to maximize his expected utility in terms of these utilities and probabilities chosen by him. . . .

But the point is that the basic claim of Bayesian theory does not lie in the suggestion that we *should* make a conscious effort to maximize our expected utility; rather, it lies in the mathematical theorem telling us that *if* we act in accordance with a few very important rationality axioms then we *shall* inevitably maximize our expected utility.

It is also made clear the subjectively perceived degree of usefulness of a prize or a lottery has no conceptual link to utility in the technical sense:

But *introspective* utility functions need not have any simple relationship of utility functions inferred from people's actual *behavior*.

It is clear that, if interpreted in this way, expected utility theory has little explanatory power. It is presupposed or observed that preferences among lotteries satisfy the axioms; hence there exists a utility function in the technical sense, which is derived from actual choices. Obviously, we would have a circle if we tried to use this utility function to explain why the agent behaves in a certain way. The most we can hope for is that *some* choices suffice for the derivation of the utility function, and that *other* choices can be explained by this function. In this case the utility function only serves as an intermediary, and the real explanation is from one set of observed behaviour to another. It cannot be used to explain why one starts to act in a certain way.

In particular, risk aversion cannot be explained by reference to a utility curve. If the observable sign of risk aversion is a tendency to prefer certain lotteries to others, this is already used in the derivation of the utility function, and the concavity of the utility curve is only a roundabout way of describing the existence of risk aversive behaviour.

On this view, the utility curve is just a definitional reformulation of given facts, i.e. given preferences, and it has no immediate relationship to utility in the sense of degree of usefulness. Therefore, it would be more correct and less inviting to misunderstandings to use a term like *indicator function* instead of utility function. I will do so in the sequel.

Expected utility theory (or Bayesian decision theory) is usually regarded as a normative theory, also from the formalistic point of view. Harsanyi, in the paper mentioned, uses the expression "a normative theory of rationality, such as Bayesian theory". The normative content lies in a proposed definition of rationality: to be rational is to satisfy the axioms, and one should be rational.

By the central theorem, to satisfy the axioms is equivalent to acting as if maximising *some* indicator function. In this sense, the prescriptive content of the theory is rather formalistic too: It only gives conditional recommendations – if you decide this way and that way, then you have to decide this way too – but never says anything substantially or specifically about a particular decision. In indicator function language: the theory is completely indifferent to the choice of indicator function.

Used as criteria of rationality, the axioms have to be defended by intuitive arguments, just like any other proposed definitional criterion. The central theorem is, of course, neither a proof of their "validity", nor of their "appropriateness". There are different opinions about this problem and it will be further discussed below.

It follows from the above that the formalistic interpretation of expected utility theory is of little use as a guide to practical action. It can never help us to get started or to make the first few decisions in a new field. Only when some decisions are already made can it tell us what we have to do in order to be coherent.

4. The realistic interpretation of expected utility theory

The idea of the realistic interpretation is that a set of decisions is not an isolated world of its own, but is depending on something outside and prior to the decisions, namely a set of values and standards. These values and standards shall enter the decision problem as parameters and shall form a reason for the decisions, and thus must not be a derived construct from the decisions themselves. Thus, while the formalistic interpretation concentrated on decision *analysis*, the realistic interpretation views itself as a tool for decision-*making* or as a guide for practical action. In the context of expected utility theory it is natural to regard the utility function as the bridge from prior values and standards to the decision problem. This function is therefore regarded as having a meaningful empirical interpretation, connected with degree of usefulness, and it is not a mere indicator function.

It is my guess that the realistic interpretation is by far the most common one in the literature, especially in texts addressed to business audiences. Yet it is difficult to find very explicit expressions of the realistic view, maybe because it is felt to be too obvious to require elaboration. The general spirit is almost always that one has to find out how much one likes different consequences, or how useful they are, and the result is then considered as a characteristic of the decision-maker or his values, which can be used to infer and explain rational decision. This is clear in, e.g., Richard Jeffrey (1965), where the word "desirability" is used instead of "utility", and in Robert Schlaifer (1969), where the relevant section is called "Quantification of likings for consequences". This is also the historically correct interpretation, since there is no doubt that what Daniel Bernoulli had in mind when he proposed to use utilities instead of monetary values as a solution to the St. Petersburg paradox was a sort of measure of the degree of satisfaction that a given sum of money could give. It is also obvious that his suggestion was intended as an explanation and not as a postfactual inference.

This immediately brings us over to a discussion of the explanatory force of expected utility theory on a realistic interpretation. It is much greater than on the formalistic view. If we accept the possibility of representing pre-existing values by a utility function, it is legitimate to refer to this function as an explanation of why a certain decision was made or a preference expressed. It is meaningful to do so, independently of whether we have any operational procedure for determining numerical values of the utility function.

The prescriptive content of the realistic view is much stronger than that of the formalistic one. It requires us to maximise the expected value of our *true* utility function, not merely of any indicator function. This, of course, implies satisfying the axioms, i.e., observing the same rationality criteria as on the formalistic view, but it amounts to much more than a requirement for a specific sort of internal structure – it is a requirement to make the decisions in accordance with pre-existing values and standards.

It is possible to fail to satisfy this rationality criterion and yet to satisfy the one of the formalistic interpretation. An agent may, e.g., start from the real usefulness of the various outcomes, but choose to aggregate these values in another way than taking the expected value (by including risk considerations, say). When this is done he may end up satisfying the expected utility axioms after all, and his preferences can therefore be described by some indicator function, which, however, can be very different from his true utility.

This also explains why the realistic interpretation is a better guide to practical action. It can help us with the first few difficult steps in a new field. We are not dependent on previous decisions about related lotteries in order to do something, but are completely free to find out one true utility in any way we choose to, which in many cases simply means by introspection.

In this connection I would like to take precautions to avoid a possible misunderstanding: My distinction between an indicator function and a real utility function is *not* the distinction between a revealed (or true) and an introspectively felt utility function. Rather, both of these latter concepts refer to what I call real utility. An indicator function summarises everything about the agent's behaviour – usefulness as well as attitude towards risk. Only if the agent knows his true utilities and makes his decisions according to the expected value of these will it agree numerically with the true utility function, although it will still be conceptually distinct from it. True utility is a measure of degree of usefulness, and like in the case of all psychophysical measurements it is possible for the agent to have introspective sensations which stand in no simple mathematical relationship to the true value. To determine the true value is a difficult problem of measurement theory, but it is a different problem from the one discussed here. Both true utility and introspectively felt utility relate to usefulness and are meaningful concepts even outside an axiomatic framework of the von Neumann-Morgenstern type.

The realistic interpretation says more than the formalistic one. It is therefore more susceptible to falsification, or, in the normative context, violation. To use insincere values or absurd standards is a violation of the realistic view, but not necessarily of the formalistic view. The relevant sort of argument, if one wants to refuse the formalistic interpretation, is therefore to try to cast some doubt on the axioms as reasonable normative principles, e.g. by pointing out absurd consequences or unnecessary restrictions on personal values. But if one wants to refuse the realistic interpretation it is in addition legitimate to refer to perceived degrees of usefulness and intuitively felt shapes of utility curves as well as to reasonable preferences in selected situations.

5. The received theory of risk aversion

A person is said to be risk averse if he prefers to have the expected monetary value of a bet rather than the bet itself. If we consider bets with small variance and if the agent's preferences can be described by a certain utility function as in expected utility theory, then he is risk averse if and only if his utility curve is concave at the mean value of the bet. And the curve is concave if the derivative is decreasing, i.e., if the second derivative is negative. So a negative second derivative is an indicator of the presence of risk aversion. From this, it seems to be a small step to use the magnitude of the second derivative, with reversed sign, as a measure of the degree of risk aversion. However, this proposed measure has the drawback that it is not invariant under affine transformations of the utility function. Since utility is not thought of as having an absolute zero or a natural unit, an affine transformation of a utility function represents the same preferences and attitudes, and should thus leave a risk aversion measure unchanged. Kenneth Arrow, therefore, in his pioneering work (1971) suggested to use the second derivative, with sign reversed, divided by the first as a *risk aversion coefficient*.

This suggestion is supported by some theoretical considerations, put forward in Pratt (1964). If we consider bets having an even chance of winning or losing *d* pounds, a risk averse agent will prefer *status quo* to accepting such a bet, and will even be willing to pay some amount of money to avoid it – a sort of insurance premium against an unwanted riskiness. This amount depends on two things: the agent's attitude towards risk, and the real (or perceived) riskiness of the bet (measured, e.g., by its variance, which is a simple function of *d*). It is usual in many other scientific situations to regard a total effect like this one as a product of a magnitude (which measures the amount of something, like the length of a piece of wire) and a propensity (which measures how much each unit of the magnitude contributes to the total effect, e.g. electrical resistivity of a wire). Another example is weight, volume and density for the total effect, the magnitude, and the propensity. If the variance of a bet is the magnitude of real riskiness, and the risk aversion coefficient the agent's propensity to convert risk into willingness to pay a risk insurance premium, then that risk aversion coefficient could be found by dividing the total risk insurance premium he is willing to pay by the variance of the bet. By using a Taylor series we can do this, and we find that the value approaches the previously defined expression (times a constant) when d approaches zero.

This expression, $-\dfrac{u''(x)}{u'(x)}$, is sometimes referred to as the *absolute* risk aversion coefficient. It measures a propensity to avoid actuarial risks, measured in absolute monetary values. But one could also be interested in *relative* risk, measured, e.g., by the quotient between the variance and the wealth. If we use this as the magnitude and regard risk aversion as a measure of how much each percent of total wealth put at risk contributes to the willingness to pay a risk insurance premium, then our previous argument will give the expression $-x\,\dfrac{u''(x)}{u'(x)}$, which is called the relative risk aversion coefficient.

Risk aversion, on this view, is obviously a property of the utility curve, and its correct interpretation is therefore dependent on how that curve is interpreted. It is impossible to go into details here, but it should be noted that the reason why Arrow became interested in defining a numerical measure of risk aversion (1971) was that he needed it to analyse and explain economic behaviour, especially people's demand for cash balances, and that therefore only a realistic interpretation would do for his purposes.

6. Risk in the expected utility framework

Our friend the gambler claimed, in his discussion with the decision analyst, that utility considerations and risk considerations were two different things and that a particular decision may be based on one or both of these. How does this fit in with our two interpretations of expected utility theory?

On the formalistic interpretation, the indicator function reflects all aspects of the preference structure, no matter on what it is based. If it should so happen, that risk and utility considerations together result in preferences which satisfy the axioms, then there exists an indicator function, not identical with the utility function, which reflects also the risk aspect.

On the realistic interpretation, preferences are determined by expected utility only, thus leaving no room for the risk aspect. Such preferences always satisfy the axioms. But someone who only subscribes to expected utility theory as a preliminary step in decision-making may go on and add risk considerations in his final decision. He may or may not end up satisfying the axioms, i.e., being in accordance with the formalistic interpretation, but in no way would he any longer be an expected utility maximiser in the realistic sense.

So it seems that we are left with the formalistic interpretation if we want to stay within the expected utility framework, but yet want to take risk into account. However, we do not, so far, know whether this position is consistent in itself. In fact, expected utility theory puts very severe restrictions on the forms that risk considerations may take – restrictions so severe that they rule out decision rules which are entirely reasonable from an intuitive point of view. The conclusion to draw from this is that expected utility, on any interpretation, essentially precludes explicit risk considerations.

It would perhaps be tempting to state this result in the form of an impossibility theorem, but I will content myself with elaborating an example, based on an actual discussion between John Watkins and John Harsanyi. The example also highlights the distinction between the formalistic and realistic views of decision theory.

Watkins is considering at what price he would be willing to sell a lottery ticket, given half a chance of £20 and half a chance of nothing:

I have a cautious gambling temperament and would pay a sizeable premium to be rid of the uncertainty. (I would actually settle for about £7.) Of course, a Bayesian could say that this only means that my marginal utility of money diminishes rather rapidly in this region; but I can promise him that it is actually pretty constant over the £0 to £20 region. (Watkins, 1977)

To this Harsanyi replies:

What is pretty constant? Is it the marginal utility of Watkins's vNM utility function? Surely not. Watkins seems to forget that this vNM utility function is actually *defined* in terms of his choices among gambles, so that his unwillingness to pay more than £7 is in fact quite conclusive evidence to show that his vNM marginal utility *is* rapidly diminishing in the relevant region – provided that Watkins's behaviour conforms to the Bayesian rationality axioms at all. (Harsanyi, 1977)

Here, both authors agree that it is possible for someone to show the preference that Watkins describes, but they disagree about how this preference is related to what they call Watkins's utility function. Harsanyi regards the expressed preference as evidence for a certain indicator function, whereas Watkins obviously thinks of his utility as something pre-existing and independent of this particular preference and admits no role for it as a determining factor of his decision, which he ascribes wholly to an aversion to risks *per se* (or to his gambling temperament, as he prefers to call it).

So far, we have two rather clear statements of the formalistic and realistic standpoints respectively, and there is little to be gained from repeating *a priori* arguments for and against. Rather, let us explore to what extent this particular decision behaviour, with its obvious risk aversion, is consistent with expected utility theory, i.e., whether there is some indicator function with which it conforms.

Of course, since there are parameters to be estimated in the expected utility model, nothing can be directly inferred from one particular decision, so we have to construe it as an instance of some more general rule. There are many dimensions along which we can vary the given example: the level of the prizes, the spread between the two prizes, the probability distribution, the number of possible outcomes, and so on. As a first attempt I will keep all the factors constant except for the first one, and I will propose to see Watkins's decision as an instance of the following rule:

R1   Whenever faced with a lottery with a 50–50 chance of two sums of money, one of which is £20 larger than the other, be prepared to sell the ticket for the lesser prize plus £7.

Is rule R1 consistent with expected utility theory? An indicator function which produces the same preferences has to satisfy the following equality for all x:

$$u(x + 7) = \tfrac{1}{2}[u(x) + u(x + 20)]$$

To simplify matters a little, I will change the amount of seven pounds to six and two thirds and to use this number as a monetary unit. The functional equation above then becomes:

$$u(x + 1) = \tfrac{1}{2}[u(x) + u(x + 3)] \tag{1}$$

One solution to this equation is

$$u(x) = 1 - a^x \quad \text{for} \quad a = \tfrac{1}{2}(\sqrt{5} - 1) \approx 0.618 \tag{2}$$

This solution is not unique, not even up to affine transformations, but I have chosen the most natural monotonous solution with $u(0) = 0$. The risk aversion coefficient is $-\log a$, which is a positive constant.

It is easy to verify that this indicator function produces the same results as the rule R1 whenever the latter is applicable. But the indicator function has a much wider scope. In fact, this embedding of R1 into expected utility theory, formalistically interpreted, produces a whole series of repercussions concerning Watkins's behaviour. Let us see if they can reasonably be accepted as normative consequences of R1.

First, we note that $u(x)$ has an upper limit – any monotonous solution to (1) has that. In fact, to accommodate the relatively large risk insurance premium for low values of x, the function has to show so sharp a decrease in marginal utility in this region that it cannot make up for it later. If we specifically look at the gain from one additional unit we find that it is

$$u(x + 1) - u(x) = a^x(1 - a) \approx 0.382\, a^x \qquad (3)$$

and the upper limit for possible further gains is

$$u(\infty) - u(x) = a^x \qquad (4)$$

The quotient between these two gains is a constant, and this means that Watkins, no matter his present wealth, should be compelled to prefer the certainty of one more unit (that is, six and two thirds of a pound) to a chance of *any* sum, however large, if that chance is less than 0.382. This is patently absurd and a strong argument against the forcing of rule *R1* into the expected utility framework.

Rule *R1* only said something about lotteries with a certain spread between the prizes. Another possibility would be to construe a rule for cases where the lesser prizes is always the *status quo* and the larger prize varies. One such rule is

R2   Whenever faced with a lottery with a 50–50 chance of nothing or a certain sum, be prepared to sell the ticket for one third of that sum.

An indicator function for this case would have to satisfy

$$u(x_0 - x) = \tfrac{1}{2}[u(x_0) + u(x_0 + 3x)] \qquad (5)$$

Again, there is no unique solution, but essentially only one which is monotonous and has $u(x_0) = 0$, namely

$$u(x) = (x - x_0)^b \quad \text{for} \quad b = \log 2/\log 3 \approx 0.631 \qquad (6)$$

The risk aversion coefficient is $(1 - b)/(x - x_0)$, which is positive and decreasing for x greater than x_0.

As before, formula (6) applies to many more cases than rule *R2*, although it agrees with it within its field of applicability, i.e., for fixed wealth x_0 (and the lesser prize being zero). But outside this range, (6) forces the agent to have preferences which are at strong variance with the intuitive rule. This is reflected in the fact that the derivative of (6) has a singularity at x_0, approaching infinity in its vicinity. Intuitively, this is needed to account for the strong risk aversion, but entails that the pattern of behaviour would be radically different for different initial wealths, even those close to x_0.

To illustrate this, recall that Watkins's willingness to sell an equal chance of nothing or three units for only one unit was a sign of strong risk aversion. A risk neutral person would not sell the ticket for less than 1.5 units. Now suppose that Watkins's initial wealth increases from x_0 to $x_0 + 3$ (i.e. by a mere £20). Then formula (6) requires him not to sell his ticket for less than 1.41 units. The small increase in initial wealth would thus turn him into near risk neutrality. This is in no way implicit in the rule *R2*, but a strange effect of the expected utility framework.

The logical structure of this example is the following: a given decision is construed as an instance of each of two more general rules, which each reflect entirely possible (and even sensible) values, and which furthermore are consistent with one another, since they deal with different "dimensions" of the decision situation. To each of these rules is then added the assumption that the agent's behaviour conforms to expected utility theory, i.e. that it can be described by an indicator function. For each of the rules individually, this assumption leads to indicator functions, which entail strongly counterintuitive behaviour for the agent in situations different from those covered by the rule. This alone casts strong doubts on the theory of expected utility. But if we in addition consider the case where the agent embraces both *R1* and *R2*, as is entirely possible, we arrive at a formal contradiction, for it is easily seen that the two functional equations have no common solution. Risk aspects are thus effectively blocked by expected utility theory under both interpretations.

7. Conjoint measurement

The general problem of conjoint measurement is to separate the effects of several factors, when only their combined effect is known or knowable. Our example in section 5 about electrical resistance of a wire is a case in point: Suppose that a number of wires of various sorts are available and that their resistances are measured. We ask if it is possible to refer these numbers back to something that is more fundamental. One possibility is to assume that resistance is proportional to length, *ceteris paribus*, and to try to turn resistance into a function of the factor length and one other factor, characterising the type of wire. Also, the assumption that willingness to pay a risk insurance premium can be seen as a product of the riskiness of the bet and a propensity to avoid risks is another example.

Obviously, there is no objective rule for splitting up a phenomenon into independent factors. Instead of regarding resistance as a result of a combination of the type of wire and length of a piece, we could have split it up into temperature and physical object, or into "length over cross-sectional area" and material, without being "wrong". Which of several possible ways one chooses depends on many more or less vague considerations, like the intuitive understanding one has of the factors, whether there already exist established measures of the separate factors, how complicated the rule for combining the factors turns out to be, and, of course, the purpose of the analysis.

Sometimes, especially in a more technical approach to conjoint measurement, this first step is taken for granted and the interest is concentrated on the mathematical properties of the measurement of the factors and the rule for combining them. But not even this more limited problem admits of a unique objective solution, but has, so to speak, too many degrees of freedom. If one, e.g., transforms the measure of one of the factors, one can compensate for this by transforming the rule of combination. If m is a measure of the first factor, n of the second, and F the rule of combination, so that $F(m, n)$ is the composite measure, a transformation of m and n into $m' = f \circ m$ and $n' = g \circ n$ through invertible functions f and g respectively, can be compensated by a shift to F', defined by $F'(x, y) = F[f^{-1}(x), g^{-1}(y)]$, thus making the triple $\langle m, n, f \rangle$ equivalent to $\langle m', n', F' \rangle$. If we have chosen to regard the thickness of a wire as a factor of its resistance, we can measure it by the cross-sectional area and use a combination rule which has this measure in a divisor position, but we could also use the radius and change the combination rule so that it divides by the square of the measure times the number π. Because of this indeterminateness, the interest has often been focused on whether there exist measures of the factors in the special case where the combination rule is simply one of addition or multiplication.

But in the case of utility of lotteries there is little reason to be interested in, e.g., additive measures. The very point of introducing utility of money was that this magnitude should not necessarily be linear with respect to the nominal amount of money. Instead I propose to look at the condition

$$\text{if } u(A) = u(B), \text{ then } u(A \circ C) = u(B \circ C) \text{ for any } C$$

instead, where the operation "$A \circ B$" is to be read "taking part in both A and B". There seem to be strong reasons for accepting this substitutivity condition for almost any type of composition rule, even though it is seldom mentioned in foundational studies of measurement. The reason for this is perhaps that it is automatically satisfied by additive measures, but even for non-additive measures it makes sense to say that if the measure "u" measures a real property, and if the composition rule "$\circ$" gives a structure with respect to that property, then what matters for $u(A \circ B)$ should be only the amount of the property in A and B and not other features of these lotteries.

The composition rule of "taking part in both" is of great importance in almost all applications of decision theory. All our every-day choices involve risk and uncertainty and we are always involved in several different types of decisions at the same time. We must always take this into consideration and make our choices depend not only on the possible outcomes of the current problem, but also on how they combine with the prospects of other future or simultaneous situations.

One's knowledge and expectations of the future are important ingredients of the background material for any decision. If a decision-maker, having taken all this into account, gives the same value to $u(A)$ and $u(B)$ and soon after finds that the decision problem C is also coming up, what then if he considers $u(A \circ C)$ to be greater than $u(B \circ C)$? I would say that there were something valuable in A which was not properly reflected by the utility function "u", some potential value which were brought into the open only by the presence of C, indicating that 'u' did not contain all the relevant information. There is nothing mystical about such potential values – they can easily be represented in a multi-dimensional theory of value, where the utility is not a single number, but a suitable n-tuple. As we shall see, the potential value may well be associated with the riskiness of the lottery.

There is only one way to deny the importance of the composition rule "playing together", and that is to say that no such rule will ever be applicable, because decision theory deals only with *complete* descriptions of the state of the world. This is indeed often claimed by theoretically-minded writers on decision theory, but a look at the examples and applications in any text-book that at all uses utility theory shows that this is not the situation for which a theory is needed.

Of course, one has to admit the consistency of the objection, but, as always, the zeal to safe-guard the theory restricts its field of applicability, so that it no longer applies to the problems it was designed to solve. If we accept the objection, then all applications and empirical tests will lack any meaning whatsoever. A practicable theory must make it possible for a gambler to decide whether to take part in a lottery without simultaneously deciding whether to take part in all possible future lotteries that may come up. And a theory must be able to help a business man to decide how much he shall bid for a job without requiring him to decide on a complete strategy for the future of his firm. The dilemma facing decision theory is thus that it can be either theoretically safe and completely inapplicable, or else open to all the difficulties connected with the composition rule 'playing together' hinted at above and deployed below.

In order to see that expected utility theory does lead to violations of the condition spelled out above, let us take a simple numerical example. I propose to use the logarithmic function for utility since it is both the historically earliest example and the one best supported by theoretical evidence. In order to have $u(0) = 0$ I will use the function $u(x) = \log(x + 100) - 2$. Nothing much depends on this particular choice – any function with decreasing risk aversion coefficient would do (the coefficient for the logarithm function is $1/x$).

Now consider a decision-maker with this utility function, who is offered a choice between three lotteries: A is getting 1500 with probability 0.25 and nothing else; B is getting 300 with probability 0.5 and nothing else; and C is the degenerate lottery of getting 100 for sure. All three lotteries have the same expected utility, viz. 0.301. Therefore, according to the expected utility model, the decision-maker is indifferent between them.

Suppose now that he is given the opportunity to take part in two of the lotteries. If they really were equally useful to him it would not matter which two he picked, but if he calculates the expected utilities for the nine possible combinations he will find the following values:

	A	B	C
A	0.545	0.536	0.533
B	0.536	0.512	0.500
C	0.533	0.500	0.477

First of all one should note that the values are not equal, but that the lotteries seem to contain some quality, which cannot be captured by the utility function. Second, this quality seems to have something to do with risk, since if the decision-maker has already decided to take part in a certain lottery, he will find the second lottery more advantageous the riskier it is.

We can expand the example by construing a continuum of lotteries, all with a given expected utility, e.g., 0.301 as above. If we define p_x as 0.301 divided by $u(x)$ and G_x as the lottery of getting x with probability p_x and nothing else, then the G_x's will form such a continuum and their riskiness will increase with x. In particular, the lotteries A, B and C above will be identical to G_{1500}, G_{300} and G_{100}. We can then easily calculate the expected utility of $G_x \circ G_y$ and find that if we keep y constant and let x approach infinity, then this expected value will approach 0.602 (i.e., twice 0.301) from below. This means that the expected utility model entails that we can come as close to additivity of utility as we want if we only chose risky enough elements from a set of what the expected utility model considers equally good lotteries.

How, then, does this argument relate to the numeric measure of risk aversion, mentioned in section 5 above? Recall that the total subjective risk, measured by the risk insurance premium someone is willing to pay, can be split up into two multiplicative components: the riskiness of the lottery (which is an objective property of the lottery and measured by, e.g., its variance), and an individual propensity to avoid risk (which is a property of the individual's utility function). In general we may want to allow this propensity to vary with wealth and it should therefore be regarded as a function of x.

For a risk averse individual, the risk premium is positive, and there-

fore the cash equivalent is smaller than the actuarial value. In particular, this is so for the lotteries G_x. When we choose the appropriate
value for p_x in G_x, we compensate the individual for his risk aversion,
and the size of the compensation, i.e., the risk premium, increases
with x because the variance increases. Because of our choice of p_x we
will have $u(G_x) = u(G_y)$ for any x and y, i.e. all the lotteries G_x will have
the same cash equivalent.

The risk premium is the actuarial value minus the cash equivalent. If
we look at the difference between the lotteries G_x and G_y with x greater
than y, we find that the greater actuarial value of G_x must be balanced
by an equally greater risk premium, since the cash equivalents are
equal. If we now consider the change from G_x and G_y to $G_x \circ A$ and
$G_y \circ A$ we have to take three things into account: the actuarial values
change, but with the same amount so that the difference is still the
same; the variances increase, but also with the same amount so that
the difference remains the same; the risk aversion coefficient is now in
general to be taken at another point at the money axis, a point further
to the right if A has positive expected return. If we now suppose that
$G_x \circ A$ is preferred to $G_y \circ A$, i.e., that the individual prefers riskier
lotteries in combinations, as in the numerical example above, then this
means that $G_x \circ A$ has a greater cash equivalent. Since the difference in
actuarial value is the same, we must conclude that the difference in risk
premiums has decreased.

The difference in risk premiums is roughly the risk aversion coefficient times the difference in variance. The latter has not changed, so
we conclude that the preference of $G_x \circ A$ over $G_y \circ A$ must depend
on a decrease in risk aversion coefficient when the lottery A was
added. We have thus established that the previously noted phenomenon of preference for risk in combinations is associated, not with
the values of the risk aversion coefficient itself, but rather with the
direction in which it changes: an individual prefers risk in combination
with favourable lotteries if his risk aversion coefficient is a decreasing
function of his wealth.

8. The various reasons for utility concavity

I now wish to conclude this paper by applying the idea of conjoint
measurement to the concept of risk aversion. In agreement with the
gambler in the introductory dialogue, I am not satisfied with the
received theory of risk aversion and with its way of conceptualising
risk. This is not to deny the usefulness of the coefficient defined in that
theory – the previous section bears witness of that – but only to
question it as a measure of the concept of risk.

What is observable is that people at times avoid actuarially fair bets.

This can be measured, e.g., by the risk insurance premium they are willing to pay. But it is a combined effect of several factors. One is the aversion to risks *per se*; another is the concavity of the *realistic* utility curve. Now suppose that we can get a grip of this latter factor: then the unknown factor which is the real focus of this investigation – the aversion to risk *per se* – can be seen as the difference between the risk premium and the factor depending on the concavity of the real utility curve. And even if we for the moment cannot find a numerical measure for this latter factor, a conceptual clarification would be an important step towards it. Therefore, the remaining part of this paper will focus on the various reasons for utility concavity.

Among the several possible reasons for the predominant concavity of utility curves, the *saturation effect* is often quoted as the most important one or even as the only one. There is little doubt that most of us feel more intensely about the first unit of something than about a later one. This feeling is more abstract as far as money is concerned (but yet obvious when we reach large enough amounts), but is very clear and easily understandable for specific kinds of goods. Money is more complicated than most other goods because its usefulness depends entirely on its being a simultaneous representative for many other goods. In this way we cannot separate the pure saturation effect from what I will later call the utility differential effect. Therefore, it is better to start with the utility of simple things which are useful because they have a value of their own, rather than a market value.

A first pair of shoes is a necessity, a couple of more pairs something rather important, even further pairs a convenience, but after a while we don't care about having another pair, and for large numbers we may even consider them a nuisance, something which is cumbersome to get rid of. The corresponding type of curve is probably rather representative for the utility of many things that pertain to our persons or to our daily needs – it is very steep at first, levels off pretty soon and stays rather flat for a while before it finally goes down a little.

Exactly how extended the different phases of such a curve are depends, besides on personal temperament and values, on how one specifies the situation. One can measure in different units – cases of beer, cans of beer and centilitres of beer – and one can make more or less strict assumptions about who is to benefit from the good – is it beer for the agent's own use, for his family's use or for the use of his friends and acquaintances? Also, the time dimension and the storability of the good plays a role – do we mean daily consumption of beer, monthly consumption of beer or simply the number of cans one has in the cellar? Numerical examples are therefore rather meaningless, yet the general shape of the curve is typical, like this:

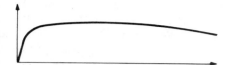

If the need for the first couple of units of the good is not so urgent, the initial steep rise and part of the plateau will probably be replaced by a rather linear increase until the same level is reached.

A particular case is when only the very first unit is useful, practically speaking, like the case of the book in the initial dialogue. This is probably the case for many household goods – stoves, lawn mowers, motor-cars for the single household, etc. *How* often it applies depends on how narrowly one defines each kind of good: my utility for screw-drivers, size 6/200, is of the "one-unit" kind, i.e. levels off after the first unit. My utility for screw-drivers in general levels off after some 5 or 10 units, my utility for tools much later and my utility for hardware articles even later. One can probably arrive at a curve arbitrarily close to the "one-unit" case if one only defines the good narrowly enough.

When it comes to money, there appears another effect, which I propose to call the *utility differential effect*. Money is not really a good in itself, but rather a vehicle for other goods. The utility of money derives from the utility of those goods that can be bought for the money, and they typically differ widely in the return they give.

This gives rise to an effect which does not exist for more basic types of goods. Given a very small budget, I first buy, among all those commodities or services that are available for purchase, those which give the highest utility per pound. The sum of these utilities is the utility of that sum of money which constitutes my budget. If my budget increases, I can buy what I bought before plus a few other things. But the new things give less utility per pound (or I would have bought them in the first case), and the average utility per pound for a budget of x pounds therefore *decreases* as x increases, i.e. the utility curve for money takes on a concave shape because of the utility differential between different types of goods.

Strictly speaking, one has to modify this conclusion a little because some commodities or services can not be divided and can therefore not contribute to the utility of sums less than their price, no matter what ratio of utility to price they can offer, but, apart from this, we here have a general type of effect which does not depend on saturation.

These two effects, or reasons for utility concavity exist quite independently of attitudes to risk *per se*. It is also possible that one can find further such effects. The sum of all such effects is a certain degree of concavity of the utility curve, and that in turn will explain why people prefer the actuarial value of a lottery to the lottery itself, without any recourse to concepts such as "risk aversion". Only when the risk premium they are willing to pay exceeds the difference between the actuarial value and the cash equivalent of the lottery (in the sense of expected utility theory) can one speak about risk aversion.

Part III
Questionable rules of rationality

This part of the volume presents some of the criticisms that have been leveled against the axioms used in deriving the MEU principle. In particular Savage's sure-thing principle will be in focus.

First, Savage presents the Allais paradox (discussed on p. 11 in the introduction to this volume). He says that when this decision problem was first presented to him, he immediately expressed the same pattern of preferences as most people do and thereby he violated the sure-thing principle. But he claims that after thorough deliberation he realizes that one of his preferences was an error which he then wanted to correct, with the result that he no longer violates the sure-thing principle.

But perhaps Savage puts the cart before the horse by using the MEU principle as a *normative* motivation underlying his judgment of error. One is then faced with the question of *why* Savage's axioms, and in particular the sure-thing principle, are normatively valid. McClennen investigates four different lines of argument that have been used to defend the normative validity of the sure-thing principle. The results of his investigation are all negative and he concludes that "the sure-thing cornerstone to the modern theory of utility and subjective probability is less secure than one would like" (p. 168 in this volume).

Kahneman and Tversky present a critique of the MEU principle as a *descriptive* model of decision making under risk. They present several classes of decision situations in which actual preferences violate the axioms of expected utility theory. Among these is a class of decision problem which is related to the Allais paradox where people systematically underweight outcomes that are merely probable in comparison with outcomes that are obtained with uncertainty and thus they violate the sure-thing principle. In addition people generally neglect outcomes that are shared by all alternatives. This means that if the outcomes are positive, people tend to be *risk averse* and prefer sure gains over a larger gain that is merely probable. On the other hand, if gains are replaced by losses in these decision problems, the preferences are reversed: In the negative domain, people exhibit *risk prone* preferences so that a large loss that is merely probable is considered better than a smaller loss that is certain.

After a review of the empirical material that conflicts with the MEU principle, Kahneman and Tversky present their alternative theory for decision making under risk, called *prospect theory*. The theory consists of two phases: First comes an *editing* phase which is a preliminary analysis of the decision situation often resulting in a simpler representation of the alternatives and the outcomes. Second, the *evaluation* phase uses values related to *gains* and *losses* rather than to final assets, and decision weights rather than probabilities. The value function, which thus replaces the utility function in the MEU principle, is generally concave for gains and convex for losses, leading to the preference patterns presented above. The decision weights that replace the probabilities in the MEU principle, are generally lower than the corresponding probabilities, except in the range of very low probabilities. By this overweighting of small probabilities prospect theory predicts that people should buy both insurance and lottery tickets.

Here Kahneman and Tversky, having a descriptive intent, deliberately ignore several problems connected with the requirement of logical coherence for prescriptive theories. Those associated with gains and losses as utility carriers were mentioned in the introduction to Part II of this book. The problems with decision weights have been discussed by Hansson (1975), who discards the idea. Apart from the more or less obvious objection that arbitrary ways of splitting up outcomes may influence the decision unduly, there is a more profound result. The weight functions with a shape that permits them to explain standard psychological facts about people's perceptions of very low and high probabilities (such as Kahneman's and Tversky's) will also give a smaller weight to an expected value of a probability than the expected value of the weight of the probability, thereby implying behavior opposite to that found in Ellsberg-like cases.

There is another problem for their theory, emphasized by a defender of the MEU principle. There are decision situations in which it is possible to make a Dutch book against the decision maker so that he will lose money no matter what happens if he follows the recommendations of prospect theory. Kahneman and Tversky discuss this problem (on pp. 183–214, this volume) and they claim that most of these worrisome situations will be detected and eliminated in the editing phase.

Mark Machina concentrates on the following rationality criterion, which, following Samuelson (1952), has become known as the independence axiom:

Independence axiom: Alternative *a* is weakly preferred (i.e., preferred or indifferent) to alternative *b* if and only if the lottery which consists of alternative *a* with probability p and alternative *c* with probability 1 − p is weakly preferred to the lottery where *b* is obtained with probability p and *c* is obtained with probability 1 − p, for all positive probabilities p and all alternatives *c*.

It is clear that this axiom entails the sure-thing principle. Thus all violations of the sure-thing principle will also be violations of the independence axiom.

Machina gives a presentation of four different types of violations of the independence axiom. Using a simple graphical description for the case when the decision situation involves three states, he presents a generalization of expected utility theory. The graphical technique helps him illustrate different assumptions about the functional structure of individual preferences. He then shows that by replacin χ the independence axiom with a different assumption (which he calls Hypothesis II) about the functional "pattern" of the indifference curves, it is possible to systematically explain and *predict* all the observed violations of the independence axiom. In contrast to the prospect theory of Kahneman and Tversky, the choices recommended by Machina's theory are never subject to Dutch books (see Machina, 1982; this volume pp. 224–225).

However Teddy Seidenfeld (1987) has shown that all decision theories which violate the independence axiom lead to *sequential incoherence* in the sense that there is some two-stage decision, with choices at successive times t_1 and t_2, where the agent is required to reverse a strict preference between two sequential alternatives a_1 and a_2: At t_1 the decision maker prefers a_1 to a_2, but he is also aware that at t_2 he will prefer a_2 to a_1 regardless of what happens at t_1.

9.   Allais' paradox

Leonard J. Savage

Introspection about certain hypothetical decision situations suggests that the sure-thing principle and, with it, the theory of utility are normatively unsatisfactory. Consider an example based on two decision situations each involving two gambles.[1]

> Situation 1. Choose between
>
> Gamble 1. $500,000 with probability 1; and
> Gamble 2. $2,500,000 with probability 0.1,
> $500,000 with probability 0.89,
>   status quo with probability 0.01.
>
> Situation 2. Choose between
>
> Gamble 3. $500,000 with probability 0.11,
>   status quo with probability 0.89; and
> Gamble 4. $2,500,000 with probability 0.1,
>   status quo with probability 0.9.

Many people prefer Gamble 1 to Gamble 2, because, speaking qualitatively, they do not find the chance of winning a *very* large fortune in place of receiving a large fortune outright adequate compensation for even a small risk of being left in the status quo. Many of the same people prefer Gamble 4 to Gamble 3; because, speaking qualitatively, the chance of winning is nearly the same in both gambles, so the one with the much larger prize seems preferable. But the intuitively accept-

---

[1] This particular example is due to Allais (1953). Another interesting example was presented somewhat earlier by Georges Morlat (see C.N.R.S., 1954).

Reprinted from *The Foundations of Statistics*, Dover Publications, Inc., New York, 1972, pp. 101–103, by kind permission of the publisher.

able pair of preferences, Gamble 1 preferred to Gamble 2 and Gamble 4 to Gamble 3, is not compatible with the utility concept or, equivalently, the sure-thing principle. Indeed that pair of preferences implies the following inequalities for any hypothetical utility function.

$$U(\$500,000) < 0.1U(\$2,500,000) + 0.89U(\$500,000) + 0.01U(\$0),$$
$$0.1U(\$2,500,000) + 0.9U(\$0) < 0.11U(\$500,000) + 0.89U(\$0); \tag{1}$$

and these are obviously incompatible.

Examples like the one cited do have a strong intuitive appeal; even if you do not personally feel a tendency to prefer Gamble 1 to Gamble 2 and simultaneously Gamble 4 to Gamble 3, I think that a few trials with other prizes and probabilities will provide you with an example appropriate to yourself.[2]

If, after thorough deliberation, anyone maintains a pair of distinct preferences that are in conflict with the sure-thing principle, he must abandon, or modify, the principle; for that kind of discrepancy seems intolerable in a normative theory. Analogous circumstances forced D. Bernoulli (1738) to abandon the theory of mathematical expectation for that of utility. In general, a person who has tentatively accepted a normative theory must conscientiously study situations in which the theory seems to lead him astray; he must decide for each by reflection – deduction will typically be of little relevance – whether to retain his initial impression of the situation or to accept the implications of the theory for it.

To illustrate, let me record my own reactions to the example with which this heading was introduced. When the two situations were first presented, I immediately expressed preference for Gamble 1 as opposed to Gamble 2 and for Gamble 4 as opposed to Gamble 3, and I still feel an intuitive attraction to those preferences. But I have since accepted the following way of looking at the two situations, which amounts to repeated use of the sure-thing principle.

One way in which Gambles 1–4 could be realized is by a lottery with a hundred numbered tickets and with prizes according to the schedule shown in Table 9.1.

Now, if one of the tickets numbered from 12 through 100 is drawn, it will not matter, in either situation, which gamble I choose. I therefore focus on the possibility that one of the tickets numbered from 1 through 11 will be drawn, in which case Situations 1 and 2 are exactly parallel. The subsidiary decision depends in both situations on whether I would sell an outright gift of $500,000 for a 10-to-1 chance to win $2,500,000 – a conclusion that I think has a claim to universality, or

---

[2] Allais has announced (but not yet published) an empirical investigation of the responses of prudent, educated people to such examples. [See also Allais (1979a, b).]

Table 9.1. *Prizes in units of $100,000 in a lottery realizing, Gambles 1–4*

		Ticket number		
		1	2–11	12–100
Situation 1	Gamble 1	5	5	5
	Gamble 2	0	25	5
Situation 2	Gamble 3	5	5	0
	Gamble 4	0	25	0

objectivity. Finally, consulting my purely personal taste, I find that I would prefer the gift of $500,000 and, accordingly, that I prefer Gamble 1 to Gamble 2 and (contrary to my initial reaction) Gamble 3 to Gamble 4.

It seems to me that in reversing my preference between Gambles 3 and 4 I have corrected an error. There is, of course, an important sense in which preferences, being entirely subjective, cannot be in error; but in a different, more subtle sense they can be. Let me illustrate by a simple example containing no reference to uncertainty. A man buying a car for $2,134.56 is tempted to order it with a radio installed, which will bring the total price to $2,228.41, feeling that the difference is trifling. But, when he reflects that, if he already had the car, he certainly would not spend $93.85 for a radio for it, he realizes that he has made an error.

10. Sure-thing doubts

Edward F. McClennen

If, after thorough deliberation, anyone maintains a pair of distinct preferences that are in conflict with the sure-thing principle, he must abandon, or modify, the principle; for that kind of discrepancy seems intolerable in a normative theory. Analogous circumstances forced D. Bernoulli to abandon the theory of mathematical expectation for that of utility. In general, a person who has tentatively accepted a normative theory must conscientiously study situations in which the theory seems to lead him astray; he must decide for each by reflection – deduction will typically be of little relevance – whether to retain his initial impression of the situation or to accept the implications of the theory for it. (*Savage, 1972, p. 102, p. 64, this volume*)

1. Introduction

Consider the problem of choosing from among the four gambles shown in Figure 10.1.

The author is greatly indebted to T. Seidenfeld, of Washington University, for many long discussions (and many heated debates) on the subject of this paper. He has also benefited greatly from conversations, over the last few years, with I. Levi, of Columbia; F. Schick, of Rutgers; K. Arrow, of Stanford; R. D. Luce, of Harvard; P. Lyon, of Cambridge; M. Machina, of University of California, San Diego; E. Freeman, of The Wharton School; and T. Rader, J. Little, and M. Meyer, all of Washington University.

Reprinted from *Foundations of Utility and Risk Theory with Applications*, ed. by B. P. Stigum and F. Wenstøp, D. Reidel Publishing Company, Dordrecht, 1983, pp. 117–136, by kind permission of the author and the publisher.

	(66) RED	(33) YELLOW	(01) BLACK
A	$2400	2500	0
B	2400	2400	2400
C	0	2500	0
D	0	2400	2400

Figure 10.1

Monetary prizes are based on the color ball drawn from an urn containing balls in the stated proportions. Recent studies show that in cases of this sort a significant number of persons will choose B in preference to A, and C in preference to D, in pairwise choice situations (see Kahneman and Tversky, 1979, Chapter 11, this volume, MacCrimmon and Larsson, 1979, and Schoemaker, 1980). Such a pattern of preference is in violation of the *sure-thing principle*, which, as an axiom of rational choice, is ubiquitous in one form or other to contemporary treatments of utility theory and the theory of subjective probability (see MacCrimmon and Larsson, 1979, and Luce and Suppes, 1965, for surveys of different versions of the axiom). For contexts in which well-defined probabilities are presupposed, one particularly clear way in which this principle can be formulated, following Samuelson, is as the *strong independence axiom*, which requires that: for all X, Y, Z, and $p > 0$,

$$X \succeq Y \text{ iff } [X, p; Z, 1 - p] \succeq [Y, p; Z, 1 - p],$$

where X, Y, Z, are lotteries (gambles), p is a probability, and $[X, p; Z, 1 - p]$ is a compound or multistage gamble in which one has p probability of playing X and $1 - p$ probability of playing Z, etc. (Samuelson, 1966, p. 133).

What is one to make of preference patterns which violate this axiom? The prevailing view is that while such violations may limit somewhat the predictive power of the sure-thing principle, the principle as *normative* for preference remains quite secure.[1] Oddly enough, one finds very little by way of systematic defense of this view. A survey of the literature does suggest, however, at least four distinct lines of argu-

---

[1] The tendency of most of the critics of expected utility theory to focus on the descriptive issue has not helped either to clarify or to resolve the normative issue. M. Allais is a notable exception. He has, from the very outset, been concerned with the theory as normative for choice. It is he who sets in motion what debate there has been. See *Econometrie* (1953) for the proceedings (in French) of the conference he organized in 1952; and Allais and Hagen (1979) for a translation of one of his early papers, for numerous relevant articles by others, and for a recent statement of his position. Hansson (1975) and Dreyfus and Dreyfus (1978) are also very valuable.

	(66)R	(33)Y	(01)B
A+	$2410	2510	10
B	2400	2400	2400
C	0	2500	0
D+	10	2410	2410

Figure 10.2

ment. The *first* adopts the strategy of a *reductio* and seeks to show that preference patterns in violation of the sure-thing principle have implications which even the persons exhibiting such preferences would find unacceptable. The *second* builds on the consideration that a gamble is properly interpreted as a "disjunctive" as distinct from a "conjunctive" bundle of goods, and is not, as such, subject to the kind of complementarity problem which frustrates the use of independence principles in the scaling of consumer preferences for (conjunctive) bundles of commodities. The *third* treats the axiom as a commutivity principle governing sequential choice and requiring the agent to be indifferent between the temporal order in which certain choice and chance events take place. The *fourth* treats the principle as a dominance principle and thereby seeks to ground it in what seems to be a secure intuition, namely, that rational choice is choice which maximizes the preferences of the agent with respect to consequences.

I propose to consider the merits of each of these arguments. My conclusions are negative. The first two arguments must, I believe, be dismissed. The first begs the question and the second involves a *non sequitur*. The third is much more substantial, but I think that when it is carefully examined, the restriction which it places on rational choice is excessive, and by no means as intuitive as its supporters have supposed. The last argument is clearly the most powerful. I shall argue, however, that the reservations expressed in regard to the third argument carry over here. I am forced to conclude, then, that the sure-thing cornerstone to the modern theory of utility and subjective probability is less secure than one would like.

2. The unacceptable implications argument

Suppose that B is chosen in preference to A and C is chosen in preference to D. It seems reasonable to assume that this preference pattern would still obtain if the two inferior options, A and D, were improved marginally, by the addition of a small constant amount, say $10, to each possible pay-off. See Figure 10.2.

	R	Y	B
B/C	[2400/ 0]	[2400/2500]	[2400/ 0]
A+/D+	[2410/ 10]	[2410/2510]	[2410/ 10]

Figure 10.3

If B is chosen in preference to A+, and C is chosen in preference to D+, then it seems reasonable to suppose that in the event one is offered the opportunity to choose between A+ and B, if a fair coin lands heads, and the opportunity to choose between D+ and C, if the same coin lands tails, the contingency plan, [Choose B, if heads; C, if tails] would be preferred to the contingency plan [Choose A+, if heads; D+, if tails]. The notion here is that the contingency plan [B/C] gives one a fifty-fifty chance to play each of the two preferred strategies, while the contingency plan [A+, D+] gives one a fifty-fifty chance to play each of the two dispreferred strategies. For the payoffs for these two strategies, as a function of what color ball is drawn from the urn, see Figure 10.3.

Here [$x/$y] is a lottery in which one has 0.5 probability of getting $x and 0.5 probability of getting $y. But preference for [B/C] over [A+/D+] now seems quite unacceptable, since in terms of final payoffs and compounded probabilities, [A+/D+] gives one the same probabilities ov 'ι each of four possible payoffs as [B/C], but the payoffs for [A+/D+] are in each case superior to the corresponding payoffs in [B/C]. Something, then, has to give. (This is the line of attack taken by Raiffa, 1970, pp. 83–84, and also 1961, pp. 690–694.)

Far from serving as an effective *reductio*, the argument simply begs the question. It goes through smoothly only on the assumption that the agent who chooses B in preference to A+ and C in preference to D+ must rank [B/C] in preference to [A+/D+]. But this assumption, as Samuelson (1952, p. 672) reminds us, simply invokes another version of the strong independence axiom. Those who want to insist on preference for B over A and C over D, and thereby reject the independence axiom, have only to be on their guard and reject the appeal to another version of the same axiom.[2]

---

[2] It would appear that Ellsberg's otherwise powerful attack on Savage's use of the sure-thing principle runs into difficulty precisely because he is unwilling to carry the argument through in this manner. While he is prepared to give up Savage's version of the axiom, he apparently is also inclined to accept the axiom as applied to the case of determinate probabilities – see here Ellsberg (1961, p. 651, footnote 7, p. 252, footnote 6, this volume), in particular, where he seems to concede a great deal to Raiffa, who proceeds in his reply (Raiffa, 1961), to catch him out on this point. It is interesting to note that in an earlier article, Ellsberg (1954) took the position that the strong indepen-

3. Independence and complementarity

Another defense of the independence axiom proceeds by reference to the concept of complementarity. Samuelson (1952, p. 672), for example, argues that while in the domain of non-stochastic goods an independence axiom would be "empirically absurd," in the domain of stochastic goods such an axiom makes sense since the possible outcomes of a lottery or gamble are *disjunctive*; just one of the possible outcomes can occur. This argument is first alluded to, as a matter of fact, by von Neumann and Morgenstern:

> By a combination of two events we mean this: Let the two events be denoted by B and C and use, for the sake of simplicity, the probability 50 percent–50 percent. Then the "combination" is the prospect of seeing B occur with probability 50 percent and (if B does not occur) C with the (remaining) probability of 50 percent. We stress that the two alternatives are mutually exclusive, so no possibility of complementarity and the like exists. (von Neumann & Morgenstern, 1953, pp. 17–18. See also Marschak, 1951, pp. 502–503.)

Although the distinction between stochastic and nonstochastic goods is obviously important, the argument hardly suffices to establish the legitimacy of the independence axiom. As it stands, the intended argument is a *non sequitur*. Since the theory of choice under conditions of uncertainty and risk is concerned with disjunctive bundles of goods, one does not expect to encounter here the sort of complementarity problem which arises in the case of conjunctive bundles of goods. But that serves only to remove one possible objection to the independence axiom; it does not show the axiom to be acceptable as a normative principle.[3]

4. The order of choice and chance

The interpretation of the independence axiom as a commutivity principle was first suggested by Rubin (1949, pp. 1–2). According to him the axiom states that:

---

dence axiom was "indubitably the most plausible" of the axioms of utility theory. Space considerations preclude my exploring these matters at greater length in this paper. Let me simply note for now that Ellsberg focuses on what are essentially a series of counterexamples to one version of the sure-thing principle. My concern, rather, has been with examining the arguments which others have offered in support of various versions of the principle. For purposes of tracing the debate initiated by Ellsberg, the exchange between Fellner and Brewer in Brewer (1963), Brewer and Fellner (1965), Fellner (1961), and Fellner (1963), is particularly useful.

[3] It is very hard to read the results in, e.g., Becker, DeGroot, and Marschak (1963), Coombs (1975), and Ellsberg (1961), without taking them as suggesting that subjects behave as if there were certain complementarity-like effects which arise within the context of stochastic goods.

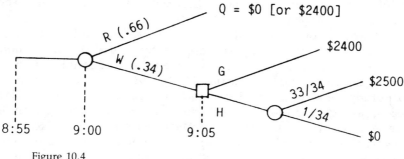

Figure 10.4

... it is immaterial in which order choice or a random event occurs, provided that a decision can be made before the random event occurs which corresponds to an arbitrary decision made afterward.

More recently, Raiffa (1970, p. 82) has made use of this interpretation as part of an analysis of a sequential version of our original decision problem (see Figure 10.4). One is to imagine a story such as this: At 9:00 A.M., a (friendly) experimenter draws a ball from an urn containing red and white balls in proportion 66:34. In the event a red ball is drawn, one gets the prize specified at point Q. In the event a white ball is drawn, one gets a choice to make, which must be executed by 9:05 A.M., between (G) receiving $2400 for certain, or (H) participating in a gamble in which one has 33/34 probability of getting $2500 and 1/34 probability of getting $0. Note that if Q is set at $2400, then H is equivalent to A (as defined in our original problem) and G is equivalent to B; alternatively, if Q is set at $0, then H is equivalent to C and G is equivalent to D.

Raiffa now poses the following questions:

1. If you obtain a white ball, would your choice between alternatives G and H depend on the detailed description of prize Q?
2. If at 8:55 A.M. the experimenter asks you to announce which alternative you will choose if you draw a white ball, will your decision depend on Q? Would it differ from the choice you would actually make at 9:05 when the chips are down?

Raiffa supposes that each of these questions should be answered in the negative. But, by inspection, those who do answer these questions in the negative will either rank A over B and C over D, or B over A and D over C, or, finally, *both* A and B indifferent to one another *and* C and D indifferent to one another: that is, they will never rank *B* over *A* and *C* over *D*.

Raiffa does not say why these questions should all be answered in the negative. What he supposes, apparently, is that the disposition of most persons to so answer them is evidence of a commitment on their part to the commutivity principle. On this view, Raiffa's task reduces to putting persons in mind of what they already (if only implicitly) accept, and getting them to revise their preferences to bring them into line with this commitment. The problem with this line of reasoning, however, is that it works only up to the point where we encounter someone who does not answer all the questions in the negative, and even Raiffa is prepared to admit that there are such persons. What argument can be offered to the holdouts? How are we to convince them that their preferences are irrational?

One argument for a negative answer to the question in (1) above is to be found in the following remark of Arrow's (offered by him as a gloss on what he terms the Principle of Conditional Preference):

> ... what might have happened under conditions that we know won't prevail should have no influence on our choice of actions. Suppose, that is, we are given information that certain states of nature are impossible. We reform our beliefs about the remaining states of nature, and on the basis of these new beliefs we form a new ordering of the actions. The Principle of Conditional Preference ... asserts that the ordering will depend only on the consequences of the actions for those states of the world not ruled out by the information. (Arrow 1972, p. 23)[4]

What about the questions posed in (2)? One is requested to "announce" at 8:55 which alternative one will choose, in the event a white ball is drawn from the urn. The request is ambiguous. It could be taken as a request to *predict* at 8:55 how one will choose at 9:05, i.e., to state in advance what one intends to do at 9:05. Alternatively, it could be a request to indicate what one's choice would be if it had to be made right then, at 8:55, prior to finding out whether a white or a red ball is drawn from the urn. The second part of the question posed in (2) above reinforces the first of these interpretations. The suggestion is that the choice is "actually" to be made at 9:05 (when the "chips are down"). Of course, on this interpretation the second part of the question in (2) begs to be answered in the negative, for it now reads: Would the choice you expect to make at 9:05 differ from the choice you would actually make at 9:05 when the chips are down? This won't do

---

[4] Chernoff (1949) argues in a similar fashion, but inexplicably treats the point, not as defense of a negative answer to what Raiffa formulates in (1), but as defense of a negative answer to the questions posed in (2). In the final, published version of Chernoff's paper (1954), a "correction" is effected by virtue of his simply leaving the comment out altogether.

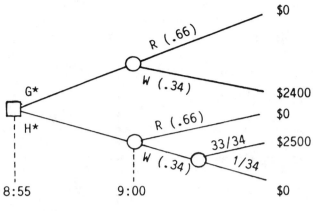

Figure 10.5

at all: commitment to Rubin's Commutivity Principle surely requires more than simply a negative answer to this question!

We must suppose, then, that what is requested in the first part of (2) is whether, in the event one must make the choice between *G* and *H* at 8:55, that choice would depend upon knowing what prize will be awarded at Q. In short, (2) is concerned with *ex ante* choice between *G* and *H*, and not simply prediction of *ex post* choice. But how are we to understand *ex ante* choice between *G* and *H*? Figure 10.4 clearly describes the case of *ex post* choice between *G* and *H*. For the *ex ante* case we need to consider the problem in Figure 10.5. Rubin's commutivity principle, then, requires that if *G* is chosen over *H* in the problem given in Figure 10.4, *G** must be chosen over *H** in the problem given in Figure 10.5.

The ambiguity noted above in the way Raiffa frames his questions, however, leaves us with a pressing problem. How can we be sure any longer that Raiffa's subjects, whom he reports are disposed to answer the questions in the negative, really do have a commitment to Rubin's Commutivity Principle? The problem, as it turns out, is quite pervasive. Markowitz, and also Kahneman and Tversky, have reported that subjects who displayed the preference patterns noted at the outset of this paper revised their rankings when presented with a sequential version of the same problem. But in each case the instructions to the subjects were ambiguous.[5] Machina, who has otherwise presented one of the most interesting reappraisals of the expected utility theory to

---

[5] See Markowitz (1959, pp. 221–222) and Kahneman and Tversky (1979, pp. 271–272, pp. 192–194, this volume). In each case, the intention seems to have been to ask the subjects how they would choose if the choice had to be made *ex ante*, in an irrevocable

appear in a long time, seems to fall into a similar trap. After formulating the independence axiom, he suggests that the following case can be made for its plausibility (adapting his argument to the notation of the present paper):

The argument for the "rationality" of this prescription [that X $\succeq$ Y *iff* (X, p; Z, 1 − p) $\succeq$ (Y, p; Z, 1 − p)] is straightforward: the choice among the latter pair of prospects is equivalent in terms of final probabilities to being presented with a coin which has 1 − p chance of landing tails (in which case you will "win" the lottery Z) and being asked *before the flip* whether you would prefer to win the lottery X or Y in the event of a head. Now, either the coin will land tails, in which case your choice won't have mattered, or else it will land heads, in which case you are in effect back to a decision between X and Y and you should clearly make the same choice as you did before. (Machina, 1981, pp. 166–167. For his critique of expected-utility theory, see Machina, 1982a.)

Machina interprets the problem as one in which the agent has both an *ex ante* and an *ex post* choice to make. If the agent, upon being requested to make a choice now at 8:55, knows that the choice will, in the event a white ball is drawn, be presented to him once again at 9:05, then it would be very odd if he were to make a choice now which was different from the one planned for later. Such a preference reversal would be subject to exploitation by others. If one preferred H^* to G^* at 8:55, but G to H at 9:05, then someone else could set up a parallel choice situation in which one would be permitted to select H^* instead of G^* only on payment of a small fee, and then, in the event a white ball is drawn, permitted to trade back H^* for G. One would then have paid out a fee to end up where one could have been from the very start, by simply opting at 8:55 to stay with G^*. In this respect, however, *ex ante* choice which is subject to reconsideration – what might be termed provisional choice – is something quite different from *ex ante* choice which is not subject to reconsideration, but rather, is irrevocable. It takes a situation in which the agent has the opportunity both

fashion. Yet the accounts which each give of what they said to subjects suggests a double, and conflicting, message. The subjects are first oriented to the problem as one in which choice clearly is to be made *ex post*, and then a redescription is presented; but it is unclear whether the redescription is of a problem in which one is to say *now* what one will choose *later*, or a problem in which one is to say *now* how one will choose *now*. I am not impressed, then, by the documented results that subjects tend to repeat in regard to the question about "*ex ante*" choice what they have already said about *ex post* choice. Note also that the subjects did have the opportunity to register their choice under a description which was unambiguously of *ex ante* choice (i.e., when presented with the original, non-sequential version of the problem), and the evidence is that a significant number chose differently when the problem was presented in that fashion. What I conclude from all this is that there isn't any very clear evidence concerning whether persons do, or do not, answer Raiffa's second question (at least as it was presumably intended to be understood) in the negative.

to choose *ex ante*, and then reconsider that choice *ex post*, for such a preference pattern to be exploited. Rubin's axiom, I suggest, is not about this kind of situation, but about the relation between *irrevocable ex ante* choice and *ex post* choice.[6]

If we fix now on the proper interpretation of the commutivity principle, what argument can be offered in favor of the recommendation that *ex ante* choice coincide with *ex post* choice? Let us refer once again to Figure 10.5, where *ex ante* choice between G and H is represented by choice between G^* and H^*, and to Figure 10.4, where the choice between G and H is *ex post*. The choice which the agent confronts at 8:55, in Figure 10.5, is *not* the same choice which he confronts at 9:05 in Figure 10.4. It is true, of course, that if one focuses on the multistaged aspect of the problem presented in Figure 10.5, one can note that the components which differentiate G^* and H^* at the second stage are what one is to choose between in the problem in Figure 10.4. But, as the modern theory of expected utility itself insists, these structural features are to be treated with the greatest caution. Savage (1972, pp. 102–103, pp. 163–165, this volume) used to insist, when confronted with the fact that even he managed to choose in violation of the independence axiom, that he was grateful to his own theory for setting him straight. The suggestion was that violators of the independence axiom are "taken in" by features of the problem which are irrelevant, from a purely rational point of view. I must confess that once the confusions I have tried to detail above are cleared up, it seems to me an open question as to who has been taken in by features of the problem which are, if not irrelevant, at least not necessarily relevant from a rational point of view. I grant that fixing on the components which differentiate G^* and H^* in Figure 10.5, asking oneself how one would choose between them, if, counterfactually, one were to be able to choose between them, and then using the answer to that question to determine how one will choose between G^* and H^*, may be a permissible way to evaluate the choice between G^* and H^*. I have been unable to isolate an argument, however, as to why one *must* evaluate the options in this manner.[7] I conclude, then, that the case has not been made for the independence axiom as a commutivity principle.

---

[6] For the distinction between irrevocable and reconsiderable *ex ante* choice, and for very clear examples of the troubles one gets into if this distinction is not preserved, I am indebted to Seidenfeld (1981). See also Levi's reply to Seidenfeld in the same volume.

[7] One way to understand the standard decision tree analysis is that it proceeds as if one problem (a problem in which choice is to be made *ex ante*) can be resolved by solving another problem (which differs in that there are choices to be made *ex post*). See, in particular, Schlaifer (1969, pp. 57–58), for a very clear statement of this. The argument of the present paper, simply put, is that this assumption is nowhere subjected to the kind of scrutiny it deserves.

5. The sure-thing or dominance argument

One can discern in the literature a line of argument quite distinct from that just discussed. It proceeds by way of postulating what has come to be known as the "sure-thing" principle, and then showing that the independence condition follows logically from the "sure-thing" principle when the latter is taken in conjunction with certain other allegedly non-controversial assumptions. Friedman and Savage introduce the argument in the following manner:

> [The independence] postulate is implied by a principle that we believe practically unique among maxims for wise action in the face of uncertainty, in the strength of its intuitive appeal. The principle is universally known and recognized.... To illustrate the principle before defining it, suppose a physician now knows that his patient has one of several diseases for each of which the physician would prescribe immediate bed rest. We assert that under this circumstance the physician should and, unless confused, will prescribe immediate bed rest....

Much more abstractly, consider a person constrained to choose between a pair of alternatives a and b, without knowing whether a particular event E does (or will) in fact obtain. Suppose that, depending on his choice and whether E does obtain, he is to receive one of four (not necessarily distinct) gambles, according to the following schedule:

	Event	
Choice	E	not-E
a	$f(a)$	$g(a)$
b	$f(b)$	$g(b)$

The principle in sufficient generality for the present purpose asserts: If the person does not prefer $f(a)$ to $f(b)$, and does not prefer $g(a)$ to $g(b)$, then he will not prefer the choice a to b. (Friedman and Savage, 1952, pp. 468–469. See also Savage, 1951 and 1972.)

If the sure-thing principle can serve to underpin the independence axiom, what can be said in support of the sure-thing principle itself? The argument proceeds along the following lines.[8] The outcome of b is,

---

[8] For discussions which connect the sure-thing principle and the independence axiom, see Friedman and Savage (1952, p. 468), Savage (1972, pp. 22–23, pp. 80–82, this volume), and also Arrow (1972, pp. 52ff). It is interesting to note that the Friedman and Savage example of the physician is of a situation in which choice can presumably be exercised now, or deferred until later (after finding out which disease the patient has). This feature plays no role in the formal characterization of the principle, however. There is a gap, then, between the example and the principle the example is supposed to explicate, a gap which needs to be filled in, and which raises, among other things, just the questions posed in Section 4 above.

by hypothesis, at least as good as the outcome of a, regardless of the turn of events (whether E or not-E takes place). So one does at least as well by choosing b as one does by choosing a, regardless of which of a set of mutually exclusive and exhaustive events takes place. But in that case, surely b is at least as good as a. The force of this sort of line of reasoning is perhaps even clearer in connection with the following version of the sure-thing principle. Suppose that every outcome of b is at least as good as the corresponding outcome of a, and at least one outcome of b is strictly preferred to the corresponding outcome of a. Then surely one must prefer b to a. To choose a over b in such a case would be to deliberately bring about an outcome that will be at best no better, and which may well turn out to be worse, than the outcome that one could have brought about had one chosen b. Such a choice, it can be argued, fails to satisfy the requirement that one choose so as to maximize one's preferences with respect to outcomes.

On first consideration this seems compelling. But it needs to be remarked that the principle as formulated is very strong, for it is framed with respect to outcomes which are defined in terms of an arbitrarily selected partition of events. The principle requires that if there exists any partition of events such that the outcomes of the one act are at least as good as the outcomes of the other act, for each event in the partition, then the first act is at least as good as the second. In particular, this means that the principle is not limited in its application to consequences or outcomes which can be characterized in non-probabilistic terms. Consider the example given above: If the doctor prescribes bed-rest, and the patient turns out to have disease X, then we need not suppose that the outcome is that the patient is cured; we need only suppose that the doctor judges that if the patient has disease X, he or she is more likely to be cured (or the probability of an earlier cure is greater), if bed-rest is prescribed than if it is not. The outcome in this, and the other corresponding cases, is, then, a lottery over various less proximate outcomes. Similar considerations apply to the decision problem given in Figure 10.5. If one chooses H^* and a white ball is drawn, then the outcome is that one confronts a lottery in which one has 33 chances out of 34 of getting \$2500 and 1 chance out of 34 of getting \$0. The principle as framed by Friedman and Savage is designed to apply to such probabilistically defined (and relatively proximate) outcomes. Thus, for example, for subjects who report that they prefer \$2400 outright to receiving the aforementioned lottery, since they are bound to get \$0 in the event a red ball is drawn, regardless of whether they choose H^* or G^*, the sure-thing principle requires a choice of G^* over H^*.

This application of the sure-thing principle to partitions whose outcomes are themselves explicitly defined in probabilistic terms raises a substantial issue. Consider the person who prefers \$2400 outright to the lottery over $2500 and $0, but who chooses H^* over G^*, contrary to the sure-thing principle. While there seems no particular bar to treating the state of affairs which results from choice of H^* and a white ball being drawn from the urn – having 33 out of 34 chances of getting 2500 and 1 chance out of 34 of getting $0 – as an "outcome" of that intersection of a choice and a chance event, it is not at all clear that this "outcome" is relevant to the "sure-thing" evaluation of G^* and H^*. Within the framework of the finer partitioning of this very outcome – which finer partitioning is an *explicit* part of the problem as defined – it is simply not true that one always does at least as well by choosing G^* as by choosing H^*, regardless of which events take place. One possible outcome of H^* is $2500, and this is, by hypothesis, strictly preferred to any outcome of G^*.[9]

I do not mean to suggest that application of the sure-thing principle can be undercut in such cases simply by displaying *some* (other) partition of events such that the outcomes which it defines fail to satisfy the antecedent conditions of that principle. Nor do I mean to suggest that application of the principle can be undercut by the consideration that the partition to which appeal is made *might* itself be refined in such a way – with the specification of even more remote outcomes or consequences – that the principle would not apply to that refinement. The case upon which I have focused is one in which the partition to which the principle does not apply is itself an *explicit refinement* of the partition to which the principle does apply. In such a case, I suggest, the claim that application of the sure-thing principle can be grounded in the notion of the maximization of preferences with respect to consequences is questionable.

The problem here is, in a sense, parallel to the problem which arises with respect to the commutivity principle. Application of the commutivity principle to the choice situation represented in Figure 10.5 seems quite plausible if we suppress an explicitly given feature of that situation, namely, that the only choice to be exercised by the agent is at 8:55, i.e. if we conflate the choice situations in Figures 10.4 and 10.5. Similarly, the application of the sure-thing principle seems plausible if we suppress the explicit refinement of the outcome associated with H^*

---

[9] Savage (1972, p. 99), considers the possibility of restricting the principle to partitions whose outcomes can be characterized in non-probabilistic terms, e.g., cash prizes. By way of rebuttal he argues that "... a cash prize is to a large extent a lottery ticket in that the uncertainty as to what will become of a person if he has a gift of a thousand dollars is not in principle different from the uncertainty about what will become of him if he holds a lottery ticket ...". This is an important point, but space considerations preclude my exploring it at the present time. Suffice it to say that a view to the effect that there is no bedrock level of certainty, that it is risk all the way down, would seem to cast *more*, not less, doubt on the attempt to ground the sure-thing principle in the idea of maximizing preferences for outcomes.

and the event of a white ball being drawn (W) – a refinement specified in terms of outcomes which are evaluatively distinct from one another (getting \$2500, getting \$0) – and thereby treat the outcome of H^* and W as if its internal structure were of no particular relevance.[10]

6. Conclusions

I do not think that the questions raised above suffice to show that the sure-thing or independence axiom must be rejected as normative for rational choice. What they show, I think, is that the arguments that have been put forward in support of the principle are less impressive than many have supposed. Some will want to argue, no doubt, that since I have not managed to show that the axiom must be rejected, and since it has proved so useful and powerful for theory building in the decision and social sciences, I must lose my case. If the issue were one concerning the use of the axiom as part of a descriptive or predictive theory of rational choice, I think the point would be well taken. But this is not the issue to which I have addressed myself in this paper. My concern has been with the axiom as *normative* for rational choice. With regard to the normative issue, I can only plead that we enter here into a matter which poses problems analogous to those which arise with regard to questions of rights and obligations. The issue is one of where the burden of proof is to lie. I believe that it should lie with those who would insist that a certain pattern of preference or choice behavior is irrational. Machina (1981, p. 167) reminds us that the economist Marschak was so taken with the independence or sure-thing principle that he came close to suggesting that it be taught in curricula along with the principles of arithmetic and logic. (For the original remarks of Marschak, see Allais & Hagen, 1979, pp. 168–172, and also Marschak, 1968, p. 49). This is worrisome. We reach in our search for norms of rationality for something which will win the assent of the widest possible group, that will hopefully transcend ideology; but the reaching is still for something normative, by means of which we will judge not only our own choices but the choices of others. This calls for great caution. If the sure-thing or independence axiom has yet to be linked in any decisive fashion to behavior which can be taken as paradigmatically rational then we would be well advised not to judge irrational those patterns which fail to conform to this principle. Such a presumption may not ideally serve the ends of theory construction, but it admirably serves the ideal of tolerance.

---

[10] There is perhaps an even deeper connection between the two arguments. Both derive their plausibility in part, I think, from a presupposition to the effect that there could be no complementarity problem in regard to the components of disjunctive prospects. I hope to address this matter directly in a separate paper.

1986 Postscript to "Sure-Thing Doubts"

Since publishing "Sure-Thing Doubts" I have become even more skeptical of the independence axiom. However, with respect to one of the four arguments examined, I am now inclined to offer a somewhat different criticism. I noted in Section 4 that Raiffa's sequential choice argument for the independence axiom turns on giving a negative answer to two questions, and observed that while I had no great quarrel with such an answer to the first question, I had not been able to isolate an argument as to why the second question must be answered in the negative. I now think that it is Raiffa's negative answer to the first question that must be challenged, and that there is, in fact, a way to defend a negative answer to the second question. All of this I have tried to explore in detail in "Dynamic Choice and Rationality," to appear in B. Munier (ed.), *Risk, Decision and Rationality: Proceedings of the Third International Conference on the Foundations and Applications of Utility, Risk and Decision Theories* (Dordrecht: D. Reidel, 1987), and in *Rationality and Dynamic Choice: Foundational Explorations*, a monograph-length work to be published by Cambridge University Press.

All of this has come about in a manner that puts me even more in debt to both T. Seidenfeld and M. Machina. The former pressed me with counterexamples that were very troublesome so long as I continued to suppose that the weakness in Raiffa's argument concerned his negative answer to the second question. The latter convinced me that one could question Raiffa's assumption that the first question must be answered in the negative.

This revised diagnosis of Raiffa's argument carries with it two advantages. In the second to last paragraph of Section 4, I noted that Machina himself interprets Raiffa's problem as if the agent has an *ex post* as well as an *ex ante* choice, and that this is what lends plausibility to an insistence on the independence axiom as normative for choice. That is, I acknowledged that if the agent faces both an *ex ante* and an *ex post* choice, he can be exploited. I suggested in response that the situation described by Raiffa concerned the relation between how one chooses *ex ante* in a case where that choice is irrevocable (i.e., there is no *ex post* choice) and how one chooses in a case in which one has only an *ex post* choice. In short, exploitation was possible only in a case distinct from the one upon which I supposed it appropriate to focus. What I failed to note at that time, however, but what T. Seidenfeld quickly pressed on me with various examples, was that, given any agent with preferences in violation of the independence axiom, there will be others willing and able to offer him sequential gambles in which he does have both *ex ante* and *ex post* choices, and hence in which he will be subject to exploitation.

But if an agent is prepared to answer Raiffa's first question affirmatively he need not be subject to exploitation. Such an agent can decide at 8:55 to trade G^* for H^*, and then choose to stay with H^* at 9:05. An agent who is capable of this will not be subject to exploitation. On the other hand, it seems plausible to suppose that such an agent will opt for G over H at 9:05, if contra the original example, he were faced with that choice *de novo*, i.e., as a simple choice between those options and not as a choice made against the background of a decision at 8:55 to execute H^*. But it is easy enough to show that an agent who treats choice at a continuation point within the tree differently from *de novo* choice, when the prospects are otherwise the same, will not satisfy Arrow's Conditional Preference Principle. Thus, in order to deal with T. Seidenfeld's counterexamples, I was led to reject my earlier assumption that the Conditional Preference Principle is controlling for preference at tree continuation points. But this new diagnosis has the advantage that it provides an account of how agents who reject the independence axiom can avoid being exploited.

The new diagnosis has an additional advantage, I believe, in that it serves also to strengthen the argument I offered in Section 5 against an appeal to dominance conditions. The proposed rejection of the Conditional Preference Principle carries with it the implication that evaluation of options at a given point within the decision tree may be sensitive to aspects of the decision tree as a whole – and hence to what from that vantage point within the decision tree merely counts as what might have been or could have been realized if other choices earlier on had been made, or if events had taken a different turn. Suppose that, contrary to the sure-thing principle, the agent prefers the combination of f(a) given E and g(a) given not-E to the combination of f(b) given E and g(b) given not-E, but does not prefer f(a) to f(b) and does not prefer g(a) to g(b), when confronted *de novo* with these simple pair-wise choices. Granted those preferences for the *de novo* simple pair-wise choices, it does not follow that, when presented with such pairs of prospects within the context of an unfolding decision tree, the agent must rank them in the same manner.

What an analysis along these lines suggests is that the dominance argument as it has typically been employed presupposes that preference at any decision node must coincide with what the agent would prefer if he were faced with a *de novo* choice. In terms perhaps more familiar to economists, the dominance argument presupposes, in effect, that our preferences at any choice point are separable from the possible prospects to be associated with earlier stages in the same decision tree. Turning this around, then, to suppose that one's evaluations are non-separable – that how we would rank various options at a given point within the decision tree may be a function of the prospects to be associated with the tree as a whole – is to suppose that one is in a setting in which the dominance principle, as usually understood, does not apply. Moreover, all of this appears to provide me with just the clue for which I had been searching, and to which I alluded in footnote 10, as how to understand the connection between Raiffa's sequential choice argument and the traditional dominance argument, for this questionable separability assumption is just what lies behind the sequential choice argument as well.

11. Prospect theory: An analysis of decision under risk

Daniel Kahneman and Amos Tversky

1. Introduction

Expected utility theory has dominated the analysis of decision making under risk. It has been generally accepted as a normative model of rational choice (Keeney and Raiffa, 1976), and widely applied as a descriptive model of economic behavior (e.g., Friedman and Savage, 1948, and Arrow, 1971). Thus, it is assumed that all reasonable people would wish to obey the axioms of the theory (von Neumann & Morgenstern, 1944, and Savage, 1954), and that most people actually do, most of the time.

The present paper describes several classes of choice problems in which preferences systematically violate the axioms of expected utility theory. In the light of these observations we argue that utility theory, as it is commonly interpreted and applied, is not an adequate descriptive model and we propose an alternative account of choice under risk.

2. Critique

Decision making under risk can be viewed as a choice between prospects or gambles. A prospect $(x_1, p_1; \ldots; x_n, p_n)$ is a contract that

This work was supported in part by grants from the Harry F. Guggenheim Foundation and from the Advanced Research Projects Agency of the Department of Defense and was monitored by Office of Naval Research under Contract N00014-78-C-0100 (ARPA Order No. 3469) under Subcontract 78-072-0722 from Decisions and Designs, Inc. to Perceptronics, Inc. We also thank the Center for Advanced Study in the Behavioral Sciences at Stanford for its support.

Reprinted from *Econometrica*, 47 (1979), pp. 263-291, by kind permission of the authors and the publisher.

yields outcome x_i with probability p_i, where $p_1 + p_2 + \cdots + p_n = 1$. To simplify notation, we omit null outcomes and use (x, p) to denote the prospect $(x, p; 0, 1 - p)$ that yields x with probability p and 0 with probability $1 - p$. The (riskless) prospect that yields x with certainty is denoted by (x). The present discussion is restricted to prospects with so-called objective or standard probabilities.

The application of expected utility theory to choices between prospects is based on the following three tenets.

i   Expectation: $U(x_1, p_1; \ldots; x_n, p_n) = p_1 u(x_1) + \cdots + p_n u(x_n)$.

That is, the overall utility of a prospect, denoted by U, is the expected utility of its outcomes.

ii   Asset Integration: $(x_1, p_1; \ldots; x_n, p_n)$ is acceptable at asset position w if $U(w + x_1, p_1; \ldots; w + x_n, p_n) > u(w)$.

That is, a prospect is acceptable if the utility resulting from integrating the prospect with one's assets exceeds the utility of those assets alone. Thus, the domain of the utility function is final states (which include one's asset position) rather than gains or losses.

Although the domain of the utility function is not limited to any particular class of consequences, most applications of the theory have been concerned with monetary outcomes. Furthermore, most economic applications introduce the following additional assumption.

iii   Risk Aversion: u is concave ($u'' < 0$).

A person is risk averse if he prefers the certain prospect (x) to any risky prospect with expected value x. In expected utility theory, risk aversion is equivalent to the concavity of the utility function. The prevalence of risk aversion is perhaps the best known generalization regarding risky choices. It led the early decision theorists of the eighteenth century to propose that utility is a concave function of money, and this idea has been retained in modern treatments (Pratt, 1964, Arrow, 1971).

In the following sections we demonstrate several phenomena which violate these tenets of expected utility theory. The demonstrations are based on the responses of students and university faculty to hypothetical choice problems. The respondents were presented with problems of the type illustrated below.

Which of the following would you prefer?

> A: 50% chance to win 1000,   B: 450 for sure.
> 50% chance to win nothing;

The outcomes refer to Israeli currency. To appreciate the significance of the amounts involved, note that the median net monthly income for a family is about 3000 Israeli pounds. The respondents were asked to imagine that they were actually faced with the choice described in the problem, and to indicate the decision they would have made in such a case. The responses were anonymous, and the instructions specified that there was no 'correct' answer to such problems, and that the aim of the study was to find out how people choose among risky prospects. The problems were presented in questionnaire form, with at most a dozen problems per booklet. Several forms of each questionnaire were constructed so that subjects were exposed to the problems in different orders. In addition, two versions of each problem were used in which the left-right position of the prospects was reversed.

The problems described in this paper are selected illustrations of a series of effects. Every effect has been observed in several problems with different outcomes and probabilities. Some of the problems have also been presented to groups of students and faculty at the University of Stockholm and at the University of Michigan. The pattern of results was essentially identical to the results obtained from Israeli subjects.

The reliance on hypothetical choices raises obvious questions regarding the validity of the method and the generalizability of the results. We are keenly aware of these problems. However, all other methods that have been used to test utility theory also suffer from severe drawbacks. Real choices can be investigated either in the field, by naturalistic or statistical observations of economic behavior, or in the laboratory. Field studies can only provide for rather crude tests of qualitative predictions, because probabilities and utilities cannot be adequately measured in such contexts. Laboratory experiments have been designed to obtain precise measures of utility and probability from actual choices, but these experimental studies typically involve contrived gambles for small stakes, and a large number of repetitions of very similar problems. These features of laboratory gambling compli-cate the interpretation of the results and restrict their generality.

By default, the method of hypothetical choices emerges as the simplest procedure by which a large number of theoretical questions can be investigated. The use of the method relies on the assumption that people often know how they would behave in actual situations of choice, and on the further assumption that the subjects have no special reason to disguise their true preferences. If people are reasonably accurate in predicting their choices, the presence of common and systematic violations of expected utility theory in hypothetical prob-lems provides presumptive evidence against that theory.

Certainty, probability, and possibility

In expected utility theory, the utilities of outcomes are weighted by their probabilities. The present section describes a series of choice problems in which people's preferences systematically violate this principle. We first show that people overweight outcomes that are considered certain, relative to outcomes which are merely probable – a phenomenon which we label the *certainty effect*.

The best known counter-example to expected utility theory which exploits the certainty effect was introduced by the French economist Maurice Allais in 1953. Allais' example has been discussed from both normative and descriptive standpoints by many authors (MacCrimmon and Larsson, 1979, and Slovic and Tversky, 1974). The following pair of choice problems is a variation of Allais' example, which differs from the original in that it refers to moderate rather than to extremely large gains. The number of respondents who answered each problem is denoted by N, and the percentage who choose each option is given in brackets.

Problem 1: Choose between
   A: 2500 with probability   .33,   B: 2400 with certainty.
     2400 with probability   .66,
     0 with probability   .01;
   N = 72   [18]   [82]*

Problem 2: Choose between
   C: 2500 with probability   .33,   D: 2400 with probability .34,
     0 with probability   .67;   0 with probability   .66.
   N = 72   [83]*   [17]

The data show that 82 percent of the subjects chose B in Problem 1, and 83 percent of the subjects chose C in Problem 2. Each of these preferences is significant at the .01 level, as denoted by the asterisk. Moreover, the analysis of individual patterns of choice indicates that a majority of respondents (61 percent) made the modal choice in both problems. This pattern of preferences violates expected utility theory in the manner originally described by Allais. According to that theory, with $u(0) = 0$, the first preference implies

$$u(2400) > .33u(2500) + .66u(2400) \text{ or } .34u(2400) > .33u(2500)$$

while the second preference implies the reverse inequality. Note that Problem 2 is obtained from Problem 1 by eliminating a .66 chance of winning 2400 from both prospects under consideration. Evidently, this change produces a greater reduction in desirability when it alters the character of the prospect from a sure gain to a probable one, than when both the original and the reduced prospects are uncertain.

A simpler demonstration of the same phenomenon, involving only two-outcome gambles is given below. This example is also based on Allais (1953).

Problem 3:

    A: (4000, .80),   or   B: (3000).

    N = 95 [20]   [80]*

Problem 4:

    C: (4000, .20),   or   D: (3000, .25).

    N = 95 [65]*   [35]

In this pair of problems as well as in all other problem-pairs in this section, over half the respondents violated expected utility theory. To show that the modal pattern of preferences in Problems 3 and 4 is not compatible with the theory, set $u(0) = 0$, and recall that the choice of B implies $u(3000)/u(4000) > 4/5$, whereas the choice of C implies the reverse inequality. Note that the prospect $C = (4000, .20)$ can be expressed as $(A, .25)$, while the prospect $D = (3000, .25)$ can be rewritten as $(B, .25)$. The substitution axiom of utility theory asserts that if B is preferred to A, then any (probability) mixture (B, p) must be preferred to the mixture (A, p). Our subjects did not obey this axiom. Apparently, reducing the probability of winning from 1.0 to .25 has a greater effect than the reduction from .8 to .2. The following pair of choice problems illustrates the certainty effect with non-monetary outcomes.

Problem 5:

A: 50% chance to win a three-week tour of England, France, and Italy;

B. A one-week tour of England, with certainty.

N = 72   [22]   [78]*

Problem 6:

C: 5% chance to win a three-week tour of England, France, and Italy;

D: 10% chance to win a one-week tour of England.

N = 72   [67]*   [33]

The certainty effect is not the only type of violation of the substitution axiom. Another situation in which this axiom fails is illustrated by the following problems.

Problem 7:

    A: (6000, .45),   B: (3000, .90).

    N = 66 [14]   [86]*

Problem 8:

    C: (6000, .001),   D: (3000, .002).

    N = 66 [73]*   [27]

Note that in Problem 7 the probabilities of winning are substantial (.90 and .45), and most people choose the prospect where winning is more probable. In Problem 8, there is a *possibility* of winning, although the probabilities of winning are minuscule (.002 and .001) in both prospects. In this situation where winning is possible but not probable, most people choose the prospect that offers the larger gain. Similar results have been reported by MacCrimmon and Larsson (1979).

The above problems illustrate common attitudes toward risk or chance that cannot be captured by the expected utility model. The results suggest the following empirical generalization concerning the manner in which the substitution axiom is violated. If (y, pq) is equivalent to (x, p), then (y, pqr) is preferred to (x, pr), $0 < p, q, r < 1$. This property is incorporated into an alternative theory, developed in the second part of the paper.

The reflection effect

The previous section discussed preferences between positive prospects, i.e., prospects that involve no losses. What happens when the signs of the outcomes are reversed so that gains are replaced by losses? The left-hand column of Table 11.1 displays four of the choice problems that were discussed in the previous section, and the right-hand column displays choice problems in which the signs of the outcomes are reversed. We use $-x$ to denote the loss of x, and $>$ to denote the prevalent preference, i.e., the choice made by the majority of subjects.

In each of the four problems in Table 11.1 the preference between negative prospects is the mirror image of the preference between positive prospects. Thus, the reflection of prospects around 0 reverses the preference order. We label this pattern the *reflection effect*.

Let us turn now to the implications of these data. First, note that the reflection effect implies that risk aversion in the positive domain is accompanied by risk seeking in the negative domain. In Problem 3′, for example, the majority of subjects were willing to accept a risk of .80 to lose 4000, in preference to a sure loss of 3000, although the gamble has a lower expected value. The occurrence of risk seeking in choices between negative prospects was noted early by Markowitz (1952). Williams (1966) reported data where a translation of outcomes produces a dramatic shift from risk aversion to risk seeking. For example, his subjects were indifferent between (100, .65; −100, .35) and (0), indicating risk aversion. They were also indifferent between (−200, .80) and (−100), indicating risk seeking. A recent review by Fishburn and Kochenberger (1979) documents the prevalence of risk seeking in choices between negative prospects.

Table 11.1 *Preferences between Positive and Negative Prospects*

	Positive prospects			Negative prospects		
Problem 3: N = 95	(4000, .80) [20]	<	(3000). [80]*	Problem 3': N = 95	(-4000, .80) [92]*	>	(-3000). [8]
Problem 4: N = 95	(4000, .20) [65]*	>	(3000, .25). [35]	Problem 4': N = 95	(-4000, .20) [42]	<	(-3000, .25) [58]
Problem 7: N = 66	(3000, .90) [86]*	>	(6000, .45). [14]	Problem 7': N = 66	(-3000, .90) [8]	<	(-6000, .45) [92]*
Problem 8: N = 66	(3000, .002) [27]	<	(6000, .001). [73]*	Problem 8': N = 66	(-3000, .002) [70]*	>	(-6000, .001). [30]

Second, recall that the preferences between the positive prospects in Table 11.1 are inconsistent with expected utility theory. The preferences between the corresponding negative prospects also violate the expectation principle in the same manner. For example, Problems 3' and 4', like Problems 3 and 4, demonstrate that outcomes which are obtained with certainty are overweighted relative to uncertain outcomes. In the positive domain, the certainty effect contributes to a risk averse preference for a sure gain over a larger gain that is merely probable. In the negative domain, the same effect leads to a risk seeking preference for a loss that is merely probable over a smaller loss that is certain. The same psychological principle – the overweighting of certainty – favors risk aversion in the domain of gains and risk seeking in the domain of losses.

Third, the reflection effect eliminates aversion for uncertainty or variability as an explanation of the certainty effect. Consider, for example, the prevalent preferences for (3000) over (4000, .80) and for (4000, .20) over (3000, .25). To resolve this apparent inconsistency one could invoke the assumption that people prefer prospects that have high expected value and small variance (see, e.g., Allais, 1953; Markowitz 1959; and Tobin 1958). Since (3000) has no variance while (4000, .80) has large variance, the former prospect could be chosen despite its lower expected value. When the prospects are reduced, however, the difference in variance between (3000, .25) and (4000, .20) may be insufficient to overcome the difference in expected value. Because (−3000) has both higher expected value and lower variance than (−4000, .80), this account entails that the sure loss should be preferred, contrary to the data. Thus, our data are incompatible with the notion that certainty is generally desirable. Rather, it appears that certainty increases the aversiveness of losses as well as the desirability of gains.

Probabilistic insurance

The prevalence of the purchase of insurance against both large and small losses has been regarded by many as strong evidence for the concavity of the utility function for money. Why otherwise would people spend so much money to purchase insurance policies at a price that exceeds the expected actuarial cost? However, an examination of the relative attractiveness of various forms of insurance does not support the notion that the utility function for money is concave everywhere. For example, people often prefer insurance programs that offer limited coverage with low or zero deductible over comparable policies that offer higher maximal coverage with higher deductibles – contrary to risk aversion (see, e.g., Fuchs, 1976). Another type of insurance problem in which people's responses are inconsistent with the concavity hypothesis may be called probabilistic insurance. To illustrate this concept, consider the following problem, which was presented to 95 Stanford University students.

Problem 9: Suppose you consider the possibility of insuring some property against damage, e.g., fire or theft. After examining the risks and the premium you find that you have no clear preference between the options of purchasing insurance or leaving the property uninsured.

It is then called to your attention that the insurance company offers a new program called *probabilistic insurance*. In this program you pay half of the regular premium. In case of damage, there is a 50 percent chance that you pay the other half of the premium and the insurance company covers all the losses; and there is a 50 percent chance that you get back your insurance payment and suffer all the losses. For example, if an accident occurs on an odd day of the month, you pay the other half of the regular premium and your losses are covered; but if the accident occurs on an even day of the month, your insurance payment is refunded and your losses are not covered.

Recall that the premium for full coverage is such that you find this insurance barely worth its cost.

Under these circumstances, would you purchase probabilistic insurance:

	Yes,	No.
N = 95	[20]	[80]*

Although Problem 9 may appear contrived, it is worth noting that probabilistic insurance represents many forms of protective action where one pays a certain cost to reduce the probability of an undesirable event – without eliminating it altogether. The installation of a burglar alarm, the replacement of old tires, and the decision to stop smoking can all be viewed as probabilistic insurance.

The responses to Problem 9 and to several other variants of the same question indicate that probabilistic insurance is generally unattractive. Apparently, reducing the probability of a loss from p to $p/2$ is less valuable than reducing the probability of that loss from $p/2$ to 0.

In contrast to these data, expected utility theory (with a concave u) implies that probabilistic insurance is superior to regular insurance. That is, if at asset position w one is just willing to pay a premium y to insure against a probability p of losing x, then one should definitely be willing to pay a smaller premium ry to reduce the probability of losing x from p to $(1 - r)p$, $0 < r < 1$. Formally, if one is indifferent between $(w - x, p; w, 1 - p)$ and $(w - y)$, then one should prefer probabilistic insurance $(w - x, (1 - r)p; w - y, rp; w - ry, 1 - p)$ over regular insurance $(w - y)$.

To prove this proposition, we show that

$$pu(w - x) + (1 - p)u(w) = u(w - y)$$

implies

$$(1 - r)pu(w - x) + rpu(w - y) + (1 - p)u(w - ry) > u(w - y)$$

Without loss of generality, we can set $u(w - x) = 0$ and $u(w) = 1$. Hence, $u(w - y) = 1 - p$, and we wish to show that

$$rp(1 - p) + (1 - p)u(w - ry) > 1 - p \quad \text{or} \quad u(w - ry) > 1 - rp$$

which holds if and only if u is concave.

This is a rather puzzling consequence of the risk aversion hypothesis of utility theory, because probabilistic insurance appears intuitively riskier than regular insurance, which entirely eliminates the element of risk. Evidently, the intuitive notion of risk is not adequately captured by the assumed concavity of the utility function for wealth.

The aversion for probabilistic insurance is particularly intriguing because all insurance is, in a sense, probabilistic. The most avid buyer of insurance remains vulnerable to many financial and other risks which his policies do not cover. There appears to be a significant difference between probabilistic insurance and what may be called contingent insurance, which provides the certainty of coverage for a specified type of risk. Compare, for example, probabilistic insurance against all forms of loss or damage to the contents of your home and contingent insurance that eliminates all risk of loss from theft, say, but does not cover other risks, e.g., fire. We conjecture that contingent insurance will be generally more attractive than probabilistic insurance when the probabilities of unprotected loss are equated. Thus, two prospects that are equivalent in probabilities and outcomes could have different values depending on their formulation. Several demonstrations of this general phenomenon are described in the next section.

The isolation effect

In order to simplify the choice between alternatives, people often disregard components that the alternatives share, and focus on the components that distinguish them (Tversky, 1972). This approach to choice problems may produce inconsistent preferences, because a pair of prospects can be decomposed into common and distinctive components in more than one way, and different decompositions sometimes lead to different preferences. We refer to this phenomenon as the *isolation effect*.

Problem 10: Consider the following two-stage game. In the first stage, there is a probability of .75 to end the game without winning anything, and a probability of .25 to move into the second stage. If you reach the second stage you have a choice between

(4000, .80)   and   (3000).

Your choice must be made before the game starts, i.e., before the outcome of the first stage is known.

Note that in this game, one has a choice between .25 × .80 = .20 chance to win 4000, and a .25 × 1.0 = .25 chance to win 3000. Thus, in terms of final outcomes and probabilities one faces a choice between (4000, .20) and (3000, .25), as in Problem 4 above. However, the dominant preferences are different in the two problems. Of 141 subjects who answered Problem 10, 78 percent chose the latter prospect, contrary to the modal preference in Problem 4. Evidently, people ignored the first stage of the game, whose outcomes are shared by both prospects, and considered Problem 10 as a choice between (3000) and (4000, .80), as in Problem 3 above.

The standard and the sequential formulations of Problem 4 are represented as decision trees in Figures 11.1 and 11.2, respectively. Following the usual convention, squares denote decision nodes and circles denote chance nodes. The essential difference between the two representations is in the location of the decision node. In the standard form (Figure 11.1), the decision maker faces a choice between two risky prospects, whereas in the sequential form (Figure 11.2) he faces a choice between a risky and a riskless prospect. This is accomplished by introducing a dependency between the prospects without changing either probabilities or outcomes. Specifically, the event "not winning 3000" is included in the event "not winning 4000" in the sequential formulation, while the two events are independent in the standard formulation. Thus, the outcome of winning 3000 has a certainty advantage in the sequential formulation, which it does not have in the standard formulation.

The reversal of preferences due to the dependency among events is particularly significant because it violates the basic supposition of a decision-theoretical analysis, that choices between prospects are determined solely by the probabilities of final states.

It is easy to think of decision problems that are most naturally represented in one of the forms above rather than in the other. For example, the choice between two different risky ventures is likely to be viewed in the standard form. On the other hand, the following problem is most likely to be represented in the sequential form. One may invest money in a venture with some probability of losing one's capital if the venture fails, and with a choice between a fixed agreed return and a percentage of earnings if it succeeds. The isolation effect implies that the contingent certainty of the fixed return enhances the

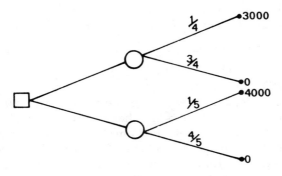

Figure 11.1 The representation of Problem 4 as a decision tree (standard formulation).

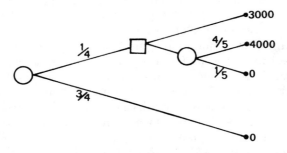

Figure 11.2 The representation of Problem 10 as a decision tree (sequential formulation).

attractiveness of this option, relative to a risky venture with the same probabilities and outcomes.

The preceding problem illustrated how preferences may be altered by different representations of probabilities. We now show how choices may be altered by varying the representation of outcomes.

Consider the following problems, which were presented to two different groups of subjects.

Problem 11: In addition to whatever you own, you have been given 1000. You are now asked to choose between

    A: (1000, .50)   and   B: (500).
    N = 70 [16]   [84]*

Problem 12: In addition to whatever you own, you have been given 2000. You are now asked to choose between

    C: (−1000, .50)   and   D: (−500).
    N = 68 [69]*   [31]

The majority of subjects chose B in the first problem and C in the second. These preferences conform to the reflection effect observed in Table 11.1, which exhibits risk aversion for positive prospects and risk seeking for negative ones. Note, however, that when viewed in terms of final states, the two choice problems are identical. Specifically,

$$A = (2000, .50; 1000, .50) = C, \quad \text{and} \quad B = (1500) = D.$$

In fact, Problem 12 is obtained from Problem 11 by adding 1000 to the initial bonus, and subtracting 1000 from all outcomes. Evidently, the subjects did not integrate the bonus with the prospects. The bonus did not enter into the comparison of prospects because it was common to both options in each problem.

The pattern of results observed in Problems 11 and 12 is clearly inconsistent with utility theory. In that theory, for example, the same utility is assigned to a wealth of $100,000, regardless of whether it was reached from a prior wealth of $95,000 or $105,000. Consequently, the choice between a total wealth of $100,000 and even chances to own $95,000 or $105,000 should be independent of whether one currently owns the smaller or the larger of these two amounts. With the added assumption of risk aversion, the theory entails that the certainty of owning $100,000 should always be preferred to the gamble. However, the responses to Problem 12 and to several of the previous questions suggest that this pattern will be obtained if the individual owns the smaller amount, but not if he owns the larger amount.

The apparent neglect of a bonus that was common to both options in Problems 11 and 12 implies that the carriers of value or utility are changes of wealth, rather than final asset positions that include current wealth. This conclusion is the cornerstone of an alternative theory of risky choice, which is described in the following sections.

3. Theory

The preceding discussion reviewed several empirical effects which appear to invalidate expected utility theory as a descriptive model. The remainder of the paper presents an alternative account of individual decision making under risk, called prospect theory. The theory is developed for simple prospects with monetary outcomes and stated probabilities, but it can be extended to more involved choices. Prospect theory distinguishes two phases in the choice process: an early phase of editing and a subsequent phase of evaluation. The editing phase consists of a preliminary analysis of the offered prospects, which often yields a simpler representation of these prospects. In the second phase, the edited prospects are evaluated and the prospect of highest value is chosen. We next outline the editing phase, and develop a formal model of the evaluation phase.

The function of the editing phase is to organize and reformulate the options so as to simplify subsequent evaluation and choice. Editing consists of the application of several operations that transform the outcomes and probabilities associated with the offered prospects. The major operations of the editing phase are described below.

Coding. The evidence discussed in the previous section shows that people normally perceive outcomes as gains and losses, rather than as final states of wealth or welfare. Gains and losses, of course, are defined relative to some neutral reference point. The reference point usually corresponds to the current asset position, in which case gains and losses coincide with the actual amounts that are received or paid. However, the location of the reference point, and the consequent coding of outcomes as gains or losses, can be affected by the formulation of the offered prospects, and by the expectations of the decision maker.

Combination. Prospects can sometimes be simplified by combining the probabilities associated with identical outcomes. For example, the prospect (200, .25; 200, .25) will be reduced to (200, .50). and evaluated in this form.

Segregation. Some prospects contain a riskless component that is segregated from the risky component in the editing phase. For example, the prospect (300, .80; 200, .20) is naturally decomposed into a sure gain of 200 and the risky prospect (100, .80). Similarly, the prospect (−400, .40; −100, .60) is readily seen to consist of a sure loss of 100 and of the prospect (−300, .40).

The preceding operations are applied to each prospect separately. The following operation is applied to a set of two or more prospects.

Cancellation. The essence of the isolation effects described earlier is the discarding of components that are shared by the offered prospects. Thus, our respondents apparently ignored the first stage of the sequential game presented in Problem 10, because this stage was common to both options, and they evaluated the prospects with respect to the results of the second stage (see Figure 11.2). Similarly, they neglected the common bonus that was added to the prospects in Problems 11 and 12. Another type of cancellation involves the discarding of common constituents, i.e., outcome-probability pairs. For example, the choice between (200, .20; 100, .50; −50, .30) and (200, .20; 150, .50; −100, .30) can be reduced by cancellation to a choice between (100, .50; −50, .30) and (150, .50; −100, .30).

Two additional operations that should be mentioned are simplification and the detection of dominance. The first refers to the simplification of prospects by rounding probabilities or outcomes. For example, the prospect (101, .49) is likely to be recoded as an even chance to win 100. A particularly important form of simplification involves the discarding of extremely unlikely outcomes. The second operation involves the scanning of offered prospects to detect dominated alternatives, which are rejected without further evaluation.

Because the editing operations facilitate the task of decision, it is assumed that they are performed whenever possible. However, some editing operations either permit or prevent the application of others. For example, (500, .20; 101, .49) will appear to dominate (500, .15; 99, .51) if the second constituents of both prospects are simplified to (100, .50). The final edited prospects could, therefore, depend on the sequence of editing operations, which is likely to vary with the structure of the offered set and with the format of the display. A detailed study of this problem is beyond the scope of the present treatment. In this paper we discuss choice problems where it is reasonable to assume either that the original formulation of the prospects leaves no room for further editing, or that the edited prospects can be specified without ambiguity.

Many anomalies of preference result from the editing of prospects. For example, the inconsistencies associated with the isolation effect result from the cancellation of common components. Some intransitivities of choice are explained by a simplification that eliminates small differences between prospects (see Tversky, 1969). More generally, the preference order between prospects need not be invariant across contexts, because the same offered prospect could be edited in different ways depending on the context in which it appears.

Following the editing phase, the decision maker is assumed to evaluate each of the edited prospects, and to choose the prospect of highest value. The overall value of an edited prospect, denoted V, is expressed in terms of two scales, π and v.

The first scale, π, associates with each probability p a decision weight $\pi(p)$, which reflects the impact of p on the over-all value of the prospect. However, π is not a probability measure, and it will be shown later that $\pi(p) + \pi(1 - p)$ is typically less than unity. The second scale, v, assigns to each outcome x a number $v(x)$, which reflects the subjective value of that outcome. Recall that outcomes are defined relative to a reference point, which serves as the zero point of the value scale. Hence, v measures the value of deviations from that reference point, i.e., gains and losses.

The present formulation is concerned with simple prospects of the form $(x, p; y, q)$, which have at most two non-zero outcomes. In such a prospect, one receives x with probability p, y with probability q, and nothing with probability $1 - p - q$, where $p + q \leq 1$. An offered prospect is strictly positive if its outcomes are all positive, i.e., if $x, y >$

0 and $p + q = 1$; it is strictly negative if its outcomes are all negative. A prospect is regular if it is neither strictly positive nor strictly negative.

The basic equation of the theory describes the manner in which π and v are combined to determine the over-all value of regular prospects.

If $(x, p; y, q)$ is a regular prospect (i.e., either $p + q < 1$, or $x \geqslant 0 \geqslant y$, or $x \leqslant 0 \leqslant y$), then

$$V(x, p; y, q) = \pi(p)v(x) + \pi(q)v(y) \tag{1}$$

where $v(0) = 0$, $\pi(0) = 0$, and $\pi(1) = 1$. As in utility theory, V is defined on prospects, while v is defined on outcomes. The two scales coincide for sure prospects, where $V(x, 1.0) = V(x) = v(x)$.

Equation (1) generalizes expected utility theory by relaxing the expectation principle. An axiomatic analysis of this representation is sketched in the Appendix, which describes conditions that ensure the existence of a unique π and a ratio-scale v satisfying equation (1).

The evaluation of strictly positive and strictly negative prospects follows a different rule. In the editing phase such prospects are segregated into two components: (i) the riskless component, i.e., the minimum gain or loss which is certain to be obtained or paid; (ii) the risky component, i.e., the additional gain or loss which is actually at stake. The evaluation of such prospects is described in the next equation.

If $p + q = 1$ and either $x > y > 0$ or $x < y < 0$, then

$$V(x, p; y, q) = v(y) + \pi(p)[v(x) - v(y)]. \tag{2}$$

That is, the value of a strictly positive or strictly negative prospect equals the value of the riskless component plus the value-difference between the outcomes, multiplied by the weight associated with the more extreme outcome. For example, $V(400, .25; 100, .75) = v(100) + \pi(.25)[v(400) - v(100)]$. The essential feature of equation (2) is that a decision weight is applied to the value-difference $v(x) - v(y)$, which represents the risky component of the prospect, but not to $v(y)$, which represents the riskless component. Note that the right-hand side of equation (2) equals $\pi(p)v(x) + [1 - \pi(p)]v(y)$. Hence, equation (2) reduces to equation (1) if $\pi(p) + \pi(1 - p) = 1$. As will be shown later, this condition is not generally satisfied.

Many elements of the evaluation model have appeared in previous attempts to modify expected utility theory. Markowitz (1952) was the first to propose that utility be defined on gains and losses rather than on final asset positions, an assumption which has been implicitly accepted in most experimental measurements of utility (see, e.g. Davidson, et al., 1957, and Mosteller and Nogee, 1951). Markowitz also noted the presence of risk seeking in preferences among positive as well as among negative prospects, and he proposed a utility function which has convex and concave regions in both the positive and the negative domains. His treatment, however, retains the expectation principle; hence it cannot account for the many violations of this principle; see, e.g., Table 11.1.

The replacement of probabilities by more general weights was proposed by Edwards (1962), and this model was investigated in several empirical studies (e.g., Anderson and Shanteau, 1970, and Tversky, 1967). Similar models were developed by Fellner (1965), who introduced the concept of decision weight to explain aversion for ambiguity, and by van Dam (1975) who attempted to scale decision weights. For other critical analyses of expected utility theory and alternative choice models, see Allais (1953), Coombs (1975), Fishburn (1977), and Hansson (1975).

The equations of prospect theory retain the general bilinear form that underlies expected utility theory. However, in order to accommodate the effects described in the first part of the paper, we are compelled to assume that values are attached to changes rather than to final states, and that decision weights do not coincide with stated probabilities. These departures from expected utility theory must lead to normatively unacceptable consequences, such as inconsistencies, intransitivities, and violations of dominance. Such anomalies of preference are normally corrected by the decision maker when he realizes that his preferences are inconsistent, intransitive, or inadmissible. In many situations, however, the decision maker does not have the opportunity to discover that his preferences could violate decision rules that he wishes to obey. In these circumstances the anomalies implied by prospect theory are expected to occur.

The value function

An essential feature of the present theory is that the carriers of value are changes in wealth or welfare, rather than final states. This assumption is compatible with basic principles of perception and judgment. Our perceptual apparatus is attuned to the evaluation of changes or differences rather than to the evaluation of absolute magnitudes. When we respond to attributes such as brightness, loudness, or temperature, the past and present context of experience defines an adaptation level, or reference point, and stimuli are perceived in relation to this reference point (Helson, 1964). Thus, an object at a given temperature may be experienced as hot or cold to the touch depending on the temperature to which one has adapted. The same principle applies to non-sensory attributes such as health, prestige, and wealth. The same level of wealth, for example, may imply abject poverty for one person and great riches for another – depending on their current assets.

The emphasis on changes as the carriers of value should not be taken to imply that the value of a particular change is independent of initial position. Strictly speaking, value should be treated as a function in two arguments: the asset position that serves as reference point, and the magnitude of the change (positive or negative) from that reference point. An individual's attitude to money, say, could be described by a book, where each page presents the value function for changes at a particular asset position. Clearly, the value functions described on different pages are not identical: they are likely to become more linear with increases in assets. However, the preference order of prospects is not greatly altered by small or even moderate variations in asset position. The certainty equivalent of the prospect (1000, .50), for example, lies between 300 and 400 for most people, in a wide range of asset positions. Consequently, the representation of value as a function in one argument generally provides a satisfactory approximation.

Many sensory and perceptual dimensions share the property that the psychological response is a concave function of the magnitude of physical change. For example, it is easier to discriminate between a change of 3° and a change of 6° in room temperature, than it is to discriminate between a change of 13° and a change of 16°. We propose that this principle applies in particular to the evaluation of monetary changes. Thus, the difference in value between a gain of 100 and a gain of 200 appears to be greater than the difference between a gain of 1100 and a gain of 1200. Similarly, the difference between a loss of 100 and a loss of 200 appears greater than the difference between a loss of 1100 and a loss of 1200, unless the larger loss is intolerable. Thus, we hypothesize that the value function for changes of wealth is normally concave above the reference point ($v''(x) < 0$, for $x > 0$) and often convex below it ($v''(x) > 0$, for $x < 0$). That is, the marginal value of both gains and losses generally decreases with their magnitude. Some support for this hypothesis has been reported by Galanter and Pliner (1974), who scaled the perceived magnitude of monetary and non-monetary gains and losses.

The above hypothesis regarding the shape of the value function was based on responses to gains and losses in a riskless context. We propose that the value function which is derived from risky choices shares the same characteristics, as illustrated in the following problems.

Problem 13:
(6000, .25),   or   (4000, .25; 2000, .25).
N = 68   [18]   [82]*

Problem 13':
(−6000, .25),   or   (−4000, .25; −2000, .25).
N = 64   [70]*   [30]

Applying equation (1) to the modal preference in these problems yields

$$\pi(.25)v(6000) < \pi(.25)[v(4000) + v(2000)] \quad \text{and}$$
$$\pi(.25)v(-6000) > \pi(.25)[v(-4000) + v(-2000)].$$

Hence, $v(6000) < v(4000) + v(2000)$ and $v(-6000) > v(-4000) + v(-2000)$. These preferences are in accord with the hypothesis that the value function is concave for gains and convex for losses.

Any discussion of the utility function for money must leave room for the effect of special circumstances on preferences. For example, the utility function of an individual who needs \$60,000 to purchase a house may reveal an exceptionally steep rise near the critical value. Similarly, an individual's aversion to losses may increase sharply near the loss that would compel him to sell his house and move to a less desirable neighborhood. Hence, the derived value (utility) function of an individual does not always reflect "pure" attitudes to money, since it could be affected by additional consequences associated with specific amounts. Such perturbations can readily produce convex regions in the value function for gains and concave regions in the value function for losses. The latter case may be more common since large losses often necessitate changes in life style.

A salient characteristic of attitudes to changes in welfare is that losses loom larger than gains. The aggravation that one experiences in losing a sum of money appears to be greater than the pleasure associated with gaining the same amount (Galanter and Pliner, 1974). Indeed, most people find symmetric bets of the form $(x, .50; -x, .50)$ distinctly unattractive. Moreover, the aversiveness of symmetric fair bets generally increases with the size of the stake. That is, if $x > y \geqslant 0$, then $(y, .50; -y, .50)$ is preferred to $(x, .50; -x, .50)$. According to equation (1), therefore,

$$v(y) + v(-y) > v(x) + v(-x) \quad \text{and} \quad v(-y) - v(-x) > v(x) - v(y).$$

Setting $y = 0$ yields $v(x) < -v(-x)$, and letting y approach x yields $v'(x) < v'(-x)$, provided v', the derivative of v, exists. Thus, the value function for losses is steeper than the value function for gains.

In summary, we have proposed that the value function is (i) defined on deviations from the reference point; (ii) generally concave for gains and commonly convex for losses; (iii) steeper for losses than for gains. A value function which satisfies these properties is displayed in Figure 11.3. Note that the proposed S-shaped value function is steepest at the reference point, in marked contrast to the utility function postulated by Markowitz (1952) which is relatively shallow in that region.

Although the present theory can be applied to derive the value function from preferences between prospects, the actual scaling is considerably more complicated than in utility theory, because of the introduction of decision weights. For example, decision weights could

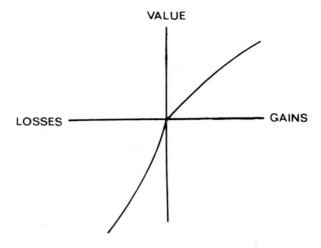

Figure 11.3   A hypothetical value function.

produce risk aversion and risk seeking even with a linear value function. Nevertheless, it is of interest that the main properties ascribed to the value function have been observed in a detailed analysis of von Neumann-Morgenstern utility functions for changes of wealth (Fishburn and Kochenberger, 1979). The functions had been obtained from thirty decision makers in various fields of business, in five independent studies (Barnes and Reinmuth, 1976; Grayson, 1960; Green, 1963; Halter & Dean, 1971; and Swalm, 1966). Most utility functions for gains were concave, most functions for losses were convex, and only three individuals exhibited risk aversion for both gains and losses. With a single exception, utility functions were considerably steeper for losses than for gains.

The weighting function

In prospect theory, the value of each outcome is multiplied by a decision weight. Decision weights are inferred from choices between prospects much as subjective probabilities are inferred from preferences in the Ramsey-Savage approach. However, decision weights are not probabilities: they do not obey the probability axioms and they should not be interpreted as measures of degree or belief.

Consider a gamble in which one can win 1000 or nothing, depending on the toss of a fair coin. For any reasonable person, the probability of winning is .50 in this situation. This can be verified in a variety of ways, e.g., by showing that the subject is indifferent between betting on heads or tails, or by his verbal report that he considers the two events equiprobable. As will be shown below, however, the decision weight $\pi(.50)$ which is derived from choices is likely to be smaller than .50. Decision weights measure the impact of events on the desirability of prospects, and not merely the perceived likelihood of these events. The two scales coincide (i.e., $\pi(p) = p$) if the expectation principle holds, but not otherwise.

The choice problems discussed in the present paper were formulated in terms of explicit numerical probabilities, and our analysis assumes that the respondents adopted the stated values of p. Furthermore, since the events were identified only by their stated probabilities, it is possible in this context to express decision weights as a function of stated probability. In general, however, the decision weight attached to an event could be influenced by other factors, e.g., ambiguity (Ellsberg, 1961, Chapter 13, this volume; Fellner, 1961).

We turn now to discuss the salient properties of the weighting function π, which relates decision weights to stated probabilities. Naturally, π is an increasing function of p, with $\pi(0) = 0$ and $\pi(1) = 1$. That is, outcomes contingent on an impossible event are ignored, and the scale is normalized so that $\pi(p)$ is the ratio of the weight associated with the probability p to the weight associated with the certain event.

We first discuss some properties of the weighting function for small probabilities. The preferences in Problems 8 and 8' suggest that for small values of p, π is a subadditive function of p, i.e., $\pi(rp) > r\pi(p)$ for $0 < r < 1$. Recall that in Problem 8, (6000, .001) is preferred to (3000, .002). Hence

$$\frac{\pi(.001)}{\pi(.002)} > \frac{v(3000)}{v(6000)} > \frac{1}{2} \quad \text{by the concavity of } v.$$

The reflected preferences in Problem 8' yield the same conclusion. The pattern of preferences in Problems 7 and 7', however, suggests that subadditivity need not hold for large values of p.

Furthermore, we propose that very low probabilities are generally overweighted, that is, $\pi(p) > p$ for small p. Consider the following choice problems.

Problem 14:
            (5000, .001),   or   (5).
    $N = 72$ [72]*   [28]

Problem 14':
            (−5000, .001),   or   (−5).
    $N = 72$ [17]   [83]*

Note that in Problem 14, people prefer what is in effect a lottery ticket over the expected value of that ticket. In Problem 14', on the other hand, they prefer a small loss, which can be viewed as the payment of an insurance premium, over a small probability of a large loss. Similar observations have been reported by Markowitz [1952]. In the present theory, the preference for the lottery in Problem 14 implies $\pi(.001)v(5000) > v(5)$, hence $\pi(.001) > v(5)/v(5000) > .001$, assuming the value function for gains is concave. The readiness to pay for insurance in Problem 14' implies the same conclusion, assuming the value function for losses is convex.

It is important to distinguish overweighting, which refers to a property of decision weights, from the overestimation that is commonly found in the assessment of the probability of rare events. Note that the issue of overestimation does not arise in the present context, where the subject is assumed to adopt the stated value of p. In many real-life situations, overestimation and overweighting may both operate to increase the impact of rare events.

Although $\pi(p) > p$ for low probabilities, there is evidence to suggest that, for all $0 < p < 1$, $\pi(p) + \pi(1 - p) < 1$. We label this property subcertainty. It is readily seen that the typical preferences in any version of Allais' example (see, e.g., Problems 1 and 2) imply subcertainty for the relevant value of p. Applying equation (1) to the prevalent preferences in Problems 1 and 2 yields, respectively,

$$v(2400) > \pi(.66)v(2400) + \pi(.33)v(2500), \quad \text{i.e.,}$$
$$[1 - \pi(.66)]v(2400) > \pi(.33)v(2500) \quad \text{and}$$
$$\pi(.33)v(2500) > \pi(.34)v(2400); \quad \text{hence,}$$
$$1 - \pi(.66) > \pi(.34), \quad \text{or} \quad \pi(.66) + \pi(.34) < 1.$$

Applying the same analysis to Allais' original example yields $\pi(.89) + \pi(.11) < 1$, and some data reported by MacCrimmon and Larsson (1979) imply subcertainty for additional values of p.

The slope of π in the interval (0, 1) can be viewed as a measure of the sensitivity of preferences to changes in probability. Subcertainty entails that π is regressive with respect to p, i.e., that preferences are generally less sensitive to variations of probability than the expectation principle would dictate. Thus, subcertainty captures an essential element of people's attitudes to uncertain events, namely that the sum of the weights associated with complementary events is typically less than the weight associated with the certain event.

Recall that the violations of the substitution axiom discussed earlier in this paper conform to the following rule: If (x, p) is equivalent to (y, pq) then (x, pr) is not preferred to (y, pqr), $0 < p, q, r \leq 1$. By equation (1),

$$\pi(p)v(x) = \pi(pq)v(y) \quad \text{implies} \quad \pi(pr)v(x) \leq \pi(pqr)v(y); \quad \text{hence,}$$
$$\frac{\pi(pq)}{\pi(p)} \leq \frac{\pi(pqr)}{\pi(pr)}.$$

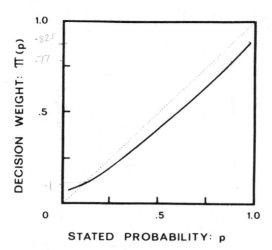

Figure 11.4 A hypothetical weighting function.

Thus, for a fixed ratio of probabilities, the ratio of the corresponding decision weights is closer to unity when the probabilities are low than when they are high. This property of π, called subproportionality, imposes considerable constraints on the shape of π: it holds if and only if log π is a convex function of log p.

It is of interest to note that subproportionality together with the overweighting of small probabilities imply that π is subadditive over that range. Formally, it can be shown that if $\pi(p) > p$ and subproportionality holds, then $\pi(rp) > r\pi(p)$, $0 < r < 1$, provided π is monotone and continuous over $(0, 1)$.

Figure 11.4 presents a hypothetical weighting function which satisfies overweighting and subadditivity for small values of p, as well as subcertainty and subproportionality. These properties entail that π is relatively shallow in the open interval and changes abruptly near the end-points where $\pi(0) = 0$ and $\pi(1) = 1$. The sharp drops or apparent discontinuities of π at the endpoints are consistent with the notion that there is a limit to how small a decision weight can be attached to an event, if it is given any weight at all. A similar quantum of doubt could impose an upper limit on any decision weight that is less than unity. This quantal effect may reflect the categorical distinction between certainty and uncertainty. On the other hand, the simplification of prospects in the editing phase can lead the individual to discard events of extremely low probability and to treat events of extremely high probability as if they were certain. Because people are limited in their ability to comprehend and evaluate extreme probabilities, highly unlikely events are either ignored or overweighted, and the difference between high probability and certainty is either neglected or exaggerated. Consequently, π is not well-behaved near the end-points.

The following example, due to Zeckhauser, illustrates the hypothesized nonlinearity of π. Suppose you are compelled to play Russian roulette, but are given the opportunity to purchase the removal of one bullet from the loaded gun. Would you pay as much to reduce the number of bullets from four to three as you would to reduce the number of bullets from one to zero? Most people feel that they would be willing to pay much more for a reduction of the probability of death from $\frac{1}{6}$ to zero than for a reduction from $\frac{4}{6}$ to $\frac{3}{6}$. Economic considerations would lead one to pay more in the latter case, where the value of money is presumably reduced by the considerable probability that one will not live to enjoy it.

An obvious objection to the assumption that $\pi(p) \neq p$ involves comparisons between prospects of the form $(x, p; x, q)$ and $(x, p'; x, q')$, where $p + q = p' + q' < 1$. Since any individual will surely be indifferent between the two prospects, it could be argued that this observation entails $\pi(p) + \pi(q) = \pi(p') + \pi(q')$, which in turn implies that π is the identity function. This argument is invalid in the present theory, which assumes that the probabilities of identical outcomes are combined in the editing of prospects. A more serious objection to the nonlinearity of π involves potential violations of dominance. Suppose $x > y > 0$, $p > p'$, and $p + q = p' + q' < 1$; hence $(x, p; y, q)$ dominates $(x, p'; y, q')$. If preference obeys dominance, then

$$\pi(p)v(x) + \pi(q)v(y) > \pi(p')v(x) + \pi(q')v(y),$$

or

$$\frac{\pi(p) - \pi(p')}{\pi(q') - \pi(q)} > \frac{v(y)}{v(x)}.$$

Hence, as y approaches, x, $\pi(p) - \pi(p')$ approaches $\pi(q') - \pi(q)$. Since $p - p' = q' - q$, π must be essentially linear, or else dominance must be violated.

Direct violations of dominance are prevented, in the present theory, by the assumption that dominated alternatives are detected and eliminated prior to the evaluation of prospects. However, the theory permits indirect violations of dominance, e.g., triples of prospects so that A is preferred to B, B is preferred to C, and C dominates A. For an example, see Raiffa (1968, p. 75).

Finally, it should be noted that the present treatment concerns the simplest decision task in which a person chooses between two available prospects. We have not treated in detail the more complicated production task (e.g., bidding) where the decision maker generates an alternative that is equal in value to a given prospect. The asymmetry between the two options in this situation could introduce systematic biases. Indeed, Lichtenstein and Slovic (1971) have constructed pairs of prospects A and B, such that people generally prefer A over B, but bid more for B than for A. This phenomenon has been confirmed in several studies, with both hypothetical and real gambles, e.g., Grether and Plott (1979). Thus, it cannot be generally assumed that the preference order of prospects can be recovered by a bidding procedure.

Because prospect theory has been proposed as a model of choice, the inconsistency of bids and choices implies that the measurement of values and decision weights should be based on choices between specified prospects rather than on bids or other production tasks. This restriction makes the assessment of v and π more difficult because production tasks are more convenient for scaling than pair comparisons.

4. Discussion

In the final section we show how prospect theory accounts for observed attitudes toward risk, discuss alternative representations of choice problems induced by shifts of reference point, and sketch several extensions of the present treatment.

Risk attitudes

The dominant pattern of preferences observed in Allais' example (Problems 1 and 2) follows from the present theory iff

$$\frac{\pi(.33)}{\pi(.34)} > \frac{v(2,400)}{v(2,500)} > \frac{\pi(.33)}{1 - \pi(.66)}.$$

Hence, the violation of the independence axiom is attributed in this case to subcertainty, and more specifically to the inequality $\pi(.34) < 1 - \pi(.66)$. This analysis shows that an Allais-type violation will occur whenever the v-ratio of the two non-zero outcomes is bounded by the corresponding π-ratios.

Problems 3 through 8 share the same structure, hence it suffices to consider one pair, say Problems 7 and 8. The observed choices in these problems are implied by the theory iff

$$\frac{\pi(.001)}{\pi(.002)} > \frac{v(3,000)}{v(6,000)} > \frac{\pi(.45)}{\pi(.90)}.$$

The violation of the substitution axiom is attributed in this case to the subproportionality of π. Expected utility theory is violated in the above manner, therefore, whenever the v-ratio of the two outcomes is bounded by the respective π-ratios. The same analysis applies to other violations of the substitution axiom, both in the positive and in the negative domain.

We next prove that the preference for regular insurance over probabilistic insurance, observed in Problem 9, follows from prospect theory – provided the probability of loss is overweighted. That is, if $(-x, p)$ is indifferent to $(-y)$, then $(-y)$ is preferred to $(-x, p/2; -y, p/2; -y/2, 1 - p)$. For simplicity, we define for $x \geq 0$, $f(x) = -v(-x)$. Since the value function for losses is convex, f is a concave function of x. Applying prospect theory, with the natural extension of equation (2), we wish to show that

$$\pi(p)f(x) = f(y) \qquad \text{implies}$$

$$f(y) \leq f(y/2) + \pi(p/2)[f(y) - f(y/2)] + \pi(p/2)[f(x) - f(y/2)]$$

$$= \pi(p/2)f(x) + \pi(p/2)f(y) + [1 - 2\pi(p/2)]f(y/2).$$

Substituting for $f(x)$ and using the concavity of f, it suffices to show that

$$f(y) \leq \frac{\pi(p/2)}{\pi(p)} f(y) + \pi(p/2)f(y) + f(y)/2 - \pi(p/2)f(y)$$

or

$$\pi(p)/2 \leq \pi(p/2), \qquad \text{which follows from the subadditivity of } \pi.$$

According to the present theory, attitudes toward risk are determined jointly by v and π, and not solely by the utility function. It is therefore instructive to examine the conditions under which risk aversion or risk seeking are expected to occur. Consider the choice between the gamble (x, p) and its expected value (px). If $x > 0$, risk seeking is implied whenever $\pi(p) > v(px)/v(x)$, which is greater than p if the value function for gains is concave. Hence, overweighting $(\pi(p) > p)$ is necessary but not sufficient for risk seeking in the domain of gains. Precisely the same condition is necessary but not sufficient for risk aversion when $x < 0$. This analysis restricts risk seeking in the domain of gains and risk aversion in the domain of losses to small probabilities, where overweighting is expected to hold. Indeed these are the typical conditions under which lottery tickets and insurance policies are sold. In prospect theory, the overweighting of small probabilities favors both gambling and insurance, while the S-shaped value function tends to inhibit both behaviors.

Although prospect theory predicts both insurance and gambling for small probabilities, we feel that the present analysis falls far short of a fully adequate account of these complex phenomena. Indeed, there is evidence from both experimental studies (Slovic, et al., 1977), survey research (Kunreuther, et al., 1978), and observations of economic behavior, e.g., service and medical insurance, that the purchase of insurance often extends to the medium range of probabilities, and that small probabilities of disaster are sometimes entirely ignored. Furthermore, the evidence suggests that minor changes in the formulation of the decision problem can have marked effects on the attractiveness of insurance (Slovic, et al., 1977). A comprehensive theory of insurance behavior should consider, in addition to pure attitudes toward uncertainty and money, such factors as the value of security, social norms of prudence, the aversiveness of a large number of small payments spread over time, information and misinformation regarding probabilities and outcomes, and many others. Some effects of these variables could be described within the present framework, e.g., as changes of reference point, transformations of the value function, or manipulations of probabilities or decision weights. Other effects may require the introduction of variables or concepts which have not been considered in this treatment.

Shifts of reference

So far in this paper, gains and losses were defined by the amounts of money that are obtained or paid when a prospect is played, and the reference point was taken to be the status quo, or one's current assets. Although this is probably true for most choice problems, there are situations in which gains and losses are coded relative to an expectation or aspiration level that differs from the status quo. For example, an unexpected tax withdrawal from a monthly pay check is experienced as a loss, not as a reduced gain. Similarly, an entrepreneur who is weathering a slump with greater success than his competitors may interpret a small loss as a gain, relative to the larger loss he had reason to expect.

The reference point in the preceding examples corresponded to an asset position that one had expected to attain. A discrepancy between the reference point and the current asset position may also arise because of recent changes in wealth to which one has not yet adapted (Markowitz, 1952). Imagine a person who is involved in a business venture, has already lost 2000 and is now facing a choice between a sure gain of 1000 and an even chance to win 2000 or nothing. If he has not yet adapted to his losses, he is likely to code the problem as a choice between $(-2000, .50)$ and (-1000) rather than as a choice between $(2000, .50)$ and (1000). As we have seen, the former representation induces more adventurous choices than the latter.

A change of reference point alters the preference order for prospects. In particular, the present theory implies that a negative translation of a choice problem, such as arises from incomplete adaptation to recent losses, increases risk seeking in some situations. Specifically, if a risky prospect $(x, p; -y, 1 - p)$ is just acceptable, then $(x - z, p; -y - z, 1 - p)$ is preferred over $(-z)$ for $x, y, z > 0$, with $x > z$.

To prove this proposition, note that

$$V(x, p; y, 1 - p) = 0 \qquad \text{iff} \qquad \pi(p)v(x) = -\pi(1 - p)v(-y).$$

Furthermore,

$$
\begin{aligned}
V(x &- z, p; -y - z, 1 - p) \\
&= \pi(p)v(x - z) + \pi(1 - p)v(-y - z) \\
&> \pi(p)v(x) - \pi(p)v(z) + \pi(1 - p)v(-y) \\
&\quad + \pi(1 - p)v(-z) \qquad \text{by the properties of } v. \\
&= -\pi(1 - p)v(-y) - \pi(p)v(z) + \pi(1 - p)v(-y) \\
&\quad + \pi(1 - p)v(-z) \qquad \text{by substitution,} \\
&= -\pi(p)v(z) + \pi(1 - p)v(-z) \\
&> v(-z)[\pi(p) + \pi(1 - p)] \qquad \text{since } v(-z) < -v(z), \\
&> v(-z) \qquad \text{by subcertainty.}
\end{aligned}
$$

This analysis suggests that a person who has not made peace with his losses is likely to accept gambles that would be unacceptable to him otherwise. The well known observation (McGlothlin, 1956) that the tendency to bet on long shots increases in the course of the betting day provides some support for the hypothesis that a failure to adapt to losses or to attain an expected gain induces risk seeking. For another example, consider an individual who expects to purchase insurance, perhaps because he has owned it in the past or because his friends do. This individual may code the decision to pay a premium y to protect against a loss x as a choice between $(-x + y, p; y, 1 - p)$ and (0) rather than as a choice between $(-x, p)$ and $(-y)$. The preceding argument entails that insurance is likely to be more attractive in the former representation than in the latter.

Another important case of a shift of reference point arises when a person formulates his decision problem in terms of final assets, as advocated in decision analysis, rather than in terms of gains and losses, as people usually do. In this case, the reference point is set to zero on the scale of wealth and the value function is likely to be concave everywhere (Spetzler, 1968). According to the present analysis, this formulation essentially eliminates risk seeking, except for gambling with low probabilities. The explicit formulation of decision problems in terms of final assets is perhaps the most effective procedure for eliminating risk seeking in the domain of losses.

Many economic decisions involve transactions in which one pays money in exchange for a desirable prospect. Current decision theories analyze such problems as comparisons between the status quo and an alternative state which includes the acquired prospect minus its cost. For example, the decision whether to pay 10 for the gamble (1000, .01) is treated as a choice between (990, .01; −10, .99) and (0). In this analysis, readiness to purchase the positive prospect is equated to willingness to accept the corresponding mixed prospect.

The prevalent failure to integrate riskless and risky prospects, dramatized in the isolation effect, suggests that people are unlikely to perform the operation of subtracting the cost from the outcomes in deciding whether to buy a gamble. Instead, we suggest that people usually evaluate the gamble and its cost separately, and decide to purchase the gamble if the combined value is positive. Thus, the gamble (1000, .01) will be purchased for a price of 10 if $\pi(.01)v(1000) + v(-10) > 0$.

If this hypothesis is correct, the decision to pay 10 for (1000, .01), for example, is no longer equivalent to the decision to accept the gamble (990, .01; −10, .99). Furthermore, prospect theory implies that if one is indifferent between $(x(1 - p), p; - px, 1 - p)$ and (0) then one will not pay px to purchase the prospect (x, p). Thus, people are expected to exhibit more risk seeking in deciding whether to accept a fair gamble than in deciding whether to purchase a gamble for a fair price. The location of the reference point, and the manner in which choice problems are coded and edited emerge as critical factors in the analysis of decisions.

Extensions

In order to encompass a wider range of decision problems, prospect theory should be extended in several directions. Some generalizations are immediate; others require further development. The extension of equations (1) and (2) to prospects with any number of outcomes is straightforward. When the number of outcomes is large, however, additional editing operations may be invoked to simplify evaluation. The manner in which complex options, e.g., compound prospects, are reduced to simpler ones is yet to be investigated.

Although the present paper has been concerned mainly with monetary outcomes, the theory is readily applicable to choices involving other attributes, e.g., quality of life or the number of lives that could be lost or saved as a consequence of a policy decision. The main properties of the proposed value function for money should apply to other attributes as well. In particular, we expect outcomes to be coded as gains or losses relative to a neutral reference point, and losses to loom larger than gains.

The theory can also be extended to the typical situation of choice, where the probabilities of outcomes are not explicitly given. In such situations, decision weights must be attached to particular events rather than to stated probabilities, but they are expected to exhibit the essential properties that were ascribed to the weighting function. For example, if A and B are complementary events and neither is certain, $\pi(A) + \pi(B)$ should be less than unity – a natural analogue to sub-certainty.

The decision weight associated with an event will depend primarily on the perceived likelihood of that event, which could be subject to major biases (Tversky & Kahneman, 1974). In addition, decision weights may be affected by other considerations, such as ambiguity or vagueness. Indeed, the work of Ellsberg (1961, Chapter 13, this volume) and Fellner (1965) implies that vagueness reduces decision weights. Consequently, subcertainty should be more pronounced for vague than for clear probabilities.

The present analysis of preference between risky options has developed two themes. The first theme concerns editing operations that determine how prospects are perceived. The second theme involves the judgmental principles that govern the evaluation of gains and losses and the weighting of uncertain outcomes. Although both themes should be developed further, they appear to provide a useful framework for the descriptive analysis of choice under risk.

Appendix[1]

In this appendix we sketch an axiomatic analysis of prospect theory. Since a complete self-contained treatment is long and tedious, we merely outline the essential steps and exhibit the key ordinal properties needed to establish the bilinear representation of equation (1). Similar methods could be extended to axiomatize equation (2).

Consider the set of all regular prospects of the form $(x, p; y, q)$ with $p + q < 1$. The extension to regular prospects with $p + q = 1$ is straightforward. Let $\geq$ denote the relation of preference between prospects that is assumed to be connected, symmetric and transitive, and let $\simeq$ denote the associated relation of indifference. Naturally, $(x, p; y, q) \simeq (y, q; x, p)$. We also assume, as is implicit in our notation, that $(x, p; 0, q) \simeq (x, p; 0, r)$, and $(x, p; y, 0) \simeq (x, p; z, 0)$. That is, the null outcome and the impossible event have the property of a multiplicative zero.

Note that the desired representation [equation (1)] is additive in the probability-outcome pairs. Hence, the theory of additive conjoint measurement can be applied to obtain a scale V which preserves the

[1] We are indebted to David H. Krantz for his help in the formulation of this section.

preference order, and interval scales f and g in two arguments such that

$$V(x, p; y, q) = f(x, p) + g(y, q).$$

The key axioms used to derive this representation are:

Independence: $(x, p; y, q) \gtrsim (x, p; y', q')$ iff $(x', p'; y, q) \gtrsim (x', p'; y', q')$.

Cancellation: If $(x, p; y', q') \gtrsim (x', p'; y, q)$ and $(x', p'; y'', q'') \gtrsim (x'', p''; y', q')$, then $(x, p; y'', q'') \gtrsim (x'', p''; y, q)$.

Solvability: If $(x, p; y, q) \gtrsim (z, r) \gtrsim (x, p; y', q')$ for some outcome z and probability r, then there exist y'', q'' such that

$$(x, p; y'' q'') \simeq (z, r).$$

It has been shown that these conditions are sufficient to construct the desired additive representation, provided the preference order is Archimedean (Debreu, 1960, and Krantz, et al., 1971). Furthermore, since $(x, p; y, q) \simeq (y, q; x, p)$, $f(x, p) + g(y, q) = f(y, q) + g(x, p)$, and letting $q = 0$ yields $f = g$.

Next, consider the set of all prospects of the form (x, p) with a single non-zero outcome. In this case, the bilinear model reduces to $V(x, p) = \pi(p)v(x)$. This is the multiplicative model, investigated in Roskies (1965) and Krantz et al. (1971). To construct the multiplicative representation we assume that the ordering of the probability-outcome pairs satisfies independence, cancellation, solvability, and the Archimedean axiom. In addition, we assume sign dependence (Krantz et al., 1971) to ensure the proper multiplication of signs. It should be noted that the solvability axiom used in Roskies (1965) and Krantz et al. (1971) must be weakened because the probability factor permits only bounded solvability.

Combining the additive and the multiplicative representations yields

$$V(x, p; y, q) = f[\pi(p)v(x)] + f[\pi(q)v(y)].$$

Finally, we impose a new distributivity axiom:

$$(x, p; y, p) \simeq (z, p) \quad \text{iff} \quad (x, q; y, q) \simeq (z, q).$$

Applying this axiom to the above representation, we obtain

$$f[\pi(p)v(x)] + f[\pi(p)v(y)] = f[\pi(p)v(z)]$$

implies

$$f[\pi(q)v(x)] + f[\pi(q)v(y)] = f[\pi(q)v(z)].$$

Assuming, with no loss of generality, that $\pi(q) < \pi(p)$, and letting $\alpha = \pi(p)v(x)$, $\beta = \pi(p)v(y)$, $\gamma = \pi(p)v(z)$, and $\theta = \pi(q)/\pi(p)$, yields $f(\alpha) + f(\beta) = f(\gamma)$ implies $f(\theta\alpha) + f(\theta\beta) = f(\theta\gamma)$ for all $0 < \theta < 1$.

Because f is strictly monotonic we can set $\gamma = f^{-1}[f(\alpha) + f(\beta)]$. Hence, $\theta\gamma = \theta f^{-1}[f(\alpha) + f(\beta)] = f^{-1}[f(\theta\alpha) + f(\theta\beta)]$.

The solution to this functional equation is $f(\alpha) = k\alpha^c$ (Aczél, 1966). Hence, $V(x, p; y, q) = k[\pi(p)v(x)]^c + k[\pi(q)v(y)]^c$, for some $k, c > 0$. The desired bilinear form is obtained by redefining the scales π, v, and V so as to absorb the constants k and c.

12. Generalized expected utility analysis and the nature of observed violations of the independence axiom

Mark Machina

1. Introduction

First expressed by Allais in the early fifties, dissatisfaction with the expected utility model of individual risk taking behavior has mushroomed in recent years, as the number of papers in this volume, its predecessor (Allais & Hagen, 1979), and elsewhere indicates (see, e.g., Chew and MacCrimmon, 1979; Fishburn, 1982b, 1983; Handa, 1977; and Kahneman and Tversky, 1979). The nature of the current debate, i.e. whether to reject a theoretically elegant and heretofore tremendously useful descriptive model in light of accumulating evidence against its underlying assumptions, is a classic one in science, and the spur to new theoretical and empirical research which it is offering cannot help but leave economists, psychologists, and others who study this area with a better understanding of individual behavior toward risk.

In terms of its logical foundations, the expected utility model may be thought of as following from four assumptions concerning the individual's ordering of probability distributions over wealth: completeness (i.e. any two distributions can be compared), transitivity of both strict and weak preference, continuity, and the so-called "independence axiom." This latter axiom, really the cornerstone of the theory, may be stated as "a risky prospect A is weakly preferred (i.e. preferred or

I am indebted to Maurice Allais, Kenneth Arrow, John Harsanyi, and Ed McClennen for discussions of this material during the Conference [Foundations of Utility and Risk Theory, Oslo, 1982], and to Beth Hayes, Joel Sobel, and Halbert White for helpful comments on the manuscript. All errors and opinions, however, are my own.

Reprinted from *Foundations of Utility and Risk Theory with Applications*, ed. by B. P. Stigum and F. Wenstøp, D. Reidel Publishing Company, Dordrecht, 1983, pp. 263–293, by kind permission of the author and the publisher.

indifferent) to a risky prospect B if and only if a $p:(1 - p)$ chance of A or C respectively is weakly preferred to a $p:(1 - p)$ chance of B or C, for arbitrary positive probability p and risky prospects A, B, and C." While the first three assumptions serve to imply that the individual's preferences may be represented by a real-valued maximand or "preference functional" defined over probability distributions, it is the independence axiom which gives the theory its main empirical content by placing a restriction on the functional form of the preference functional, implying that it (or some monotonic transformation of it) must be "linear in the probabilities" and hence representable as the mathematical expectation of some von Neumann-Morgenstern utility index defined over the set of pure outcomes.

Although the normative validity of the independence axiom has often been questioned in the past (see for example Allais, 1952/1979a, Tversky, 1975, Wold, 1952, and the examples offered in Dreze, 1974 and Machina, 1981), the primary form of attack on the expected utility hypothesis has been on the empirical validity of the independence axiom. Beginning with the famous example of Allais (discussed in detail below), the empirical/experimental research on the independence axiom has uncovered four types of *systematic* violations of the axiom: the "common consequence effect," the "common ratio effect" (which includes the "Bergen Paradox" and "certainty effect" as special cases), "oversensitivity to changes in small probabilities," and the "utility evaluation effect" (described below). While defenders of the expected utility model have claimed that such violations, systematic or otherwise, would disappear once the nature of such "errors" had been pointed out to subjects (e.g., Raiffa, 1968, pp. 80–86, Savage, 1972, pp. 102–103; pp. 163–165, this volume), empirical tests of this assertion (MacCrimmon, 1968, pp. 9–11, Slovic and Tversky, 1974) have fairly convincingly refuted it, and it is now generally acknowledged that, as a descriptive hypothesis, the independence axiom is not able to stand up to the data.

Accordingly, the defense of the expected utility model has shifted to the other two *sine qua non*'s of a useful theory, namely analytic power and the ability to generate refutable predictions and policy implications in a wide variety of situations.[1] Expected utility supporters have pointed out that descriptive models are like lifeboats in that "you don't abandon a leaky one until something better comes along," and insist that a mere ability to rationalize "aberrant" observations is not enough for an alternative model to replace expected utility – to be acceptable, the alternative must at least approximate the analytic power and versatility of expected utility analysis. On the whole they have been

---

[1] Of course, any comparison of the refutable implications of two competing models should be followed immediately by a discussion of which of these implications have and have not in fact been refuted.

correct in so arguing, as many of the alternatives which have been offered have had little predictive power, and various ones have been restricted to only pairwise choice, have implied intransitive behavior, were able to accommodate only discrete probability distributions, or even possessed the property that the individual can be led into "making book against him/herself!"

The purpose of this chapter is to describe an alternative to expected utility analysis (in fact, a generalization of it) which is designed to possess the high analytic power of expected utility as well as to parsimoniously capture the nature of observed departures from the independence axiom. On the one hand, this technique, termed "generalized expected utility analysis," allows us to apply the major concepts, tools, and results of expected utility theory to the analysis of almost completely general preferences (specifically, any set of preferences which is complete, transitive, and "smooth" in the sense described below). On the other hand, however, this technique is capable of simply characterizing any additional behavioral restrictions we might feel are warranted, such as general risk aversion, declining risk aversion, comparative risk aversion between individuals, and in particular, a simple condition on preferences which serves to generate all four of the above mentioned systematic violations of the independence axiom. In addition, because of the very weak assumptions required, it turns out that many of the other alternatives and generalizations of expected utility theory which have been offered are special cases of the present analysis, which can therefore be used to derive further results in these special cases.

The following section offers a brief overview of those aspects of expected utility theory which will be relevant for the present purposes. Section 3 offers a simple graphical and algebraic description of generalized expected utility analysis, including extensions of the expected utility concepts of the "risk averse concave utility function" and the Arrow-Pratt measure of risk aversion to the general case of "smooth" preferences.[2] Section 4 offers a survey of the four known types of systematic violations of the independence axiom, as well as a description and discussion of the simple condition on preferences which serves to generate each of these four types of behavior. Section 5 offers a brief conclusion.

2. The expected utility model

In this and the following sections, we adopt the standard choice-theoretic approach of assuming that the individual has a complete,

---

[2] For a more complete and rigorous treatment of much of the material in Sections 3 and 4, see Machina (1982a, 1982b, 1984).

transitive preference ordering over the set $D[0, M]$ of all cumulative distribution functions $F(.)$ over the wealth interval $[0, M]$. As in standard consumer theory (see, for example, Debreu, 1959, chap. 4), completeness, transitivity, and continuity are sufficient to imply that we can represent the individual's ranking by some real-valued preference functional $V(.)$ over $D[0, M]$, so that the probability distribution $F^*(.)$ is weakly preferred to $F(.)$ if and only if $V(F^*) \geq V(F)$. In those cases when we find it useful to consider the subset $D\{x_1, \ldots, x_n\}$ of probability distributions over the payoffs $x_1 < \cdots < x_n$, we shall represent the typical distribution in $D\{x_1, \ldots, x_n\}$ by the vector of corresponding probabilities $(p_1, \ldots, p_n)$ and represent the restriction of $V(.)$ to $D\{x_1, \ldots, x_n\}$ by $V(p_1, \ldots, p_n)$.

Now, if we in addition assume that the individual satisfies the independence axiom, it follows (see, e.g., Herstein and Milnor, 1953) that $V(.)$, or some monotonic transformation of $V(.)$, will possess the functional form $V(F) \equiv \int U(x)dF(x)$ [or in the discrete case, $V(p_1, \ldots, p_n)$ $\equiv \Sigma U(x_i)p_i$], i.e., the mathematical expectation of the von Neumann-Morgenstern utility function $U(.)$ with respect to $F(.)$ [or $(p_1, \ldots, p_n)$]. In other words, $V(.)$ can be represented as a linear functional of $F(.)$ [or in the discrete case, as a linear function of $(p_1, \ldots, p_n)$], hence the phrase that the preferences of an expected utility maximizer are "linear in the probabilities." In this case it is also clear that the distribution $F^*(.)$ will be weakly preferred to $F(.)$ if and only if $\int U(x)dF^*(x) \geq \int U(x)dF(x)$, or equivalently, if and only if

$$\int U(x)[dF^*(x) - dF(x)] \geq 0. \tag{1}$$

For purposes of illustration, it is useful to consider the subset $D\{x_1, x_2, x_3\}$ of all probability distributions over the wealth levels $x_1 < x_2 < x_3$ in $[0, M]$, which may be represented by the points in the unit triangle in the (p_1, p_3) plane, as in Figure 12.1 (with p_2 defined by $p_2 = 1 - p_1 - p_3$). Because of the "linearity" property of expected utility maximizers, such individuals' indifference curves in this space (the solid lines in Figure 12.1) will be parallel straight lines, with preferred indifference curves lying to the northwest.[3] The dashed lines in the figure are what may be termed "iso-expected *value* loci," i.e., loci of probability distributions with the same mean. Northeast movements along such loci, since they represent changes in the distribution which preserve the mean but increase the probability of the worst and best outcomes (i.e. increase p_1 and p_3 at the expense of p_2), are seen to be precisely the set of "mean preserving spreads" in the sense of Rothschild and Stiglitz

---

[3] The indifference curves here are the loci of solutions to the equation $p_1U(x_1) + (1 - p_1 - p_3)U(x_2) + p_3U(x_3) = k$ for different values of the constant k. Northwest movements make the individual better off since they consist of either increases in p_3 at the expense of p_2, increases in p_2 at the expense of p_1, or a combination of the two.

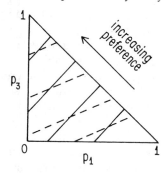

Figure 12.1

(1970). Thus, if the indifference curves are steeper than these loci, as in Figure 12.1, mean preserving spreads will always make the individual worse off, or in other words, the individual is risk averse. Conversely, if the indifference curves are flatter than the iso-expected value loci, the individual will be risk loving in the sense that mean preserving spreads will be preferred.

In fact, there is even a stronger sense in which the steepness of the indifference curves provides a measure of risk aversion. Solving the equation in footnote 3, we obtain that the slope of these indifference curves is equal to

$$-\frac{(U(x_3) - U(x_2)) - (U(x_2) - U(x_1))}{U(x_3) - U(x_2)} + 1. = \frac{u(x_2) - u(x_1)}{u(x_3) - u(x_2)} \quad (2)$$

Neglecting the addition of the constant 1, this expression (negative the ratio of a second difference of utility to a first difference) may be thought of as the discrete analogue of the Arrow-Pratt measure $-U''(x)/U'(x)$, and indeed, Pratt (1964, Thm. 1) has shown that they are related in that the more concave the utility function, the greater the value of expression (2) for fixed x_1, x_2, and x_3. Thus, given two expected utility maximizers, the one with the steeper indifference curves will be the more risk averse over $D\{x_1, x_2, x_3\}$.

3. Generalized expected utility analysis: A brief overview

Although there certainly have been studies which have found individual preferences over uncertain and certain prospects which violate both transitivity and completeness (or at least stability) of preferences, (see, e.g., Kahneman and Tversky, 1979, pp. 271–273, pp. 192–195, this volume; Tversky 1969, 1975; Grether, 1978; and Grether and Plott, 1979) by far the largest and most systematic body of empirical results are those revealing systematic violations of the independence axiom. Of the three, it is in some sense fortunate that it is independence and not the other two which is most frequently violated – while dropping either transitivity or completeness/stability would lead to a fundamental break with the traditional theory of choice, dropping independence [i.e. linearity of $V(.)$] amounts to simply changing the functional form of the preference functional, something which is done frequently in economic theory and econometrics.

One of the virtues of generalized expected utility analysis is that it can be developed with extremely weak assumptions on the functional form of the preference functional. Specifically, we need only assume that $V(.)$ is a differentiable functional of $F(.)$ (i.e. "smooth in the probabilities"), which is equivalent to assuming that indifference curves in $D\{x_1, x_2, x_3\}$ (or more generally, indifference hypersurfaces in $D[0, M]$) are smooth and vary smoothly in the appropriate sense. Differentiability or smoothness of preferences is considered to be an extremely weak assumption in standard choice theory, and it is sufficiently weak so that many (though not all) of the functional forms which have been offered to replace expected utility are special cases of it (see below).

Algebraically, the assumption that the preference functional $V(.)$ is differentiable in $F(.)$ means that we can take the usual first order Taylor expansion of $V(.)$ about any point in its domain, i.e. about any distribution $F_o(.)$ in $D[0, M]$, so that for each $F_o(.)$ in $D[0, M]$ there will exist some linear functional $\psi(.; F_o)$ (linear in its first argument) such that

$$V(F) - V(F_o) = \psi(F - F_o; F_o) + o(\|F - F_o\|), \tag{3}$$

where, as in standard calculus, $o(.)$ denotes a function of higher order than its argument, and $\|.\|$ is the L^1 norm, a standard measure of the "distance" between two functions.

Because $\psi(F - F_o; F_o)$ is linear in its first argument, it can be represented as the expectation of some function with respect to $F(.) - F_o(.)$, so that we may rewrite (3) as

$$V(F) - V(F_o) = \int U(x; F_o)[dF(x) - dF_o(x)] + o(\|F - F_o\|), \tag{4}$$

where the notation $U(.; F_o)$ is used to denote the dependence of $\psi(.; F_o)$, and hence its integral representation, upon the function $F_o(.)$, i.e. upon the point in the domain about which we are taking the Taylor expansion. As in standard calculus, we know that for differential movements about the domain of $V(.)$, [i.e., for changes from $F_o(.)$ to some "very close" $F(.)$], the first order or linear term in (4) will dominate the higher order term, so that an individual with preference functional $V(.)$ will rank differential shifts from $F_o(.)$ according to the sign of the term $\int U(x; F_o)[dF(x) - dF_o(x)]$. Recalling expression (1), however, we see that this is *precisely* the same ranking that would be used by an expected utility maximizer with a utility function $U(.; F_o)$. Of course in some sense this is no surprise: preferences which are "smooth" (i.e. differentiable) are locally linear, and we know that in ranking probability distributions, linearity is equivalent to expected utility maximization.

Thus, even though an individual with a smooth preference functional $V(.)$ will not necessarily satisfy the independence axiom and possesses no "global" von Neumann-Morgenstern utility function, we see that at each distribution $F_o(.)$ in $D[0, M]$ there will exist a "local utility function" $U(.; F_o)$ over $[0, M]$ which represents the individual's preferences at $F_o(.)$. Because of the analogy between equations (1) and (4), it is clear that if $U(x; F_o)$ is increasing in x then the individual will prefer all differential first order stochastically dominating shifts from $F_o(.)$, (see Hadar and Russell, 1969, for the definition of first order stochastic dominance) and $U(x; F_o)$ will be concave in x if and only if the individual is made worse off by all differential mean preserving spreads about $F_o(.)$ [i.e., is locally risk averse in the neighborhood of $F_o(.)$].

Of course, as with any linear approximation to a differentiable function, the ranking determined by the first order linear term [i.e., by the local utility function $U(.; F_o)$] will typically not correspond exactly to the ranking determined by $V(.)$ over any open neighborhood of $F_o(.)$ in $D[0, M]$. However, and again by analogy with standard calculus, it is possible to completely and exactly reconstruct the preference functional from knowledge of what its linear approximations (i.e., derivatives) look like at every point in the domain, by use of the Fundamental Theorem of Integral Calculus. To do this, we take any path of the form $\{F(.; \alpha) \mid \alpha \in [0, 1]\}$ from $F_o(.)$ to $F(.)$ [not necessarily "near" $F_o(.)$], so that $F(.; 0) = F_o(.)$ and $F(.; 1) = F(.)$, and use the fact that $V(F) - V(F_o)$ will be simply the integral of $dV(F(.; \alpha))/d\alpha$ as α runs from 0 to 1. In the case of the "straight line" path $F(.; \alpha) \equiv \alpha F(.) + (1 - \alpha) F_o(.)$, for example, we have

$$V(F) - V(F_o) = \int_0^1 \frac{dV(F(.; \alpha))}{d\alpha} \, d\alpha =$$
$$= \int_0^1 \{\textstyle\int U(x; F(.; \alpha)) \, [dF(x) - dF_o(x)]\} d\alpha, \tag{5}$$

since the derivative of the higher order term in (4) as α increases will be zero (see Machina, 1982a for details).

Besides yielding a way to completely reconstruct the preference functional $V(.)$ from knowledge of the local utility functions, equation (5) yields insight on how generalized expected utility analysis may be used to obtain *global* characterizations of behavior in terms of "expected utility" type conditions on the local utility functions. For example, say that $F(.)$ differs from $F_o(.)$ by a "large" mean preserving spread. If the local utility functions $U(.; F(.; \alpha)$ are concave in x at each $F(.; \alpha)$, then it follows that the term in curled brackets in (5) will be nonpositive for each α, so that $V(.)$ will weakly prefer $F(.)$ to $F_o(.)$. Indeed, it is shown formally in Machina (1982a) that the "expected utility" condition of concavity of (all) the local utility functions is *equivalent* to the individual being averse to all mean preserving spreads, or in other words, to the individual being globally risk averse.

A similar method was used in Machina (1982a) to prove two other extensions of "expected utility" analysis to the case of individuals with preference functionals which do not necessarily satisfy the independence axiom. Using straight line paths as in the previous paragraph, it is straightforward to show that an individual's preferences will exhibit "monotonicity," i.e., preference for first order stochastically dominating distributions, if and only if all the local utility functions are increasing in x. The second result extends the well known "Arrow-Pratt theorem" of comparative risk aversion: if we form the natural analogue to the Arrow-Pratt measure in our more general setting, i.e., $-U_{11}(x; F)/U_1(x; F)$ (where subscripts denote successive partial derivatives with respect to x), we have that one individual will be everywhere more risk averse than another in the standard behavioral senses (see Machina, 1982a) if and only if the "generalized Arrow-Pratt term" of the first individual is everywhere higher than that of the second, or equivalently, if and only if the first individual's local utility functions are everywhere more concave than the second's.

Note that while these types of extended expected utility theorems might seem "more complex" than those of expected utility theory since they involve checking all the local utility functions rather than a single von Neumann-Morgenstern utility function, they are in fact "less complex" in that the expected utility theorems may be thought of as derived from the more general theorems with the *additional* restriction that all of the local utility functions are identical.

The above algebraic arguments admit of a nice graphical interpretation in terms of the triangle diagram of Section 2 above. Since we are now considering preferences over the subset $D\{x_1, x_2, x_3\}$ of $D[0, M]$, we shall use the symbol $P_o = (p_{1,o}, p_{2,o}, p_{3,o})$ instead of $F_o(.)$ to denote the probability distribution about which we expand the preference functional. Figure 12.2 illustrates the general principle that if preferences (and hence indifference curves) are smooth, then there will exist a "tangent" (i.e., linear approximating) expected utility preference field to the individual's indifference curves at each distribution, as illustrated by the parallel straight lines which are tangent to the individual's actual (nonlinear) indifference curves at P_o. Figure 12.3

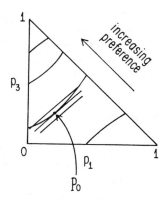

Figure 12.2

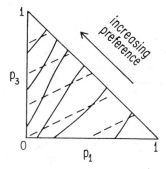

Figure 12.3

illustrates the above result that global risk aversion is equivalent to all the local utility functions being concave. Graphically, it is clear that what is necessary and sufficient for all mean preserving spreads (i.e., all northeast movements along iso-expected value lines) to make the individual worse off is not that the indifference curves necessarily be linear, but rather that they be everywhere steeper than the (dashed) iso-expected value lines. Of course, this is equivalent to the condition that the tangents to the indifference curves be everywhere steeper, which from the analysis of Section 2 is seen to be equivalent to the condition that all the *local* utility functions are concave in x. Finally, we could illustrate the above generalized Arrow-Pratt theorem on comparative risk aversion by a pair of nonlinear preference fields, one of whose indifference curves always intersected the other's from below (i.e., were everywhere steeper).

Having developed the above results for the case of general differentiable preference functionals, it is useful to see how they might be applied to specific special cases, i.e., to specific nonlinear functional forms. Pursuing the Taylor expansion analogy further, we see that the simplest generalization of "linearity in the probabilities" is "quadratic in the probabilities," or in other words, a functional form such as

$$V(F) \equiv \int R(x)dF(x) + \tfrac{1}{2}[\int S(x)dF(x)]^2 \qquad (6)$$

whose local utility function can be calculated to be

$$U(x; F) = R(x) + S(x)[\int S(z)dF(z)]. \qquad (7)$$

Thus, if $R(.)$ and $S(.)$ are both positive, increasing, and concave it follows that $V(.)$ will exhibit both monotonicity and global risk aversion, and conditions under which one preference functional of this form was everywhere more risk averse than another could similarly be determined. Table 12.1 presents several specific functional forms which have been suggested by researchers which are examples of smooth preference functionals, together with their calculated local utility functions.

It is clear that many more generalizations of "expected utility" type results to non-expected utility maximizers can be derived: for some examples, the reader is referred to Machina (1982a, 1982b, 1984). We conclude this section with remarks on two issues which seem to have caused a lot of confusion in the "expected utility vs. non-expected utility" debate, namely whether non-expected utility maximizers can necessarily be tricked into "making book against themselves," and the nature of "cardinality vs. ordinality of preferences" in the context of expected utility vs. non-expected utility maximization.

There are two senses in which non-expected utility maximizers might make book against themselves (i.e., violate a preference for first order stochastic dominance in either a single choice or a sequence of choices). The first is that for *certain types* of non-expected utility functional forms, most notably the "subjective expected utility" or "prospect theory" form $\Sigma u(x_i)f(p_i)$ (e.g., Edwards, 1955; Kahneman & Tversky, 1979, Chapter 11, this volume), it is *necessarily* true that the functional form will prefer some prospects to others which stochastically dominate them. Such a property is clearly undesirable, and in the present author's view, makes such forms unacceptable as the basis of descriptive theories (it is straightforward to show that this form is not a special case of a general differentiable preference functional). The second sense is that if an individual has a differentiable preference functional and the local utility functions are not all increasing, then the individual will prefer some distributions to others which stochastically dominate them. Of course, the analogous result is also true of expected utility maximizers: to achieve a preference for first order stochastic dominance, we must posit utility functions, von Neumann-Morgenstern or local, which are increasing in x. It is clear that the *real* issue is whether there can exist non-expected utility maximizing individuals who will not make book against themselves, or whether making book against oneself is an *intrinsic* property of non-expected utility maximizers. The answer is easy – we know from above that individuals with increasing local utility functions always prefer stochastically dominating distributions in pairwise choices, and the transitivity which follows from the maximization of $V(.)$ ensures that such individuals will never violate stochastic dominance preference in a sequence of choices either.

The final issue is the apparent confusion that going from expected utility to non-expected utility involves going from "cardinal" preferences to "ordinal" preferences. This is not true. There are two related, though distinct, functions for the expected utility maximizer: the preference functional $V(.)$ over $D[0, M]$ (which happens to be linear) and the von Neumann-Morgenstern utility function $U(.)$ over $[0, M]$. The first of these is ordinal in that any monotonic transformation of $V(.)$ will represent the same preference ranking over $D[0, M]$, and the second is cardinal in that another von Neumann-Morgenstern utility function $U^*(.)$ will represent the individual's preferences if and only if $U^*(x) \equiv aU(x) + b$ $(a > 0)$. *Precisely the same* is true of non-expected utility maximizers: clearly the preference functional $V(.)$ of a non-expected utility maximizer is ordinal, and in Machina (1982a) it was shown that the *local* utility functions $U(.; F)$ are cardinal in that another set of local utility functions will represent the same preferences if and only if they are a positive linear transformation of the original set. Thus, the *preference functionals* of all individuals, expected utility maximizing or otherwise, are always ordinal, and the *utility functions*, von Neumann-Morgenstern or local, are always cardinal. Whether or not the independence axiom is satisfied is irrelevant.

4. The nature of systematic violations of the independence axiom

One of the most important points made by the defenders of expected utility theory is that dropping the independence axiom (i.e. linearity) and retaining only transitivity and completeness (and possibly smoothness) results in a model which possesses almost no predictive power. We have seen in the previous section how generalized expected utility analysis, while not *requiring* strong behavioral assumptions in order to apply, nevertheless still admits of refutable hypotheses such as mono-

Table 12.1 *Local utility functions for various functional forms of $V(.)$*

Mathematical form	Reference*	$V(F)$	$U(x; F)$
Linear (i.e. expected utility)	von Neumann & Morgenstern (1944)	$\int U(x)dF(x)$	$U(x)$
Mean & variance of utility (special case of simple & general quadratic)	Allais (1952, p. 108)	$\bar{u} - \lambda\int(U(x) - \bar{u})^2 dF(x)$ ($\bar{u} = \int U(z)dF(z)$)	$U(x) - \lambda U(x)^2 + 2\lambda U(x)\bar{u}$
Simple quadratic (special case of general quadratic)	Machina (1982a, p. 295)	$\int R(x)dF(x) \pm \frac{1}{2}[\int S(x)dF(x)]^2$	$R(x) \pm S(x)\int S(z)dF(z)$
General quadratic	Machina (1982a, n. 45)	$\int\int T(x,z)dF(x)dF(z)$ ($T(x,z) = T(z,x)$)	$2\int T(x,z)dF(z)$
First three moments of utility	Hagen (1979, p. 272)	$\bar{u} + f(s^2, m^3)$ ($\bar{u} = \int U(z)dF(z)$, $s^2 = \int(U(z)-\bar{u})^2 dF(z)$, $m^3 = \int(U(z)-\bar{u})^3 dF(z)$)	$U(x) + f_1 \cdot [U(x)^2 - 2U(x)\bar{u}]$ $+ f_2 \cdot U(x)[U(x)^2 - 3U(x)\bar{u}$ $+ 3\bar{u}^2 - 3s^2]$
Rational (i.e. ratio of two linear forms)	Chew & MacCrimmon (1979) Fishburn (1983)	$\dfrac{\int w(x)dF(x)}{\int \alpha(x)dF(x)}$	$\dfrac{w(x) - V(F)\ \alpha(x)^{**}}{\int\alpha(z)dF(z)}$

* The reference cited for each functional form is not necessarily the first appearance of that form, nor should it be inferred that the respective author necessarily "prefers" that form over others they may have presented. In some instances I have slightly changed the exact form as given in the reference for greater simplicity.

** I am indebted to Chew Soo Hong and Kenneth MacCrimmon (private correspondence) for the derivation of the local utility function of the rational form. The expression in the table differs from theirs due to a difference in notation.

tonicity and risk aversion, via assumptions on the local utility functions which are analogous to the expected utility conditions. In the present section we review the evidence on the four known types of systematic violations of the independence axiom, and show that they will all follow from a *single* assumption on the shape of the individual preference functional $V(.)$, which we term "Hypothesis II."[4] Thus, in addition to the usual hypotheses of monotonicity and risk aversion, generalized expected utility analysis admits of an evidently quite powerful refutable hypothesis on precisely how individuals violate the independence axiom, and one which has been substantially confirmed by the evidence so far.

4.1 The common consequence effect

As an example of the first type of systematic violation of the axiom, the common consequence effect, we shall consider the first, and still most famous, specific example of this effect, namely the so-called "Allais Paradox" (see Allais, 1952/1979a, p. 89; Morrison, 1967; Moskowitz, 1974; Raiffa, 1968; and Slovic and Tversky, 1974, for example). First proposed by Allais in 1952, this example consists of obtaining the subject's preference ranking over the two pairs of risky prospects

$$a_1: \{100\% \text{ chance of } \$1M \quad \text{versus} \quad a_2: \begin{cases} 10\% \text{ chance of } \$5M \\ 89\% \text{ chance of } \$1M \\ 1\% \text{ chance of } \$ \ 0 \end{cases}$$

and

$$a_3: \begin{cases} 10\% \text{ chance of } \$5M \\ 90\% \text{ chance of } \$ \ 0 \end{cases} \quad \text{versus} \quad a_4: \begin{cases} 11\% \text{ chance of } \$1M \\ 89\% \text{ chance of } \$ \ 0 \end{cases}$$

where $\$1M = \$1,000,000$. While it is easy to show that an expected utility maximizer would prefer either a_1 and a_4 (if $[.10U(5M) - .11U(1M) + .01U(0)] < 0$) or else a_2 and a_3 (if $[.10U(5M) - .11U(1M) + .01U(0)] > 0$), experimenters such as those listed above have repeatedly found that the modal if not majority choice of subjects has been a_1 and a_3, which violates the independence axiom.

The common consequence effect is really a generalization of the type of violation exhibited in the Allais Paradox, and involves choices between pairs of prospects of the form:

---

[4] "Hypothesis I" is a separate hypothesis on the typical shape of the local utility function which, in conjunction with Hypothesis II, serves to generate behavior of the ‚type observed by Friedman & Savage (1948) and Markowitz (1952) (see Machina, 1982a).

Prospect \ Probability	p	$1 - p$
b_1	k	C^*
b_2	a^*	C^*
b_3	k	c^*
b_4	a^*	c^*

where a^*, C^*, and c^* are (possibly) random prospects with C^* stochastically dominating c^*, and k is a sure outcome lying between the highest and lowest outcomes of a^*, so that, for example, b_2 is a prospect with the same ultimate probabilities as a compound prospect yielding a p chance of a^* and a $1 - p$ chance of C^*. It is clear that an individual satisfying the independence axiom would rank b_1 and b_2 the same as b_3 and b_4: whether the "common consequence" was C^* (as in the first pair) or c^* (as in the second) would be "irrelevant." However, researchers such as Kahneman and Tversky (1979), MacCrimmon (1968), and MacCrimmon and Larsson (1979) as well as the five listed previously have found a tendency for individuals to violate the independence axiom by preferring b_1 to b_2 and b_4 to b_3 in problems of this type (this is the same type of behavior as exhibited in the Allais Paradox, since the prospects a_1, a_2, a_3, and a_4 there correspond to b_1, b_2, b_4 and b_3, respectively, with $k = C^* = \$1M$, $c^* = \$0$, and a^* a 10/11:1/11 chance of $\$5M$ or $\$0$). In other words, the better (in the sense of stochastic dominance) the "common consequence," the more risk averse the choice (since a^* is riskier than k).

4.2 The common ratio effect

A second type of systematic violation of the independence axiom, the so-called "common ratio effect," also follows from an early example of Allais' (Allais, 1952/1979a, p. 91) and includes the "Bergen Paradox" of Hagen (1979) and the "certainty effect" of Kahneman and Tversky (1979) [pp. 185–187, this volume] as special cases. This effect involves rankings over pairs of prospects of the form:

$$c_1: \begin{cases} p & \text{chance of } \$X \\ 1 - p & \text{chance of } \$0 \end{cases} \quad \text{versus} \quad c_2: \begin{cases} q & \text{chance of } \$Y \\ 1 - q & \text{chance of } \$0 \end{cases}$$

and

$$c_3: \begin{cases} \alpha p & \text{chance of } \$X \\ 1 - \alpha p & \text{chance of } \$0 \end{cases} \quad \text{versus} \quad c_4: \begin{cases} \alpha q & \text{chance of } \$Y \\ 1 - \alpha q & \text{chance of } \$0 \end{cases}$$

where $p > q$, $X < Y$, and $0 < \alpha < 1$ [the term "common ratio" derives from the equality of prob(X)/prob(Y) in c_1 vs. c_2 and c_3 vs. c_4]. Once again, it is clear that an individual satisfying the independence axiom would rank c_1 and c_2 the same as c_3 and c_4; however, researchers have found a systematic tendency for subjects to depart from the independence axiom by preferring c_1 to c_2 and c_4 to c_3. Thus, Kahneman and Tversky (1979) found, for example, that while 86% of their subjects preferred a .90:.10 chance of $3000 or $0 to a .45:.55 chance of $6000 or $0, 73% preferred a .001:.999 chance of $6000 or $0 to a .002:.998 chance of $3000 or $0 [p. 187, Problems 7 and 8, this volume]. Besides Kahneman and Tversky, other researchers who have found this effect are Hagen (1979, pp. 285–296), MacCrimmon and Larsson (1979, pp. 350–359), and Tversky (1975).

4.3 Oversensitivity to changes in small probability/outlying events

A third type of systematic violation of the independence axiom is that, relative to the "linearity" property of expected utility, individuals tend to exhibit what may be termed an "oversensitivity to changes in the probabilities of small probability/outlying events." While the formalization of this notion requires both a precise definition of what it means for an individual to become "more sensitive" to changes in the probability of an event (relative to changes in the probabilities of certain other events) as well as what it means for an event to become "more outlying" relative to other events, we begin with an intuitive discussion of this notion, using the Allais Paradox of Section 4.1 as an example.

Note that, in the Allais example, the changes from prospects a_1 to a_2 and from a_4 to a_3 both consist of a (beneficial) shift of .10 units of probability mass from the outcome $1M to the outcome $5M and a (detrimental) shift of .01 units from $1M to $0. Since the typical individual prefers a_1 to a_2, we see that when the initial distribution is a_1, i.e., when the outcome $0 is a low probability event, the increase in its probability (at the expense of the preferred outcome $1M) is not compensated for by the beneficial shift of mass up to $5M. However, when the initial distribution is a_4, we see that the event $0 is no longer such a low probability/outlying event (since its probability is now .89) and we find that the individual is no longer as sensitive to the increase in its probability, in the sense that the beneficial shift from $1M to $5M is now enough to compensate and the change to a_3 is preferred. In other words, when the initial distribution changed in a manner which made the outcome $0 "less outlying," the individual became less sensitive to changes in its probability relative to changes in the probabilities of $5M and $1M.

There is an alternative way to view this example which helps bring out another aspect of the notion of "outlyingness." Note that the change in the initial distribution from a_1 to a_4 may be thought of as making the event \$5M "more outlying" relative to the events \$1M and \$0 since, although the probability of the event \$5M itself hasn't changed, the bulk of the distribution has moved *farther away* from the event \$5M. And in response, the individual has become more sensitive to changes in the probability of \$5M, since the beneficial increase in its probability (at the expense of \$1M) which was not enough to outweigh the detrimental shift when the initial distribution was a_1 is now enough to outweigh it when the initial distribution is a_4.

The above discussion serves as motivation for our formalizations of the notions of "changes in sensitivity" and "outlyingness." Noting that any change in a probability distribution must consist of one or more "shifts" of probability mass from one outcome to another, we define the marginal rate of substitution $MRS(x_2 \to x_3, x_2 \to x_1; F)$ as the amount of probability mass which must be shifted from payoff level x_2 to x_3 per unit amount shifted from x_2 to x_1 in order to leave the individual indifferent, when the amounts shifted are infinitesimally small and the initial distribution is $F(.)$ (in the following discussion, we assume $x_1 < x_2 < x_3$). Then, the notion of increased sensitivity in the above discussion of the Allais Paradox may be formalized by saying that a change in the initial distribution $F(.)$ makes the individual *more sensitive to changes in the probability of* x_1 *versus changes in the probabilities of* x_2 *and* x_3 *(and equivalently, less sensitive to changes in the probability of* x_3 *relative to changes in the probabilities of* x_1 *and* x_2*)* if the change serves to raise the value of $MRS(x_2 \to x_3, x_2 \to x_1; F)$ (i.e., the individual is more sensitive to changes in the probability of x_1 if a shift of probability mass from the intermediate value x_2 to x_1 now requires more of a compensating shift of mass from x_2 up to x_3, and similarly for the case of a decreased sensitivity to changes in the probability of x_3 relative to changes in the probabilities of x_1 and x_2).

Again using the discussion of the Allais Paradox as motivation, we will say that any rightward shift of mass within the interval $[x_2, \infty)$ serves to change the initial distribution in a manner which makes the event x_3 *less outlying relative to events* x_1 *and* x_2, since rightward shifts of mass within the interval $[x_2, x_3]$ clearly move the distribution away from x_1 and x_2 and toward x_3, and rightward shifts within the interval $[x_3, \infty)$ also serve to make x_3 less of a "large" outcome relative to the bulk of the distribution, since they result in x_3 being farther from the "right edge" of the distribution. Similarly, leftward shifts of mass within the interval $(-\infty, x_2]$ serve to make the event x_1 *less outlying relative to the events* x_2 *and* x_3. Thus, our formalization of the "over-sensitivity condition" is:

any change in the initial distribution which serves to make an event more (less) outlying relative to a pair of other events serves to change the relevant marginal rate of substitution so as to make the individual more (less) sensitive to changes in the probability of that event relative to changes in the probabilities of the other two events.

While using a notion (the marginal rate of substitution) which is not typically seen in the analysis of preferences over probability distributions, the above condition is very much in the spirit of the Hicks-Allen "diminishing marginal rate of substitution" assumption of non-stochastic demand theory, in that it relates changes in a fundamental marginal rate of substitution to changes in the "current consumption bundle" (in this case, the initial distribution). Furthermore, this condition may be shown to be equivalent to the common consequence effect and to imply the common ratio effect (see Machina, 1982a), and in Section 4.5 below will be shown to possess a nice graphical interpretation in terms of the indifference curves in the unit triangle diagram.

4.4 The utility evaluation effect

The final type of systematic violation of the independence axiom may be termed the "utility evaluation effect." It is well known that there are several ways of evaluating or "assessing" the von Neumann-Morgenstern utility function of an expected utility maximizer, all of which, according to the theory, should yield the same function subject to positive linear transformations (see, for example, Farquhar, 1982). However, in actual practice different techniques have "recovered" utility functions from the same individual which differ in systematic ways.

One of the most frequently used assessment methods is termed the "fractile method" (see McCord and de Neufville, 1982). This method begins by arbitrarily defining $U(0) = 0$ and $U(M) = 1$ for some positive M, and picking some fixed probability $\bar{p}$ between zero and unity. The first step in the method then consists of determining the individual's certainty equivalent of a $\bar{p}: 1 - \bar{p}$ chance of M or 0. If we term this certainty equivalent c_1, it follows from the equation $U(c_1) = \bar{p}U(M) + (1 - \bar{p})U(0)$ that $U(c_1)$ will have the value $\bar{p}$. The second and third step consist of finding the certainty equivalent c_2 of a $\bar{p}: 1 - \bar{p}$ chance of c_1 and 0 [so that $U(c_2) = \bar{p}U(c_1) + (1 - \bar{p})U(0) = \bar{p}^2$] and the certainty equivalent c_3 of a $\bar{p}: 1 - \bar{p}$ chance of M or c_1 [so that $U(c_3) = \bar{p}U(M) + (1 - \bar{p})U(c_1) = \bar{p} + (1 - \bar{p})\bar{p}$]. Further points on the utility curve are determined by finding the certainty equivalents of a $\bar{p}: 1 - \bar{p}$ chance of c_2 or 0, a $\bar{p}: 1 - \bar{p}$ chance of c_1 or c_2, a $\bar{p}: 1 - \bar{p}$ chance of c_3 or c_1, a $\bar{p}: 1 - \bar{p}$ chance of M or c_3, etc., always interpolating by letting $\bar{p}$ be the probability of the higher of the two payoffs. Thus, if $\bar{p} = \frac{1}{2}$, the first step would find that monetary value whose utility was $\frac{1}{2}$, the second and third steps would find the values with utility levels $\frac{1}{4}$ and $\frac{3}{4}$, and so on through $\frac{1}{8}$, $\frac{3}{8}$, $\frac{5}{8}$, $\frac{7}{8}$, $\frac{1}{16}$, $\frac{3}{16}$, etc. Let $U^{\bar{p}}(.)$ denote the utility function derived in this way, for a given value of $\bar{p}$.

Of course, if the individual is an expected utility maximizer, this method ought to recover the same utility function for each value of $\bar{p}$ used, i.e. the functions $U^{1/2}(.)$ and $U^{1/3}(.)$ ought to be identical, since both would have the same normalization $U(0) = 0$ and $U(M) = 1$. However, Karmarkar (1974) discovered an almost universal tendency for the recovered $U^{\bar{p}}(.)$ curve to lie above the $U^{\bar{p}*}(.)$ curve whenever $\bar{p}$ was higher than p^*.[5] This same effect was found (though less markedly) by McCord and de Neufville (1982) and can also be recovered from the experimental data presented by Allais (1979b).[6] Once again, individuals are seen to be evidently departing from the expected utility hypothesis of linearity in a systematic manner.

4.5 Hypothesis II

The previous subsections have presented four types of systematic violations of the independence axiom that have been found by empirical researchers. Needless to say, if these four types of behavior were entirely unrelated (or even mutually contradictory), then supporters of expected utility theory would have a valid point in maintaining that any generalization of expected utility designed to accommodate them would be nothing more than an *ad hoc* extension of the model in each of these four directions.

However, it turns out that not only are each of the above four aspects of behavior compatible, but they all follow from a *single* assumption on the shape of the preference functional $V(.)$. Thus, the data are telling us that not only do individuals' preferences depart

---

[5] Of Karmarkar's four subjects, three exhibited fitted $U^{p}(.)$ curves which strictly and markedly increased with p. The fourth ("Subject B") exhibited $U^{9/10}(.)$ and $U^{3/4}(.)$ curves which were both above the $U^{1/2}(.)$ curve, but which crossed each other at one point. Since the curves of this subject were much closer to each other than the curves of the other subjects, it is possible that this crossing is due to the slightly random character of responses which is typically found in studies of this type.

[6] McCord & de Neufville found that the great majority of their subjects exhibited $U^{1/4}(.)$ curves which were below their $U^{1/2}(.)$ curves in the region where the curves had a value of $\frac{1}{4}$. However, an equal number of their subjects had $U^{3/4}(.)$ curves above and below their $U^{1/2}(.)$ curves, indicating no average departure from linearity in either direction in this region. McCord and de Neufville also found that whether the $U^{1/4}(.)$ and $U^{3/4}(.)$ curves lay above or below the $U^{1/2}(.)$ curve seemed to be correlated with the subject's degree of risk aversion, with the $U^{1/2}(.)$ curve typically lying higher relative to the other curves for risk averters and lower for risk lovers. However, since their method of classifying individuals as risk averse or risk loving was based on the concavity or convexity, and hence *height*, of the $U^{1/2}(.)$ curve, this finding may in part be a statistical artifact introduced by their method of categorizing the observations. Finally, since Allais' method of constructing his "$B_{1/2}$" curves differed slightly from the fractile method, his data may only be used to compare $U^{1/2}(.)$ with $U^{p}(.)$ for $p < \frac{1}{2}$, where it exhibits the utility evaluation effect described in this section (see Allais, 1979b, pp. 611–654).

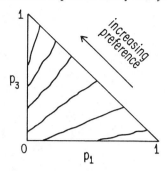

Figure 12.4

from linearity, but they do so in a single systematic manner, which in addition may be modelled quite easily and which (expected utility theorists note:) leads to further refutable restrictions on behavior.

As in standard calculus, one particularly compact way of specifying the nature of a nonlinearity in a preference functional is to specify how the derivative (i.e. the local utility function) of the functional varies as we move about the domain $D[0, M]$. Our formal hypothesis, termed "Hypothesis II," (see note 4) basically states that as we move from one probability distribution in $D[0, M]$ to another which (first order) stochastically dominates it, the local utility function becomes more concave at each point x, or stated formally in terms of the Arrow-Pratt ratio $-U_{11}(x; F)/U_1(x; F)$:

Hypothesis II: If the distribution $F^*(.)$ first order stochastically dominates $F(.)$, then

$$-U_{11}(x; F^*)/U_1(x; F^*) \geq -U_{11}(x; F)/U_1(x; F)$$
for all $x \, \varepsilon \, [0, M]$.

Hypothesis II possesses a straightforward graphical interpretation in terms of the indifference curves in the unit triangle diagram. Note first that the set of all probability distributions in the triangle which stochastically dominate a given distribution correspond to all the points which are northwest of the point representing the distribution.[7] According to Hypothesis II, therefore, the local utility functions at these northwest distributions will be more concave. However, we know from Section 3 that the more concave the (von Neumann-Morgenstern or local) utility function, the steeper the slope of the indifference curves through the point. Accordingly, Hypothesis II

---

[7] Stochastically dominating shifts in $D\{x_1, x_2, x_3\}$ are shifts which increase p_3 at the expense of p_2 and/or increase p_2 at the expense of p_1, which correspond respectively to upward and/or leftward (i.e., northward and/or westward) shifts in the unit triangle diagram.

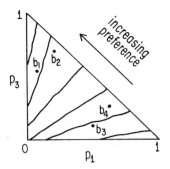

Figure 12.5

implies that indifference curves in the unit triangle "fan out" as in Figure 12.4, with steeper curves lying to the northwest and flatter curves lying to the southeast.

To get an idea of how Hypothesis II implies the common consequence effect, let us refer back to its general formulation (the table in Section 4.1) and consider the special case when the value k and the payoff levels of the prospects c^*, C^*, and a^* are all elements of $\{x_1, x_2, x_3\}$ for some $x_1 < x_2 < x_3$, so that the prospects b_1, b_2, b_3, and b_4 are all in the set $D\{x_1, x_2, x_3\}$ and hence may be plotted in the unit triangle diagram. In such a case it is straightforward to show that the four prospects will always form a parallelogram with b_2 and b_4 to the northeast of b_1 and b_3 respectively, and the segment $\overline{b_1 b_2}$ parallel to and to the north and/or west of $\overline{b_3 b_4}$, e.g., as shown in Figure 12.5. In this case it is easy to see how the "fanning out" property of indifference curves implied by Hypothesis II would lead an individual to violate the independence axiom by preferring b_1 to b_2 and b_4 to b_3, which is precisely the common consequence effect. In Machina (1982a) it was shown that Hypothesis II is in fact *equivalent* to the common consequence effect in the more general case when c^*, C^*, and a^* may be arbitrary (possibly continuous) prospects.

A similar graphical analysis demonstrates now Hypothesis II implies the second type of systematic violation of the independence axiom, namely the common ratio effect. Letting $x_1 = 0$, $x_2 = X$, and $x_3 = Y$ in the formulation of Section 4.2 and plotting the prospects c_1, c_2, c_3, c_4 in the unit triangle diagram, we once again find that c_2 and c_4 are northeast of c_1 and c_3 respectively and that $\overline{c_1 c_2}$ is parallel to and northwest of $\overline{c_3 c_4}$, as seen in Figure 12.6. And similarly, it is clear how the "fanning out" property implied by Hypothesis II would lead the individual to violate the independence axiom by preferring c_1 to c_2 and c_4 to c_3, i.e., exhibit the common ratio effect.

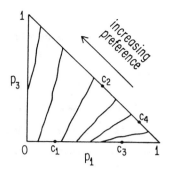

Figure 12.6

Hypothesis II's implication that the individual will be systematically oversensitive to changes in the probabilities of low probability-outlying events may be seen quite simply from Figure 12.4 above. We begin by noting that, just as in non-stochastic demand theory, the marginal rate of substitution $MRS(x_2 \rightarrow x_3, x_2 \rightarrow x_1; F)$ is precisely equal to the slope of the indifference curve through the point corresponding to the distribution $F(.)$ in the diagram, since rightward and upward movements in the diagram correspond to the shifts $x_2 \rightarrow x_3$ and $x_2 \rightarrow x_1$ respectively. Under the fanning out implication of Hypothesis II, we find that the individual is most sensitive to changes in the probability of x_1 relative to changes in the probabilities of x_2 and x_3 [i.e., $MRS(x_2 \rightarrow x_3, x_2 \rightarrow x_1; F)$ is the highest] near the left edge of the triangle, or in other words precisely when x_1 is a low probability event (i.e., p_1 is low). Note also that moving straight up in the triangle, which does not change p_1 but increases p_3 at the expense of p_2, also serves to make the event x_1 more outlying (since it moves probability mass further away from x_1) and indeed is seen to also increase the individual's sensitivity to changes in p_1, as measured by the slope of the indifference curves. An analogous argument applies to the individual's sensitivity to changes in p_3 relative to changes in p_1 and p_2.

Finally, we may also use the triangle diagram to illustrate how Hypothesis II implies the utility evaluation effect. If we were to take an individual satisfying Hypothesis II and try to "evaluate" his or her $U^{1/2}(.)$ curve, the first step (as in Section 4.4) would be to determine the certainty equivalent c_1 of a $\frac{1}{2}:\frac{1}{2}$ chance of M or 0. Consider now Figure 12.7, where we pick $x_1 = 0$, $x_2 = c_1$, and $x_3 = M$, so that the origin (i.e., the sure prospect c_1) is seen to lie on the same indifference curve as the prospect which offers a $\frac{1}{2}:\frac{1}{2}$ chance of M or 0. We then find the sure amount c_2 which is indifferent to a $\frac{1}{2}:\frac{1}{2}$ chance of c_1 or 0, and the amount c_3 which is indifferent to a $\frac{1}{2}:\frac{1}{2}$ chance of M and c_1

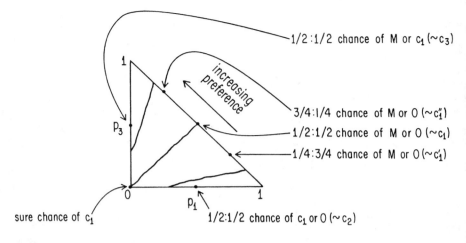

("~" denotes indifference)

Figure 12.7

(see Figure 12.7). These three points, with their associated $U^{1/2}(.)$ values of ¼, ½, and ¾, are plotted in Figure 12.8 as points on the $U^{1/2}(.)$ curve.

Now, to evaluate the first point on the $U^{1/4}(.)$ curve, we find the certainty equivalent c_1' of a ¼:¾ chance of M or 0. However, if we note where this latter prospect lies in Figure 12.7, we see that it will be preferred to a ½:½ chance of c_1' or 0, so that its certainty equivalent c_1' will be higher than c_2. This of course implies that $U^{1/2}(.)$ will attain a value of ¼ before $U^{1/4}(.)$ does, so that $U^{1/2}(.)$ lies above $U^{1/4}(.)$ in this region. Similarly, the first point on the $U^{3/4}(.)$ curve will be the value c_1'' which is indifferent to a ¾:¼ chance of M or 0, and again it is seen from Figure 12.7 that since this prospect will be less preferred than a ½:½ chance of M or c_1, c_1'' must be less than c_3, which implies that $U^{3/4}(.)$ lies above $U^{1/2}(.)$ in this range (see Figure 12.8). This analysis may be extended to a further evaluation and comparison of the three "evaluated utility functions" in a manner which continues to exhibit the utility evaluation effect.

Accordingly, it is *not* true, as some expected utility defenders might suppose, that the violations of the independence axiom which researchers have found are random and unsystematic departures from expected utility, but rather, individuals have been found to depart from expected utility in a systematic and unified manner, as captured by Hypothesis II in general and by the "fanning out" property in the special case of preferences over three-outcome distributions.

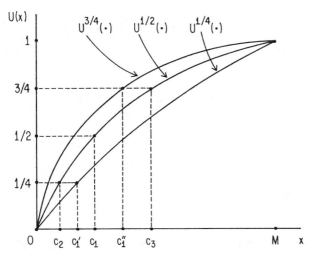

Figure 12.8

4.6. Further predictions and policy implications of Hypothesis II

It is easy to see that Hypothesis II possesses that final required property of any replacement of the expected utility hypothesis, namely the ability to generate further refutable predictions and policy implications. Of course, since each of the four types of systematic violations of expected utility discussed above is a general principle rather than a specific example, each admits of an infinite number of specific examples which serve as refutable predictions. As a new type of example, I would like to consider a problem posed by Professor Kenneth Arrow in his superb and thought provoking Plenary Talk in this Conference.[8] Arrow noted that one of the canonical problems in choice under uncertainty involves the tradeoff between the probability and the outcome value of an unfortunate event, and offered the specific example of an individual with initial wealth W facing a p probability of a loss of X (with a $1 - p$ probability of no loss). A natural question to ask here is how does the individual's marginal rate of substitution between p and X depend upon their existing values. Defining expected utility $\phi(p, X; W) \equiv pU(W - X) + (1 - p)U(W)$, we get that this marginal rate of substitution is

$$MRS_{p,X} = \left.\frac{dp}{dX}\right|_\phi = \frac{-pU'(W - X)}{U(W) - U(W - X)} \tag{8}$$

[8] This paper was originally presented at the First International Conference on Foundations of Utility and Risk Theory, Oslo, June 1982.

In his talk, Arrow noted that this expected utility formulation implied a possibly quite useful restriction on behavior, namely that, fixing X and W, the marginal rate of substitution between p and X is proportional to p, i.e., to the probability of the unfortunate event. He quite rightly noted that it would be possible to exploit this property to make important predictions *as well as policy suggestions*, say in determining the tradeoff between the probability and severity of a nuclear accident, and also noted that any acceptable alternative to expected utility would have to possess this same type of ability.

To see how generalized expected utility analysis, and more particularly Hypothesis II, might be applied to this problem, we replace the expected utility maximand $\phi(p, X; W)$ with the more general maximand $V(pG_{W-X} + (1 - p)G_W)$, where G_c stands for the distribution with unit mass at c, so that $pG_{W-X} + (1 - p) G_W$ represents the distribution in question. We then have from equation (4) that

$$MRS_{p,X} = \frac{dp}{dX}\bigg|_V = \frac{-pU_1(W - X; pG_{W-X} + (1 - p)G_W)}{U(W; pG_{W-X} + (1 - p)G_W) - U(W - X; pG_{W-X} + (1 - p)G_W)} =$$

(after some manipulation)   (9)

$$= -p\left[\int_{W-X}^W \exp\left[-\int_{W-X}^z \left\{-\frac{U_{11}(\omega; pG_{W-X} + (1 - p)G_W)}{U_1(\omega; pG_{W-X} + (1 - p)G_W)}\right\}d\omega\right]dz\right]^{-1}.$$

As usual in generalized expected utility analysis, we see the formal analogy with the expected utility case: the marginal rate of substitution in (9) is identical to that in (8) with the von Neumann-Morgenstern utility function $U(.)$ replaced by the local utility function $U(.; F)$ when $F = pG_{W-X} + (1 - p)G_W$. However, since the local utility function in (9) now depends on the precise distribution $pG_{W-X} + (1 - p)G_W$, the marginal rate of substitution is no longer strictly proportional to p as before. However, this is not to say that Hypothesis II is without implications in this case. Noting that an increase in p induces a first order stochastic worsening of the distribution $pG_{W-X} + (1 - p)G_W$, we see that under Hypothesis II an increase in p will lower the term in curled brackets in (9) (the Arrow-Pratt term) for each value of ω, so that Hypothesis II implies that the marginal rate of substitution between p and X varies less than proportionately with p. The replacement of the expected utility prediction of exact proportionality with a weak inequality on proportionality reflects the fact that Hypothesis II is a weak inequality which includes the expected utility case (i.e. the independence axiom) as a borderline case, just as, geometrically, "fanning out" includes parallel linear indifference curves as a borderline case. Nevertheless, weak inequalities are still refutable restrictions on behavior (we use them all the time in economics) and this result is clearly not without policy implications which, if not as strong as the ones generated by expected utility, are at least more accurately tied to what we have observed about individuals' actual preferences. While this is just a single example, it should be clear that Hypothesis II can be used to derive other important behavioral predictions and policy implications.

5. Conclusion

Defenders of the expected utility approach are quite correct in insisting that any alternative to expected utility not only be consistent with the data, but also be at least on the order of elegance of the expected utility theory, and capable of easily derived behavioral restrictions and implications for policy analysis. The technique of generalized expected utility analysis seems to fit these requirements. Specifically,

while making virtually no requisite assumptions on preferences other than completeness, transitivity, and smoothness, it allows us to retain the elegant set of concepts, tools, and techniques of expected utility analysis,

it admits of refutable restrictions on preferences and hence on behavior, with the concepts of monotonicity and risk aversion, for example, modelled almost exactly as in expected utility analysis, and

it admits of a restriction (Hypothesis II) which implies the four known types of observed systematic violations of the independence axiom, and which generates both additional refutable behavioral predictions as well as policy implications.

Whether the future will yield empirical observations which contradict Hypothesis II, or even the underlying assumption of smooth preferences, is really not the issue at hand (see, e.g., Kahneman and Tversky, 1979, pp. 271–273, pp. 192–195, this volume; Tversky, 1969, 1975; Grether, 1978; and Grether and Plott, 1979). The present point is that generalized expected utility analysis seems to offer a theoretically powerful and empirically supported generalization of the expected utility model. Indeed, if generalized expected utility analysis and other related models lead to the type of empirical work which will require still newer models to replace them, they will have served us well.

Part IV
Unreliable probabilities

One of the fundamental assumptions for the classical form of Bayesianism is that the decision maker's beliefs can be represented by a unique probability distribution. This assumption was named A3 in the introduction. The best known argument in favor of it is the so-called Dutch book theorem. However, there are also a number of strong arguments against this assumption.

Some of the examples discussed in Part II of this volume seem to indicate that there are decision situations which are identical in all aspects relevant to a Bayesian decision maker, but which nevertheless motivate different decisions. Even if the probabilities involved are the same, the *reliability* of the information affects the choices. We decide differently depending on the type of information available. There is a clear difference between situations where we have scanty information about the events involved (e.g., the event of a bus strike in Verona, Italy, next month) and situations where we have more or less complete knowledge of the random processes involved (e.g., the event of drawing a ball from an urn).

It is important to bear in mind that the unreliability of probability information is not connected with the problem of how exactly to determine the subjective probabilities. The second problem, which was discussed in Hacking's article in Part II of this volume, is a *measurement* problem dealing with the clarity of probability perception; the unreliability problem is *epistemological* dealing with the quality of information underlying the probability assessments. For a discussion of this distinction, see Levi (1985).

Perhaps the clearest illustration of the effect of the unreliability of probabilities is Ellsberg's paradox (discussed in the introduction). In the first paper in this part, Daniel Ellsberg presents a development of the Bayesian theory that avoids the problems created by the paradox. Among other things, he introduces the notion of "ambiguity" [what we call unreliability] which exists when a decision maker does not "know enough about the problem to rule out a number of possible distributions . . ." (p. 258 in this volume).

Besides theoretical arguments against the assumption A3 there are a number of empirical results suggesting that the assumption has no descriptive value (e.g., Becker and Brownson, 1964; Yates and Zukowski, 1976; Larson, 1980; Goldsmith and Sahlin, 1982; and Einhorn and Hogart, 1985). To give one example of the findings, Becker and Brownson (1964) in their experimental study submit that "the *ambiguity* associated with a given alternative is determined by the nature of the distribution on the probabilities of future events relevant to that action" (p. 64). In their experiment, the subjects' beliefs about the states were determined by limited information about the contents of a number of urns containing white and black balls. They were informed that each urn contained a minimum of white balls and a minimum of black balls among a total of 100, but otherwise they were given no information about the distribution of the colors. For example, the only information the subjects had about Urn V was that it contained at least 40 white balls and at least 40 black balls among the total 100.

Becker and Brownson defined the ambiguity associated with an urn as the range of the possible distributions. In Urn V, for example, the range is 20, i.e., the difference between 60 and 40. In the experiment it was found that the amount of money the subjects were willing to pay to avoid the ambiguity was consistently related to the degree of ambiguity (as they had defined it).

The central problem for assumption A3 is that it neglects the differences in the quality of information available in different decision situations, since such differences cannot be captured by a single probability distribution. As Kyburg (1968) suggests in his article (this volume p. 117), one way to resolve this problem would be to represent beliefs by probability *intervals*. If one has detailed information about an event, its associated probability interval would be narrow. In the case of perfect information it would be reduced to a point, whereas if the information available is scanty, the associated probability interval would be wider (in the case of complete ignorance it would be the total interval from 0 to 1). In Kyburg (1983a), this idea is developed into a complete decision theory. Similarly, Frederic Schick (1979) (this volume pp. 283–295) shows how Ramsey's theory can be extended to handle problems concerning the varying reliability of information. The extension of the traditional theory proposed by Schick also provides the probability of an event with upper and lower bounds. Theories related to Kyburg's and Schick's are also presented by Smith (1961, 1965), Good (1962), and Dempster (1967). (Savage mentions probability intervals as a possible solution to the problem of 'unsure' opinions, footnote, p. 58, second edition, 1954/1972.)

However, there are some conceptual drawbacks connected with representing beliefs by probability intervals. Most important, it is difficult to model how such intervals should be *updated* in the light of new evidence. This has led some authors to suggest that beliefs should be represented by a *set* of probability distributions (rather than a single distribution as in the traditional theory). The updating problem can then be handled by updating each of the distributions in the set. The intended interpretation of such a set of probability distributions is that, even if the information available to the decision maker does not allow him to pick out a unique distribution, he can determine the class of distributions consistent with his beliefs. The most prominent decision theory based on this kind of representation of beliefs has been developed by Isaac Levi (1974, Chapter 15, this volume, 1980, and 1982).

In Gärdenfors and Sahlin (1982, Chapter 16, this volume, and 1983) a further development of this idea is carried out. Apart from a set of probability distributions, a measure of the epistemic *reliability* of these distributions is introduced. The idea is that even if several probability distributions are possible from the decision maker's point of view, some of them are backed up by more information than others and are thus more reliable. On the basis of the decision theory suggested in the paper, it can be shown that the different degrees of reliability will influence the decision maker's risk behavior.

The type of decision theory suggested in this article and in Levi's article avoids the criticism directed against assumption A3 and it can explain many of the experimental findings mentioned above. Related theories are presented by Einhorn and Hogarth (1985), who develop a psychological theory and present a number of experimental findings, and Berger (1982), who discusses similar problems in statistics.

13. Risk, ambiguity, and the Savage axioms

Daniel Ellsberg

1. Are there uncertainties that are not risks?

There has always been a good deal of skepticism about the behavioral significance of Frank Knight's distinction between "measurable uncertainty" or "risk," which may be represented by numerical probabilities, and "unmeasurable uncertainty" which cannot. Knight maintained that the latter "uncertainty" prevailed – and hence that numerical probabilities were inapplicable – in situations when the decision-maker was ignorant of the statistical frequencies of events relevant to his decision; or when a priori calculations were impossible; or when the relevant events were in some sense unique; or when an important, once-and-for-all decision was concerned.[1]

Yet the feeling has persisted that, even in these situations, people tend to *behave* "as though" they assigned numerical probabilities, or "degrees of belief," to the events impinging on their actions. However, it is hard either to confirm or to deny such a proposition in the absence

Research for this paper was done as a member of the Society of Fellows, Harvard University, 1957. It was delivered in essentially its present form, except for Section III, at the December meetings of the Econometric Society, St Louis, 1960. In the recent revision of Section III, I have been particularly stimulated by discussions with A. Madansky, T. Schelling, L. Shapley, and S. Winter.
Reprinted from *Quarterly Journal of Economics*, 75 (1961), pp. 643–669, by kind permission of the publisher.

[1] F. H. Knight (1921). But see Arrow's comment: "In brief, Knight's uncertainties seem to have surprisingly many of the properties of ordinary probabilities, and it is not clear how much is gained by the distinction.... Actually, his uncertainties produce about the same reactions in individuals as other writers ascribe to risk." K. J. Arrow (1951b, pp. 417, 426).

of precisely-defined procedures for *measuring* these alleged "degrees of belief."

What might it mean operationally, in terms of refutable predictions about observable phenomena, to say that someone behaves "as if" he assigned quantitative likelihoods to events: or to say that he does not? An intuitive answer may emerge if we consider an example proposed by Shackle, who takes an extreme form of the Knightian position that statistical information on frequencies within a large, repetitive class of events is strictly irrelevant to a decision whose outcome depends on a single trial. Shackle not only rejects numerical probabilities for representing the uncertainty in this situation; he maintains that in situations where all the potential outcomes seem "perfectly possible" in the sense that they would not violate accepted laws and thus cause "surprise," it is impossible to distinguish meaningfully (i.e., in terms of a person's behavior, or any other observations) between the relative "likelihoods" of these outcomes. In throwing a die, for instance, it would not surprise us at all if an ace came up on a single trial, nor if, on the other hand, some other number came up. So Shackle concludes:

Suppose the captains in a Test Match have agreed that instead of tossing a coin for a choice of innings they will decide the matter by this next throw of a die, and that if it shows an ace Australia shall bat first, if any other number, then England shall bat first. Can we now give any meaningful answer whatever to the question, "Who will bat first?" except "We do not know?"[2]

Most of us might think we could give better answers than that. We could say, "England will bat first," or more cautiously: "I think England will probably bat first." And if Shackle challenges us as to what we "mean" by that statement, it is quite natural to reply: "We'll *bet* on England; and we'll give you good odds."

It so happens that in this case statistical information (on the behavior of dice) is available and does seem relevant even to a "single shot" decision, our bet; it will affect the odds we offer. As Damon Runyon once said, "The race is not always to the swift nor the battle to the strong, but that's the way to bet." However, it is our bet itself, and not the reasoning and evidence that lies behind it, that gives operational meaning to our statement that we find one outcome "more likely" than another. And we may be willing to place bets – thus revealing "degrees of belief" in a quantitative form – about events for which there is no

---

[2] G. L. S. Shackle (1955, p. 8). If this example were not typical of a number of Shackle's works, it would seem almost unfair to cite it, since it appears so transparently inconsistent with commonly-observed behavior. Can Shackle really believe that an Australian captain who cared about batting first would be *indifferent* between staking this outcome on "heads" or on an ace?

statistical information at all, or regarding which statistical information seems in principle unobtainable. If our pattern of bets were suitably orderly – if it satisfied certain postulated constraints – it would be possible to infer for ourselves numerical subjective probabilities for events, in terms of which some future decisions could be predicted or described. Thus a good deal – perhaps all – of Knight's class of "un-measurable uncertainties" would have succumbed to measurement, and "risk" would prevail instead of "uncertainty."

A number of sets of constraints on choice-behavior under uncer-tainty have now been proposed, all more or less equivalent or closely similar in spirit, having the implication that – for a "rational" man – *all* uncertainties can be reduced to *risks*.[3] Their flavor is suggested by Ramsay's early notions that, "The degree of a belief is ... the extent to which we are prepared to act upon it," and "The probability of ⅓ is clearly related to the kind of belief which would lead to a bet of 2 to 1" (Ramsey, 1931, p. 171, p.29, this volume). Starting from the notion that gambling choices are influenced by, or "reflect," differing degrees of belief, this approach sets out to *infer* those beliefs from the actual choices. Of course, in general those choices reveal not only the person's relative expectations but his relative preferences for outcomes; there is a problem of distinguishing between these. But if one picks the right choices to observe, and if the Savage postulates or some equivalent set are found to be satisfied, this distinction can be made unambiguously, and either qualitative, or, ideally, numerical probabili-ties can be determined. The propounders of these axioms tend to be hopeful that the rules *will* be commonly satisfied, at least roughly and most of the time, because they regard these postulates as normative maxims, widely-acceptable principles of rational behavior. In other words, people should tend to behave in the postulated fashion, because that is the way they would *want* to behave. At the least, these axioms are believed to predict certain choices that people will make when they take plenty of time to reflect over their decision, in the light of the postulates.

In considering only deliberate decisions, then, does this leave any room at all for "unmeasurable uncertainty": for uncertainties *not* re-ducible to "risks," to quantitative or qualitative probabilities?

A side effect of the axiomatic approach is that it supplies, at last (as Knight did not), a useful operational meaning to the proposition that

---

[3] Ramsey (1931) [Chapter 2, this volume]; Savage (1954); de Finetti (1951, pp. 217–226); Suppes, Davidson, and Siegel (1957). Closely related approaches, in which individual choice behavior is presumed to be stochastic, have been developed by Luce (1959) and Chipman (1960). Although the argument in this paper applies equally well to these latter stochastic axiom systems, they will not be discussed explicitly.

people do *not* always assign, or act "as though" they assigned, probabilities to uncertain events. The meaning would be that with respect to certain events they did not obey, nor did they wish to obey – *even on reflection* – Savage's postulates or equivalent rules. One could emphasize here either that the postulates failed to be acceptable in those circumstances as normative rules, or that they failed to predict reflective choices; I tend to be more interested in the latter aspect, Savage no doubt in the former. (A third inference, which H. Raiffa favors, could be that people need more drill on the importance of conforming to the Savage axioms.) But from either point of view, it would follow that *there would be simply no way to infer meaningful probabilities for those events from their choices*, and theories which purported to describe their uncertainty in terms of probabilities would be quite inapplicable in that area (unless quite different operations for measuring probability were devised). Moreover, such people could not be described as maximizing the mathematical expectation of utility on the basis of numerical probabilities for those events derived on *any* basis. Nor would it be possible to derive numerical "von Neumann-Morgenstern" utilities from their choices among gambles involving those events.

I propose to indicate a class of choice-situations in which many otherwise reasonable people neither wish nor tend to conform to the Savage postulates, nor to the other axiom sets that have been devised. But the implications of such a finding, if true, are not wholly destructive. First, both the predictive and normative use of the Savage or equivalent postulates might be improved by avoiding attempts to apply them in certain, specifiable circumstances where they do not seem acceptable. Second, we might hope that it is precisely in such circumstances that certain proposals for alternative decision rules and nonprobabilistic descriptions of uncertainty (e.g., by Knight, Shackle, Hurwicz, and Hodges and Lehmann) might prove fruitful. I believe, in fact, that this is the case.

2. Uncertainties that are not risks

Which of two events, α, β, does an individual consider "more likely"? In the Ramsey-Savage approach, the basic test is: *On which event would he prefer to stake a prize, or to place a given bet?* By the phrase, "to offer a bet *on* α" we shall mean: to make available an action with consequence *a* if α occurs (or, as Savage puts it, if α "obtains") and *b* if α does not occur (i.e., if ᾱ, or "not-α" occurs), where *a* is preferable to *b*.

Suppose, then, that we offer a subject alternative bets "on" α and "on" β (α, β need not be either mutually exclusive or exhaustive, but for convenience we shall assume in all illustrations that they are mutually exclusive).

Events

		α	β	$\bar{\alpha} \cap \bar{\beta}$
Gambles	I	a	b	b
	II	b	a	b

The Ramsey-Savage proposal is to interpret the person's preference between I and II as revealing the relative likelihood he assigns to α and β. If he does not definitely prefer II to I, it is to be inferred that he regards α as "not less probable than" β, which we will write: $\alpha \gtrapprox \beta$.

For example, in the case of Shackle's illustration, we might be allowed to bet either that England will bat first or that Australia will (these two events being complementary), staking a $10 prize in either case:

	England first	Australia first
I	$10	$0
II	$0	$10

If the event were to be determined by the toss of a die, England to bat first if any number but an ace turned up, I would strongly prefer gamble I (and if Shackle should really claim indifference between I and II, I would be anxious to make a side bet with him). If, on the other hand, the captains were to toss a coin, I would be indifferent between the two bets. In the first case an observer might infer, on the basis of the Ramsey-Savage axioms, that I regarded England as more likely to bat first than Australia (or, an ace as less likely than not to come up); in the second case, that I regarded heads and tails as "equally likely."

That inference would, in fact, be a little hasty. My indifference in the second case would indeed indicate that I assigned equal probabilities to heads and tails, *if I assigned any probabilities at all to those events*; but the latter condition would remain to be proved, and it would take further choices to prove it. I might, for example, be a "minimaxer," whose indifference between the two bets merely reflected the fact that their respective "worst outcomes" were identical. To rule out such possibilities, it would be necessary to examine my pattern of preferences in a number of well-chosen cases, in the light of certain axiomatic constraints.

In order for any relationship ⊜ among events to have the properties of a "qualitative probability relationship," it must be true that:

(a) ⊜ is a complete ordering over events; for any two events α, β either α is "not less probable than" β, or β is "not less probable than" α, and if α ⊜ β and β ⊜ γ, then α ⩾ γ.

(b) If α is more probable than β, then "not-α" (or, ᾱ) is less probable than not-β (β̄); if α is equally probable to ᾱ, and β is equally probable to β, then α is equally probably to β.

(c) If α and γ are mutually exclusive, and so are β and γ (i.e., if α ∩ γ = β ∩ γ = 0), and if α is more probable than β, then the union (α ∪ γ) is more probable than (β ∪ γ).

Savage proves that the relationship ⊜ among events, inferred as above from choices among gambles, will have the above properties if the individual's pattern of choices obeys certain postulates. To indicate some of these briefly:

P1: Complete ordering of gambles, or "actions." In the example below either I is preferred to II, II is preferred to I, or I and II are indifferent. If I is preferred to II, and II is preferred or indifferent to III, then I is preferred to III (not shown).

	α	β	ᾱ ∩ β̄
I	a	b	b
II	b	a	b

P2: The choice between two actions must be unaffected by the value of pay-offs corresponding to events for which both actions have the *same* pay-off (i.e., by the value of pay-offs in a constant column). Thus, if the subject preferred I to II in the example above, he should prefer III to IV, below, when *a* and *b* are unchanged and *c* takes any value:

	α	β	ᾱ ∩ β̄
III	a	b	c
IV	b	a	c

This corresponds to Savage's Postulate 2, which he calls the "Sure-thing Principle" and which bears great weight in the analysis. One rationale for it amounts to the following: Suppose that a person would not prefer IV to III if he *knew* that the third column would not "obtain";

if, on the other hand, he knew that the third column *would* obtain, he would still not prefer IV to III, since the pay-offs (whatever they are) are equal. So, since he would not prefer IV to III "in either event," he should not prefer IV when he does not know *whether or not* the third column will obtain.

"Except possibly for the assumption of simple ordering," Savage asserts, "I know of no other extralogical principle governing decisions that finds such ready acceptance." [p. 80, this volume].[4]

P4: The choice in the above example must be independent of the values of a and b, given their ordering. Thus, preferring I to II, the subject should prefer V to VI below, when $d > e$:

	α	β	$\bar{\alpha} \cap \bar{\beta}$
V	d	e	e
VI	e	d	e

This is Savage's Postulate 4, the independence of probabilities and pay-offs. Roughly, it specifies that the choice of event on which a person prefers to stake a prize should not be affected by the size of the prize.

In combination with a "noncontroversial" Postulate P3 (corresponding to "admissibility," the rejection of dominated actions), these four postulates, if generally satisfied by the individual's choices, imply that his preference for I over II (or III over IV, or V over VI) may safely be interpreted as sufficient evidence that he regards α as "not less probable than" β; the relationship "not less probable than" *thus operationally defined*, will have all the properties of a "qualitative probability relationship." (Other postulates, which will not be considered here, are necessary in order to establish *numerical* probabilities.) In general, as one ponders these postulates and tests them introspectively in a varie-

---

[4] Savage notes that the principle, in the form of the rationale above, "cannot appropriately be accepted as a postulate in the sense that P1 is, because it would introduce new undefined technical terms referring to knowledge and possibility that would render it mathematically useless without still more postulates governing these terms." [pp. 80–81, this volume]. He substitutes for it a postulate corresponding to P2 above as expressing the same intuitive constraint. Savage's P2 corresponds closely to "Rubin's Postulate" (Luce & Raiffa, 1957, p. 290) [this volume p. 63] or Milnor's "Column Linearity" postulate, (Luce & Raiffa, 1957, p. 297) [this volume p. 71], which implies that adding a constant to a column of pay-offs should not change the preference ordering among acts.

If numerical probabilities were assumed known, so that the subject were dealing explicitly with known "risks," these postulates would amount to Samuelson's (1952) "Special Independence Assumption", on which Samuelson relies heavily in his derivation of "von Neumann-Morgenstern utilities."

ty of hypothetical situations, they do indeed appear plausible. That is to say that they do seem to have wide validity as normative criteria (for me, as well as for Savage); they are probably[5] roughly accurate in predicting certain aspects of actual choice behavior in many situations and better yet in predicting reflective behavior in those situations. To the extent this is true, it should be possible to infer from certain gambling choices in those situations at least a qualitative probability relationship over events, corresponding to a given person's "degrees of belief."

Let us now consider some situations in which the Savage axioms do *not* seem so plausible: circumstances in which none of the above conclusions may appear valid.

Consider the following hypothetical experiment. Let us suppose that you confront two urns containing red and black balls, from one of which a ball will be drawn at random. To "bet on Red$_I$" will mean that you choose to draw from Urn I; and that you will receive a prize a (say $100) if you draw a red ball ("if Red$_I$ occurs") and a smaller amount b (say, $0) if you draw a black ("if not-Red$_I$ occurs").

You have the following information. Urn I contains 100 red and black balls, but in a ratio entirely unknown to you; there may be from 0 to 100 red balls. In Urn II, you confirm that there are exactly 50 red and 50 black balls. An observer – who, let us say, is ignorant of the state of your information about the urns – sets out to measure your subjective probabilities by interrogating you as to your preferences in the following pairs of gambles:

1. "Which do you prefer to bet on, Red$_I$ or Black$_I$: or are you indifferent?" That is, drawing a ball from Urn I, on which "event" do you prefer the $100 stake, red or black: or do you care?
2. "Which would you prefer to bet on, Red$_{II}$ or Black$_{II}$?"
3. "Which do you prefer to bet on, Red$_I$ or Red$_{II}$?"[6]
4. "Which do you prefer to bet on, Black$_I$ or Black$_{II}$?"[7]

Let us suppose that in both the first case and the second case, you are indifferent (the typical response).[8] Judging from a large number of

---

[5] I bet.

[6] Note that in no case are you invited to choose both a color and an urn freely; nor are you given any indication beforehand as to the full set of gambles that will be offered. If these conditions were altered (as in some of H. Raiffa's experiments with students), you could employ randomized strategies, such as flipping a coin to determine what color to bet on in Urn I, which might affect your choices.

[7] See immediately preceding note.

[8] Here we see the advantages of purely hypothetical experiments. In "real life," you would probably turn out to have a profound color preference that would invalidate the whole first set of trials, and various other biases that would show up one by one as the experimentation progressed inconclusively. [*continued on p. 253*]

responses, under absolutely nonexperimental conditions, your answers to the last two questions are likely to fall into one of three groups. You may still be indifferent within each pair of options. (If so, you may sit back now and watch for awhile.) But if you are in the majority, you will report that you prefer to bet on Red$_{II}$ rather than Red$_I$, and Black$_{II}$ rather than Black$_I$. The preferences of a small minority run the other way, preferring bets on Red$_I$ to Red$_{II}$, and Black$_I$ to Black$_{II}$.

If you are in either of these latter groups, you are now in trouble with the Savage axioms.

Suppose that, betting on red, you preferred to draw out of Urn II. An observer, applying the basic rule of the Ramsey-Savage approach, would infer tentatively that you regarded Red$_{II}$ as "more probable than" Red$_I$. He then observes that you also prefer to bet on Black$_{II}$ rather than Black$_I$. Since he cannot conclude that you regard Red$_{II}$ as more probable than Red$_I$ and, at the same time, not-Red$_{II}$ as more probable than not-Red$_I$ – this being inconsistent with the essential properties of probability relationships – he must conclude that your choices are not revealing judgments of "probability" at all. So far as these events are concerned, it is *impossible* to infer probabilities from your choices; you must inevitably be violating some of the Savage axioms (specifically, P1 and P2, complete ordering of actions or the Sure-thing Principle).[9]

However, the results in Chipman's almost identical experiment (1960, pp. 87–88) do give strong support to this finding; Chipman's explanatory hypothesis differs from that proposed below.

[9] In order to relate these clearly to the postulates, let us change the experimental setting slightly. Let us assume that the balls in Urn I are each marked with a I, and the balls in Urn II with a II; the contents of both urns are then dumped into a single urn, which then contains 50 Red$_{II}$ balls, 50 Black$_{II}$ balls, and 100 Red$_I$ and Black$_I$ balls in unknown proportion (or in a proportion indicated only by a small random sample, say, one red and one black). The following actions are to be considered:

	100		50	50
	R$_I$	B$_I$	R$_{II}$	B$_{II}$
I	a	b	b	b
II	b	a	b	b
III	b	b	a	b
IV	b	b	b	a
V	a	a	b	b
VI	b	b	a	a

Let us assume that a person is indifferent between I and II (between betting on R$_I$ or B$_I$), between III and IV and between V and VI. It would then follow from Postulates 1 and 2, the assumption of a complete ordering of actions and the Sure-thing Principle, that, I, II, III and IV are all indifferent to each other.

To indicate the nature of the proof, suppose that I is preferred to III (the person prefers to bet on R$_I$ rather than R$_{II}$). Postulates 1 and 2 imply that certain transformations can be performed on this pair of actions *without affecting their preference ordering*; specifically, one action can be replaced by an action indifferent to it (P1 – complete

The same applies if you preferred to bet on Red_I and $Black_I$ rather than Red_{II} or $Black_{II}$. Moreover, harking back to your earlier (hypothetical) replies, any *one* of these preferences involves you in conflict with the axioms. For if one is to interpret from your answers to the first two questions that Red_I is "equally likely" to not-Red_I, and Red_{II} is equally likely to not-Red_{II}, then Red_I (or $Black_I$) should be equally likely to Red_{II} (or to $Black_{II}$), and any preference for drawing from one urn over the other leads to a contradiction.[10]

It might be objected that the assumed total ignorance of the ratio of red and black balls in Urn I is an unrealistic condition, leading to erratic decisions. Let us suppose instead that you have been allowed to draw a random sample of two balls from Urn I, and that you have drawn one red and one black. Or a sample of four: two red and two black. Such conditions do not seem to change the observed pattern of choices appreciably (although the reluctance to draw from Urn I goes down somewhat, as shown for example, by the amount a subject will pay to draw from Urn I; this still remains well below what he will pay for Urn II). The same conflicts with the axioms appear.

Long after beginning these observations, I discovered recently that Knight had postulated an identical comparison, between a man who knows that there are red and black balls in an urn but is ignorant of the numbers of each, and another who knows their exact proportion. The results indicated above directly contradict Knight's own intuition about the situation: "It must be admitted that practically, if any decision as to conduct is involved, such as a wager, the first man would have to act

ordering) and the value of a constant column can be changed (P2 – Sure-thing Principle).

Thus starting with I and III and performing such "admissible transformations" it would follow from P1 and P2 that the *first* action in *each* of the following pairs should be preferred:

	R_I	B_I	R_{II}	B_{II}	
I	a	b	b	b	
III	b	b	a	b	
I'	a	b	b	a	P2
III'	b	b	a	a	
I''	a	b	b	a	P1
III''	a	a	b	b	
I'''	b	b	b	a	P2
III'''	b	a	b	b	
I''''	b	b	a	b	P1
III''''	a	b	b	b	

Contradiction: I preferred to III, and I'''' (equivalent to III) preferred to III'''' (equivalent to I).

[10] See immediately preceding note.

on the supposition that the chances are equal." (Knight, 1921, p. 219). If indeed people were compelled to act on the basis of some Principle of Insufficient Reason when they lacked statistical information, there would be little interest in Knight's own distinctions between risk and uncertainty so far as conduct was involved. But as many people predict their own conduct in such hypothetical situations, they do *not* feel obliged to act "as if" they assigned probabilities at all, equal or not, in this state of ignorance.

Another example yields a direct test of one of the Savage postulates. Imagine an urn known to contain 30 red balls and 60 black and yellow balls, the latter in unknown proportion. (Alternatively, imagine that a sample of two drawn from the 60 black and yellow balls has resulted in one black and one yellow.) One ball is to be drawn at random from the urn; the following actions are considered:

| | 30 | 60 | |
	Red	Black	Yellow
I	$100	$0	$0
II	$0	$100	$0

Action I is "a bet on red," II is "a bet on black." Which do you prefer?

Now consider the following two actions, under the same circumstances:

| | 30 | 60 | |
	Red	Black	Yellow
III	$100	$0	$100
IV	$0	$100	$100

Action III is a "bet on red or yellow"; IV is a "bet on black or yellow." Which of these do you prefer? Take your time!

A very frequent pattern of response is: action I preferred to II, and IV preferred to III. Less frequent is: II preferred to I, and III preferred to IV. Both of these, of course, violate the Sure-thing Principle, which requires the ordering of I to II to be preserved in III and IV (since the two pairs differ only in their third column, constant for each pair).[11]

---

[11] Kenneth Arrow has suggested the following example, in the spirit of the above one:

| | 100 | | 50 | 50 |
	R_I	B_I	R_{II}	B_{II}
I	a	a	b	b
II	a	b	a	b
III	b	a	b	a
IV	b	b	a	a

The first pattern, for example, implies that the subject prefers to bet "on" red rather than "on" black; and he also prefers to bet "against" red rather than "against" black. A relationship "more likely than" inferred from his choices would fail condition (b) above of a "qualitative probability relationship," since it would indicate that he regarded red as more likely than black, but also "not-red" as more likely than "not-black." Moreover, he would be acting "as though" he regarded "red or yellow" as less likely than "black or yellow," although red were more likely than black, and red, yellow and black were mutually exclusive, thus violating condition (c) above.

Once again, it is impossible, on the basis of such choices, to infer even qualitative probabilities for the events in question (specifically, for events that include yellow or black, but not both). Moreover, for any values of the pay-offs, it is impossible to find probability numbers in terms of which these choices could be described – even roughly or approximately – as maximizing the mathematical expectation of utility.[12]

You might now pause to reconsider your replies. If you should repent of your violations – if you should decide that your choices implying conflicts with the axioms were "mistakes" and that your "real" preferences, upon reflection, involve no such inconsistencies – you confirm that the Savage postulates are, if not descriptive rules for you, your *normative criteria* in these situations. But this is by no means a universal reaction; on the contrary, it would be exceptional.

Responses do vary. There are those who do *not* violate the axioms, or say they won't, even in these situations (e.g., G. Debreu, R. Schlaiffer, P. Samuelson); such subjects tend to apply the axioms rather than their intuition, and when in doubt, to apply some form of the Principle of Insufficient Reason. Some violate the axioms cheerfully, even with gusto (J. Marschak, N. Dalkey); others sadly but persistently, having looked into their hearts, found conflicts with the axioms and decided, in Samuelson's phrase,[13] to satisfy their preferences and let the axioms

---

Assume that I is indifferent to IV, II is indifferent to III. Suppose that I is preferred to II; what is the ordering of III and IV? If III is not preferred to IV, P2, the Sure-thing Principle is violated. If IV is not preferred to III, P1, complete ordering of actions, is violated. (If III is indifferent to IV, both P1 and P2 are violated.)

[12] Let the utility pay-offs corresponding to $100 and $0 be 1, 0; let P_1, P_2, P_3 be the probabilities corresponding to red, yellow, black. The expected value to action I is then P_1; to II, P_2; to III, $P_1 + P_3$; to IV, $P_2 + P_3$. But there are no P's, $P_i \geq 0$, $\Sigma P_i = 1$, such that $P_1 > P_2$ and $P_1 + P_3 < P_2 + P_3$.

[13] P. Samuelson (1950).

To test the predictive effectiveness of the axioms (or of the alternative decision rule to be proposed in the next section) in these situations, controlled experimentation is in order. [See Chipman's ingenious experiment (1960).] But, as Savage remarks (p. 28), the mode of interrogation implied here and in Savage's book, asking "the person not how he feels, but what he would do in such and such a situation" and giving him ample opportunity to ponder the implications of his replies, seems quite appropriate

satisfy themselves. Still others (H. Raiffa) tend, intuitively, to violate the axioms but feel guilty about it and go back into further analysis.

The important finding is that, after rethinking all their "offending" decisions in the light of the axioms, a number of people who are not only sophisticated but reasonable decide that they wish to persist in their choices. This includes people who previously felt a "first-order commitment" to the axioms, many of them surprised and some dismayed to find that they wished, in these situations, to violate the Sure-thing Principle. Since this group included L. J. Savage, when last tested by me (I have been reluctant to try him again), it seems to deserve respectful consideration.

3. Why are some uncertainties not risks?

Individuals who would choose I over II and IV over III in the example above (or, II over I and III over IV) are simply not acting "as though" they assigned numerical or even qualitative probabilities to the events in question. There are, it turns out, other ways for them to act. But what *are* they doing?

Even with so few observations, it is possible to say some other things they are *not* doing. They are not "minimaxing"; nor are they applying a "Hurwicz criterion," maximizing a weighted average of minimum pay-off and maximum for each strategy. If they were following any such rules they would have been indifferent between each pair of gambles, since all have identical minima and maxima. Moreover, they are not "minimaxing regret," since in terms of "regrets" the pairs I–II and III–IV are identical.[14]

Thus, *none* of the familiar criteria for predicting or prescribing decision-making under uncertainty corresponds to this pattern of choices. Yet the choices themselves do not appear to be careless or random. They are persistent, reportedly deliberate, and they seem to predominate empirically; many of the people who take them are eminently reasonable, and they insist that they *want* to behave this way, even though they may be generally respectful of the Savage axioms. There are strong indications, in other words, not merely of the existence of reliable patterns of blind behavior but of the operation of

in weighing "the theory's more important normative interpretation." Moreover, these nonexperimental observations can have at least negative empirical implications, since there is a presumption that people whose instinctive choices violate the Savage axioms, and who claim upon further reflection that they do not *want* to obey them, do not *tend* to obey them normally in such situations.

[14] No one whose decisions were based on "regrets" could violate the Sure-thing Principle, since all constant columns of pay-offs would transform to a column of 0's in terms of "regret"; on the other hand, such a person would violate P1, complete ordering of strategies.

definite normative criteria, different from and conflicting with the familiar ones, to which these people are trying to conform. If we are talking about you, among others, we might call on your introspection once again. What did *you* think you were doing? What were you trying to do?

One thing to be explained is the fact that you probably would *not* violate the axioms in certain other situations. In the urn example, although a person's choices may not allow us to infer a probability for yellow, or for (red or black), we may be able to deduce quite definitely that he regards (yellow or black) as "more likely than" red; in fact, we might be able to arrive at quite precise numerical estimates for his probabilities, approximating $2/3$, $1/3$. What is the difference between these uncertainties, that leads to such different behavior?

Responses from confessed violators indicate that the difference is not to be found in terms of the two factors commonly used to determine a choice situation, the relative desirability of the possible pay-offs and the relative likelihood of the events affecting them, but in a third dimension of the problem of choice: the nature of one's information concerning the relative likelihood of events. What is at issue might be called the *ambiguity* of this information, a quality depending on the amount, type, reliability and "unanimity" of information, and giving rise to one's degree of "confidence" in an estimate of relative likelihoods.

Such rules as minimaxing, maximaxing, Hurwicz criteria or minimaxing regret are usually prescribed for situations of "complete ignorance," in which a decision-maker lacks any information whatever on relative likelihoods. This would be the case in our urn example if a subject had no basis for considering any of the possible probability distributions over red, yellow, black – such as $(1,0,0)$, $(0,1,0)$, $(0,0,1)$ – as a better estimate, or basis for decision, than any other. On the other hand, the Savage axioms, and the general "Bayesian" approach, are unquestionably appropriate when a subject is willing to base his decisions on a definite and precise choice of a particular distribution: his uncertainty in such a situation is unequivocally in the form of "risk."

But the state of information in our urn example can be characterized neither as "ignorance" nor "risk" in these senses. Each subject does know enough about the problem to *rule out* a number of possible distributions, including all three mentioned above. He *knows* (by the terms of the experiment) that there are red balls in the urn; in fact, he knows that exactly $1/3$ of the balls are red. Thus, in his "choice" of a subjective probability distribution over red, yellow, black – if he wanted such an estimate as a basis for decision – he is limited to the set of potential distributions between $(1/3, 2/3, 0)$ and $(1/3, 0, 2/3)$; i.e., to the infinite set $(1/3, \lambda, 2/3 - \lambda)$, $0 \leq \lambda \leq 2/3$. Lacking any observations on the number of yellow or black balls, he may have little or no information indicating that one of the remaining, infinite set of distributions is more "likely," more worthy of attention than any other. If he should accumulate some observations, in the form of small sample distributions, this set of "reasonable" distributions would diminish, and a particular distribution might gather increasing strength as a candidate; but so long as the samples remain small, he may be far from able to select one from a number of distributions, or one composite distribution, as a unique basis for decision.

In some situations where two or more probability distributions over the states of nature seem reasonable, or possible, it may still be possible to draw on different sorts of evidence, establishing probability weights in turn to these different distributions to arrive at a final, composite distribution. Even in our examples, it would be misleading to place much emphasis on the notion that a subject has *no* information about the contents of an urn on which no observations have been made. The subject can always ask himself: "What is the likelihood that the experimenter has rigged this urn? Assuming that he has, what proportion of red balls did he probably set? If he is trying to trick me, how is he going about it? What other bets is he going to offer me? What sort of results is he after?" If he has had a lot of experience with psychological tests before, he may be able to bring to bear a good deal of information and intuition that seems relevant to the problem of weighting the different hypotheses, the alternative reasonable probability distributions. In the end, these weights, and the resulting composite probabilities, may or may not be equal for the different possibilities. In our examples, actual subjects do tend to be indifferent between betting on red or black in the unobserved urn, in the first case, or between betting on yellow or black in the second. This need not at all mean that they felt "completely ignorant" or that they could think of no reason to favor one or the other; it does indicate that the reasons, if any, to favor one or the other balanced out subjectively so that the possibilities entered into their final decisions weighted equivalently.

Let us assume, for purposes of discussion, that an individual can always assign relative weights to alternative probability distributions reflecting the relative support given by his information, experience and intuition to these rival hypotheses. This implies that he can always assign relative likelihoods to the states of nature. But how does he *act* in the presence of his uncertainty? The answer to that may depend on another sort of judgment, about the reliability, credibility, or adequacy of his information (including his relevant experience, advice and intuition) as a whole: not about the relative support it may give to one hypothesis as opposed to another, but about its ability to lend support to any hypothesis at all.

If all the information about the events in a set of gambles were in the form of sample-distributions, then ambiguity might be closely related, inversely, to the size of the sample.[15] But sample-size is not a universally useful index of this factor. Information about many events cannot be conveniently described in terms of a sample distribution; moreover, sample-size seems to focus mainly on the quantity of information. "Ambiguity" may be high (and the confidence in any particular estimate of probabilities low) even where there is ample quantity of information, when there are questions of reliability and relevance of information, and particularly where there is *conflicting* opinion and evidence.

This judgment of the ambiguity of one's information, of the over-all credibility of one's composite estimates, of one's confidence in them, cannot be expressed in terms of relative likelihoods of events (if it could, it would simply affect the final, compound probabilities). Any scrap of evidence bearing on relative likelihood should already be represented in those estimates. But having exploited knowledge, guess, rumor, assumption, advice, to arrive at a final judgment that one event is more likely than another or that they are equally likely, one can still stand back from this process and ask: "How much, in the end, is all this worth? How much do I really know about the problem? How firm a basis for choice, for appropriate decision and action, do I have?" The answer, "I don't know very much, and I can't rely on that," may sound rather familiar, even in connection with markedly unequal estimates of relative likelihood. If "complete ignorance" is rare or nonexistent, "considerable" ignorance is surely not.

Savage himself alludes to this sort of judgment and notes as a difficulty with his approach that no recognition is given to it:

there seem to be some probability relations about which we feel relatively "sure" as compared with others.... The notion of "sure" and "unsure" introduced here is vague, and my complaint is precisely that neither the theory of personal probability, as it is developed in this book, nor any other device known to me renders the notion less vague.... A second difficulty, perhaps

---

[15] See Chipman (1960, pp. 75, 93). Chipman's important work in this area, done independently and largely prior to mine, is not discussed here since it embodies a stochastic theory of choice; its spirit is otherwise closely similar to that of the present approach, and his experimental results are both pertinent and favorable to the hypotheses below (though Chipman's inferences are somewhat different).

See also the comments by N. Georgescu-Roegen on notion of "credibility," a concept identical to "ambiguity" in (1958, pp. 24–26) and (1954). These highly pertinent articles came to my attention only after this paper had gone to the printer, allowing no space for comment here.

closely associated with the first one, stems from the vagueness associated with judgments of the magnitude of personal probability.[16]

Knight asserts what Savage's approach tacitly denies, that such over-all judgments may influence decision:

The action which follows upon an opinion depends as much upon the amount of confidence in that opinion as it does upon the favorableness of the opinion itself.... Fidelity to the actual psychology of the situation requires, we must insist, recognition of these two separate exercises of judgment, the formation of an estimate and the estimation of its value. (Knight, 1921, p. 227)

Let us imagine a situation in which so many of the probability judgments an individual can bring to bear upon a particular problem are either "vague" or "unsure" that his confidence in a particular assignment of probabilities, as opposed to some other of a set of "reasonable" distributions, is very low. We may define this as a situation of high ambiguity. The general proposition to be explored below is that it is precisely in situations of this sort that self-consistent behavior violating the Savage axioms may commonly occur.

Ambiguity is a subjective variable, but it should be possible to identify "objectively" some situations likely to present high ambiguity, by noting situations where available information is scanty or obviously unreliable or highly conflicting; or where expressed expectations of different individuals differ widely; or where expressed confidence in estimates tends to be low. Thus, as compared with the effects of familiar production decisions or well-known random processes (like coin flipping or roulette), the results of Research and Development, or the performance of a new President, or the tactics of an unfamiliar opponent are all likely to appear ambiguous. This would suggest a broad field of application for the proposition above.

In terms of Shackle's cricket example: Imagine an American observer who had never heard of cricket, knew none of the rules or the method of scoring, and had no clue as to the past record or present prospects of England or Australia. If he were confronted with a set of side bets as to whether England would bat first – this to depend on the throw of a die or a coin – I expect (unlike Shackle) that he would be found to obey Savage's axioms pretty closely, or at least, to want to obey them if any discrepancies were pointed out. Yet I should not be surprised by quite different behavior, at odds with the axioms, if that particular observer

---

[16] Savage (1954, pp. 57, 58, 59). Savage later goes so far as to suggest that the "aura of vagueness" attached to many judgments of personal probability might lead to systematic violations of his axioms although the decision rule he discusses as alternative – minimaxing regret – cannot, as mentioned in footnote 14, account for the behavior in our examples.

were forced to gamble heavily on the proposition that England would *win the match*.

Let us suppose that an individual must choose among a certain set of actions, to whose possible consequences we can assign "von Neumann-Morgenstern utilities" (reflecting the fact that in choosing among *some* set of "unambiguous" gambles involving other events and these same outcomes, he obeys the Savage axioms). We shall suppose that by compounding various probability judgments of varying degrees of reliability he can eliminate certain probability distributions over the states of nature as "unreasonable," assign weights to others and arrive at a composite "estimated" distribution y^o that represents all his available information on relative likelihoods. But let us further suppose that the situation is ambiguous for him. Out of the set Y of all possible distributions there remains a set Y^o of distributions that still seem "reasonable," reflecting judgments that he "might almost as well" have made, or that his information – perceived as scanty, unreliable, *ambiguous* – does not permit him confidently to rule out.

In choosing between two actions, I and II, he can compute their expected utilities in terms of their pay-offs and the "estimated" probability distribution y^o. If the likelihoods of the events in question were as unambiguous as those in the situations in which his von Neumann-Morgenstern utilities were originally measured, this would be the end of the matter; these pay-offs embody all his attitudes toward "risk," and expected values will correspond to his actual preferences among "risky" gambles. But in this case, where his final assignment of probabilities is less confident, that calculation may leave him uneasy. "So I has a lower expectation than II, on the basis of *these* estimates of probabilities," he may reflect; "How much does that tell me? That's not much of a reason to choose II."

In this state of mind, searching for additional grounds for choice, he may try new criteria, ask new questions. For any of the probability distributions in the "reasonably possible" set Y^o, he can compute an expected value for each of his actions. It might now occur to him to ask: "What might happen to me if my best estimates of likelihood don't apply? What is the *worst* of the reasonable distributions of pay-off that I might associate with action I? With action II?" He might find he could answer this question about the lower limit of the reasonable expectations for a given action much more confidently than he could arrive at a single, "best guess" expectation; the latter estimate, he might suspect, might vary almost hourly with his mood, whereas the former might look much more solid, almost a "fact," a piece of evidence definitely worth considering in making his choice. In almost no cases (excluding "complete ignorance" as unrealistic) will the *only* fact worth noting about a prospective action be its "security level": the "worst" of the expectations associated with reasonably possible probability distributions. To choose on a "maximin" criterion alone would be to ignore entirely those probability judgments for which there is evidence. But in situations of high ambiguity, such a criterion may appeal to a conservative person as deserving *some* weight, when interrogation of his own subjective estimates of likelihood has failed to disclose a set of estimates that compel exclusive attention in his decision-making.

If, in the end, such a person chooses action I, he may explain:

In terms of my best estimates of probabilities, action I has almost as high an expectation as action II. But if my best guesses should be rotten, which wouldn't surprise me, action I gives me better protection; the worst expectation that looks reasonably possible isn't much worse than the "best guess" expectation, whereas with action II it looks possible that my expectation could really be terrible.

An advocate of the Savage axioms as normative criteria, foreseeing where such reasoning will lead, may interject in exasperation:

Why are you double-counting the "worst" possibilities? They're already taken into account in your over-all estimates of likelihoods, weighted in a reasoned, realistic way that represents – by your own claim – your best judgment. Once you've arrived at a probability distribution that reflects everything you know that's relevant, don't fiddle around with it, use it. Stop asking irrelevant questions and whining about how little you really know.

But this may evoke the calm reply:

It's no use bullying me into taking action II by flattering my "best judgment." *I* know how little that's based on; I'd back it if we were betting with pennies, but I want to know some other things if the stakes are important, and "How much might I expect to lose, without being unreasonable?" just strikes me as one of those things. As for the reasonableness of giving extra weight to the "bad" likelihoods, my test for that is pragmatic; in situations where I really can't judge confidently among a whole range of possible distributions, this rule steers me toward actions whose expected values are relatively *insensitive* to the particular distribution in that range, without giving up too much in terms of the "best guess" distribution. That strikes me as a sensible, conservative rule to follow. What's wrong with it?

"What's wrong with it" is that it will lead to violations of Savage's Postulate 2, and will make it impossible for an observer to describe the subject's choices as though he were maximizing a linear combination of pay-offs and probabilities over events. Neither of these considerations, even on reflection, may pose to our conservative subject overwhelming imperatives to change his behavior. It will *not* be true that this behavior is erratic or unpredictable (we shall formalize it in terms of a decision rule below), or exhibits intransitivities, or amounts to "throwing away utility" (as would be true, for example, if it led him occasionally to choose strategies that were strongly "dominated" by others). There is, in fact, no obvious basis for asserting that it will lead him in the long run to worse outcomes than he could expect if he reversed some of his preferences to conform to the Savage axioms.

Another person, or this same person in a different situation, might have turned instead or in addition to some other criteria for guidance. One might ask, in an ambiguous situation: "What is the *best* expectation I might associate with this action, without being unreasonable?" Or: "What is its average expectation, giving all the reasonably possible distributions equal weight?" The latter consideration would not, as it happens, lead to behavior violating the Savage axioms. The former would, in the same fashion though in the opposite direction as the "maximin" criterion discussed above; indeed, this "maximaxing" consideration could generate the minority behavior of those who, in our urn example, prefer II to I and III to IV. Both these patterns of behavior could be described by a decision rule similar to the one below, and their respective rationales might be similar to that given above. But let us continue to focus on the particular pattern discussed above, because it seems to predominate empirically (at least, with respect to our examples) and because it most frequently corresponds to *advice* to be found on decision-making in ambiguous situations.

In reaching his decision, the relative weight that a conservative person will give to the question, "What is the worst expectation that might appear reasonable?" will depend on his confidence in the judgments that go into his estimated probability distribution. The less confident he is, the more he will sacrifice in terms of estimated expected pay-off to achieve a given increase in "security level"; the more confident, the greater increase in "security level" he would demand to compensate for a given drop in estimated expectation. This implies that "trades" are possible between security level and estimated expectation in his preferences, and that does seem to correspond to observed responses. Many subjects will still prefer to bet on R_{II} than R_I in our first example even when the proportion of red to black in Urn II is lowered to 49:51, or will prefer to bet on red than on yellow in the second example even when one red ball is removed from the urn. But at *some* point, as the "unambiguous" likelihood becomes increasingly unfavorable, their choices will switch.[17]

Assuming, purely for simplicity, that these factors enter into his

[17] This contradicts the assertions by Chipman (1960, p. 88) and Georgescu-Roegen (1954, pp. 527–530, 1958, p. 25) that individuals order uncertainty-situations lexicographically in terms of estimated expectation and "credibility" (ambiguity); ambiguity appears to influence choice even when estimated expectations are not equivalent.

decision rule in linear combination, we can denote by ρ his degree of confidence, in a given state of information or ambiguity, in the estimated distribution y^o, which in turn reflects all of his judgments on the relative likelihood of distributions, including judgments of equal likelihood. Let min_x be the minimum expected pay-off to an act x as the probability distribution ranges over the set Y^o; let est_x be the expected pay-off to the act x corresponding to the estimated distribution y^o.

The simplest decision rule reflecting the above considerations would be:[18] *Associate with each x the index:*

$$\rho \cdot est_x + (1 - \rho) \cdot min_x$$

Choose that act with the highest index.

An equivalent formulation would be the following, where y^o is the estimated probability vector, y_z^{min} the probability vector in Y^o corresponding to min_x for action x and (X) is the vector of payoffs for action x: *Associate with each x the index:*

$$[\rho \cdot y^o + (1 - \rho) y_x^{min}] \, (x)$$

Choose that act with the highest index.

In the case of the red, yellow, and black balls, supposing no samples and on explicit information except that ⅓ of the balls are red, many subjects might lean toward an estimated distribution of (⅓, ⅓, ⅓): if not from "ignorance," then from counterbalancing considerations. But many of these would find the situation ambiguous; for them the "reasonable" distributions Y^o might be all those between (⅓, ⅔, 0) and (⅓, 0, ⅔). Assuming for purposes of illustration $\rho = $ ¼ (Y^o, y^o, X and ρ are all subjective data to be inferred by an observer or supplied by the individual, depending on whether the criterion is being used descriptively or for convenient decision-making), the formula for the index would be:

$$¼ \cdot est_x + ¾ \, min_x.$$

The relevant data (assigning arbitrary utility values of 6 and 0 to the money outcomes \$100 and \$0) would be:

---

[18] This rule is based upon the concept of a "restricted Bayes solution" developed by J. L. Hodges, Jr., and E. L. Lehmann (1952). The discussion throughout Section III of this paper derives heavily from the Hodges and Lehmann argument, although their approach is motivated and rationalized somewhat differently.

See also, L. Hurwicz (1951). This deals with the same sort of problem and presents a "generalized Bayes-minimax principle" equivalent, in more general form, to the decision rule I proposed in an earlier presentation of this paper (December, 1960), but both of these lacked the crucial notions developed in the Hodges and Lehmann approach of a "best estimate" distribution y^o and a "confidence" parameter ρ.

	Red	Yellow	Black	Min_x	Est_x	Index
I	6	0	0	2	2	2
II	0	6	0	0	2	.5
III	6	0	6	2	4	2.5
IV	0	6	6	4	4	4

A person conforming to this rule with these values would prefer I to II and IV to III, in violation of the Sure-thing Principle: as do most people queried. In justifying this pattern of behavior he might reproduce the rationale quoted above (*q.v.*); but most verbal explanations, somewhat less articulately, tend to be along these lines:

The expected pay-off for action I is definite: 2. The risks under action II may be no greater, but I know what the risk is under action I and I don't under action II. The expectation for action II is ambiguous, it might be better or it might be worse, anything from 0 to 4. To be on the safe side, I'll assume that it's closer to 0; so action I looks better. By the same token, IV looks better than III; I *know* that my expected pay-off with IV is 4, whereas with III it might be as low as 2 (which isn't compensated by the chance that it could be 6). In fact, I know the whole *probability distribution of payoffs* (though not the distribution over *events*) for I and IV, but I don't for II and III. I know that a payoff of 6 is *twice as likely* as 0 under IV, whereas 6 *may* be only half as likely as 0 under III.

Leaving the advocate of the Savage axioms, if he is still around to hear this, to renew his complaints about the silliness and irrelevance of such considerations, let us note a practical consequence of the decision rule which the above comment brings into focus. It has already been mentioned that the rule will favor – other things (such as the estimated expectation) being roughly equal – actions whose expected value is less sensitive to variation of the probability distribution within the range of ambiguity. Such actions may frequently be those definable as "status quo" or "present behavior" strategies. For these, ρ may be high, the range of Y_o small.

A familiar, ongoing pattern of activity may be subject to considerable uncertainty, but this uncertainty is more apt to appear in the form of "risk"; the relation between given states of nature is known precisely, and although the random variation in the state of nature which "obtains" may be considerable, its stochastic properties are often known confidently and in detail. (Actually, this confidence may be self-deceptive, based on ignoring some treacherous possibilities; nevertheless, it commonly exists.) In contrast, the ambiguities surrounding the outcome of a proposed *innovation*, a departure from current strategy, may be much more noticeable. Different sorts of events are relevant to its outcome, and their likelihoods must now be estimated, often with little evidence or prior expertise; and the effect of a given state of nature upon the outcome of the new action may itself be in question. Its variance may not appear any higher than that of the familiar action when computed on the basis of "best estimates" of the probabilities involved, yet the meaningfulness of this calculation may be subject to doubt. The decision rule discussed will not preclude choosing such an act, but it will definitely bias the choice away from such ambiguous ventures and toward the strategy with "*known* risks." Thus the rule is "conservative" in a sense more familiar to everyday conversation than to statistical decision theory; it may often favor traditional or current strategies, even perhaps at high risk, over innovations whose consequences are undeniably ambiguous. This property may recommend it to some, discredit it with others (some of whom might prefer to reverse the rule, to emphasize the more hopeful possibilities in ambiguous situations); it does not seem irrelevant to one's attitude toward the behavior.

In the equivalent formulation in terms of $y_x^{\min}$ and y^o, the subject above could be described "as though" he were assigning weights to the respective pay-offs of actions II and III, whose expected values are ambiguous, as follows (assuming $y^o = (\frac{1}{3}, \frac{1}{3}, \frac{1}{3})$ in each case):

	$y_x^{\min}$	$\rho \cdot y^o + (1 - \rho)y_x^{\min}$
II	$(\frac{1}{3}, 0, \frac{2}{3})$	$(\frac{1}{3}, \frac{1}{12}, \frac{7}{12})$
III	$(\frac{1}{3}, \frac{2}{3}, 0)$	$(\frac{1}{3}, \frac{7}{12}, \frac{1}{12})$

Although the final set of weights for each set of pay-offs resemble probabilities (they are positive, sum to unity, and represent a linear combination of two probability distributions), they differ for each action, since $y_x^{\min}$ will depend on the pay-offs for x and will vary for different actions. If these weights were interpreted as "probabilities," we would have to regard the subject's subjective probabilities as being dependent upon his pay-offs, his evaluation of the outcomes. Thus, this model would be appropriate to represent cases of true pessimism, or optimism or wishfulness (with $y_x^{\min}$ substituting for $y_x^{\min}$). However, in this case we are assuming *conservatism*, not pessimism; our subject does not actually expect the worst, but he chooses to act "*as though*" the *worst were somewhat more likely than his best estimates of likelihood would indicate.* In either case, he violates the Savage axioms; it is impossible to infer from the resulting behavior a set of probabilities *for events* independent of his payoffs. In effect, he "distorts" his best estimates of likelihood, in the direction of increased emphasis on the less favorable outcomes and to a degree depending on ρ, his confidence in his best estimate.[19]

Not only does this decision model account for "deviant" behavior in a particular, ambiguous situation, but it covers the observed shift in a subject's behavior as ambiguity decreases. Suppose that a sample is drawn from the urn, strengthening the confidence in the best estimates of likelihood, so that ρ increases, say, to ¾. The weights for the pay-offs to actions II and III would now be:

$$\rho \cdot y^o + (1 - \rho)\, y_x^{\min}$$

II $\quad (\frac{1}{3}, \frac{1}{4}, \frac{5}{12})$

III $\quad (\frac{1}{3}, \frac{5}{12}, \frac{1}{4})$

and the over-all index would be:

	Index
I	2
II	1.5
III	3.5
IV	4

In other words, the relative influence of the consideration, "What is the worst to be expected?" upon the comparison of actions is lessened. The final weights approach closer to the "best estimate" values, and I and II approach closer to indifference, as do III and IV. This latter aspect might show up behaviorally in the amount a subject is willing to pay for a given bet on yellow, or on (red or black), in the two situations.

---

[19] This interpretation of the behavior pattern contrasts to the hypothesis or decision rule advanced by Fellner (1961). Fellner seems unmistakably to be dealing with the same phenomena discussed here, and his proposed technique of measuring a person's subjective probabilities and utilities in relatively "unambiguous" situations and then using these measurements to calibrate his uncertainty in more ambiguous environments seems to me a most valuable source of new data and hypotheses. Moreover, his descriptive data and intuitive conjectures lend encouraging support to the findings reported here. However, his solution to the problem supposes a single set of weights determined independently of pay-offs (presumably corresponding to the "best estimates" here) and a "correction factor," reflecting the degree of ambiguity or confidence, which operates on these weights in a manner independent of the structure of pay-offs. I am not entirely clear on the behavioral implications of Fellner's model or the decision rule it implies, but in view of these properties I am doubtful whether it can account adequately for all the behavior discussed above.

In the limit, as ambiguity diminishes for one reason or another and ρ approaches 1, the estimated distribution will come increasingly to dominate decision. With confidence in the best estimates high, behavior on the basis of the proposed decision rule will roughly conform to the Savage axioms, and it would be possible to infer the estimated probabilities from observed choices. But prior to this, a large number of information states, distinguishable from each other and all far removed from "complete ignorance," might all be sufficiently ambiguous as to lead many decision-makers to conform to the above decision rule with ρ < 1, in clear violation of the axioms.

Are they foolish? It is not the object of this paper to judge that. I have been concerned rather to advance the testable propositions: (1) certain information states can be meaningfully identified as highly ambiguous; (2) in these states, many reasonable people tend to violate the Savage axioms with respect to certain choices; (3) their behavior is deliberate and not readily reversed upon reflection; (4) certain patterns of "violating" behavior can be distinguished and described in terms of a specified decision rule.

If these propositions should prove valid, the question of the optimality of this behavior would gain more interest. The mere fact that it conflicts with certain axioms of choice that at first glance appear reasonable does not seem to me to foreclose this question; empirical research, and even preliminary speculation, about the nature of actual or "successful" decision-making under uncertainty is still too young to give us confidence that these axioms are not abstracting away from vital considerations. It would seem incautious to rule peremptorily that the people in question should not allow their perception of ambiguity, their unease with their best estimates of probability, to influence their decision: or to assert that the manner in which they respond to it is against their long-run interest and that they would be in some sense better off if they should go against their deep-felt preferences. If their rationale for their decision behavior is not uniquely compelling (and recent discussions with T. Schelling have raised questions in my mind about it), neither, it seems to me, are the counterarguments. Indeed, it seems out of the question summarily to judge their behavior as irrational: I am included among them.

In any case, it follows from the propositions above that for their behavior in the situations in question, the Bayesian or Savage approach gives wrong predictions and, by their lights, bad advice. They act in conflict with the axioms deliberately, without apology, because it seems to them the sensible way to behave. Are they clearly mistaken?

14. Self-knowledge, uncertainty, and choice

Frederic Schick

This paper takes up some problems in the theory of choice, problems arising from questions of where and when we can choose at all, or see any point in choosing. Some of these problems have a long history, others have come up only lately. The older problems have outstayed their time. They need to be reminded of this and pressed again to retire. The others are elusive and can be troubling. I will try to come to grips with these.

A meta-problem here is how to tie all this together. I shall draw on the fact that it has been done before. The British economist G. L. S. Shackle has linked up much the same issues. Shackle discusses three problems. He holds that some radical steps must be taken in order to get around them. In this last, I think he is wrong, but we needn't stop for that now. We shall set out what follows under Shackle's three problem-headings, and start each time with Shackle himself.

1. Suppose that everything must happen exactly as it does. "... [S]uppose that history is a book already written, whose pages the hand of fate is merely turning, not composing" (Shackle, 1966, p. 72). What we shall be doing at every moment in our lives is fully noted in the book. In a world of the sort described, can we choose what we do? Shackle says that we can't. In such a world, what people think their choices "... can be no more than the clicking of the machine as it works, and ... is ... something wholly different from that explosion of essential novelty which they seem to be to the person whose tense thought and feeling give them birth" (1969, p. 3). In a fully pre-set world, choice is *illusory*.

Reprinted from *The British Journal for the Philosophy of Science*, **30** (1979), pp. 235–252, by kind permission of the author and the publisher.

Shackle takes the world of the book to be determined throughout. He sees determinism as ruling out freedom, and so – since he holds that a choice must be free – as also ruling out choices. The problem here is ancient. It goes back at least to Lucretius: either we reject determinism or we give up thinking we choose. Along with many other humanists, Shackle rejects determinism.

We are asked to decide between determinism and choice. I will be very brief here, for I follow Hume in thinking we can have them both. Hume went at it in two ways. First, he set out determinism as compatible with freedom. Granted that all events are determined, for all events are caused, those that involve us as agents included. But where we are agents, we do what we do because we want to do it. This is what makes us agents there. It also makes what we do there free. My taking a shower, and then making some coffee and going for a walk are all caused, and so they are determined, but they are caused by my wanting to do them, and so they are also free. Determinism allows for freedom. And so it allows for choices.

It may be said that how we will choose is also already in the book. On the determinist doctrine, a choice must always have been what it was. It is never open to us not to want what we wind up doing. How then is the outcome up to us? Hume's second (and basic) line challenges the idea of *must*ness. All events are caused, but this means only that each is preceded by another on an invariant pattern. Causality is a matter of regularity, not of any sort of compulsion. It has to do with sequences only, not with pushes and shoves (in Hume's terms: with *conjunction*, not with *connection*). On a Humean view, necessity is no part of the picture. We might always have chosen otherwise. Alternatives were always open.

There are of course the familiar issues. What is a causal regularity? What distinguishes a causal sequence from a mere coincidence? Or go back to the preceding paragraph: what sets off my doing something *because* I want to do it from my doing it *while* (or *right after*) wanting to? These are dreary questions, but better, I think, to slog along here than to have to settle for either determinism or choice. Humean determinists are not in the clear, but at least they needn't believe that choices are all illusory.

None of this discredits the possibility of Shackle's great book of history. Indeed, the whole question of determinism can be skirted. Even if determinism were false, the book of history might still exist – the future need not be determined in order to be foreseen. Here we approach a second problem, the problem of the foreknowledge of our future. There are in fact two separate issues under this new heading. One, again, will not hold us long. The other is harder to get at, and also more rewarding.

The first of these issues has an ancient lineage. Some of the early Christian thinkers agonised long and hard over it. For them, the problem arose from their assumption of the omniscience of God. If I am going to do x, then God, since He knows everything, knows that I will do it. This means that it is true beforehand that I will do x, for no one (not even God) can know what is not true. Since it is true beforehand, it will be true at the point of action. It will be true necessarily there, for it can't possibly be true at one time and not at another. So I will have to do x – will necessarily have to. This works out without a word about how I am going to choose. But then, what choice will I have?

Or look directly at choices. God knows beforehand whether or not I will do x. Being omniscient, He also knows whether I will *choose* to do it. If He knows that I will choose this, then I must of course choose it. If I must, it's out of my hands. Again, what sort of a choice is that? "If God foresees all things and cannot in anything be mistaken, then that which His Providence foresees must result; wherefore if It knows beforehand not only men's deeds but even their designs and wishes, there can be no freedom of judgment" (Boethius, *Cons. Phil.*, bk. 5).

Unbelievers ought not to smirk. The same predicament can also be got to without bringing in God. Suppose that *you* knew yesterday that I would do x today. Again we can work our way to the conclusion that I have no choice about it. Indeed the problem is simpler still – no one at all need foreknow the fact. Aristotle saw this clearly (*De Interp.*, chap. 9). His form of the argument is this: if I am to do x today, then it was always true in the past (even if no one knew it then) that I will now do x, and so I must now do it. My doing x is a *fait accompli*, and thus there is ". . . no need to deliberate or to take trouble."

Aristotle avoids this conclusion by rejecting the first step toward it. He holds that a proposition about an event is never true before that event, nor is its negation false. This makes trouble for the theory of knowledge. A second, less costly, line is Ryle's (1956, chap. 2). Ryle notes that the argument involves a sleight-of-hand with the terms of necessity. Granted, if it was true yesterday that I would do x today, then it is necessarily true that I will do it. But parse this conditional properly, and there is no problem. The conditional is not: if it was true yesterday that I would do x today, then it-is-necessarily-true-that-I-will-do-it. Rather, it is: if it was true yesterday that I would do x today, then-it-is-necessarily-true-that I will do x today. Formally, what will be granted is not: if (yesterday) p, then necessarily-p; but only: if (yesterday) p, then-necessarily p. The necessity operator qualifies not the consequent but the connective. It serves only to indicate that the if-then structure is analytic. It has no force left over to establish fatalism.

Let us move to the second, deeper issue of foreknowledge. This takes the foreknowing party to be the agent himself. The agent here is not omniscient. He does know, however, how he will choose. This is the form of the foreknowledge problem considered by Shackle, though he brings out only a special case. Suppose that the agent

> has in mind what he accepts as a complete list of all the relevantly distinct actions open to him, and that for each of these available actions he sees one and only one possible sequel differing from that of every other of the available actions.... If we now further assume that these sequels can be completely ordered by him so that, when he compares any two of them, one is preferred to the other, it is plain his available actions will be similarly ordered, and that the question "Which action shall I take?" will call for nothing which can be called judgment.... I shall say that, in such a case, decision is *empty* (1966, pp. 73–4).

We might put it this way. If we foresee our choice-situation (our preferences, our options, their sequels and the rest), then we can work out how we will choose. Suppose we expect to act as we choose – call this the assumption of *effectiveness*. Then if we foresee our situation, we can see also how we will act. There is nothing left to consider, no occasion for judgment: perfect foresight empties choice.

No doubt this too has a history. There must have been some medieval who asked whether t.ie omniscience of God extended to His own future, and if so, how come His choices were not empty. But no record is left of this thinker. The recent discussions start from scratch – see Ginet (1962), Pears (1968), and Goldman (1970, chap. 6). Note that the Rylean response to the two-person problem (one person foreknowing the other's choice) gets us nowhere in the one-person case. This issue can't be reduced to something independent of a foreknower. What here empties a person's choice is not its being true beforehand that he will choose that way, but his own knowing that he will. If he knows how he will choose, he also knows how he will act. But then he faces no issue, for knowing *that* he will do x, he cannot consider *whether* to do it. And where he faces no issue, he has no choice to make.

Shackle presents the problem in a very narrow form. He speaks only of someone's foreknowledge emptying his choice where this person "... supposes himself to know precisely, completely, and for certain what consequences for himself would flow from any given one of the available acts" (1969, p. 4). This agent faces no risk. Foreknowledge empties choice in the wider world too. Richard Jeffrey (1977) brings out a more inclusive problem.

Let o^x be the agent's option of doing x, and T be any logical truth. It is a thesis of utility theory that the utility of a proposition w is the weighted average of w's coming out true in this contingency or in that – given that these are exclusive and exhaustive – the weights being the probabilities (conditional on w) of these contingencies. [A special case appears as (3) in part 2 below.] Jeffrey presents his problem as follows.

Suppose you [the agent] think it within your power to make o^x true if you wish, that you prefer o^x to T, and that you are convinced that o^x is preferable to every other one of your options. Then the probability of o^x is 1, for you know you will make o^x true. But then, [setting $v = o^x$ and $w = T$ in (3)] ... your utility for o^x is equal to your utility for T, so that ... you are indifferent between o^x and T after all, in contradiction to the assumption that you prefer o^x to T (1977, p. 136; this has been touched up slightly to square with our notation here).

I shall lay out the larger problem in a way closer to Shackle's thinking and hold off Jeffrey till later. Suppose that the agent knows what the sequels of his options would be in each of various contingencies. (Shackle's special case is that in which the agent sees one contingency only.) Reflecting on the possibilities, the agent may realise that his doing x – again, this option is o^x – offers him more than anything else. Aware of his general rationality, he may infer that he will choose o^x, and so, assuming effectiveness, that o^x will come out. Can this person choose here? He knows that he will do x, so he has no issue of whether he will do it. What is there left for him to choose?

I have been speaking of this as a problem of the foreknowledge of choices, but knowledge comes into it only indirectly. Basically, it is a problem of fore*belief*. Let me put it more fully. Suppose that the agent thinks that his issue will be made up of options $o_1, o_2, \ldots o_m$, and that $s_{11}, s_{12}, \ldots s_{1n}, s_{21}, s_{22}, \ldots s_{2n}, \ldots s_{m1}, s_{m2}, \ldots s_{mn}$ would be the sequels of these options in the contingencies he sees. (An *option* here is a *live* option: not just anything the agent might do, but one of a set of possible actions that, conjointly, raise an issue for him.) Let the agent also think that his over-all preference ranking will be R. If R meets certain conditions of specificity, it follows (I show how in the second part) that the agent's options will have determinate expected utilities. Let the agent believe that he will choose rationally – that he will choose an option whose expected utility won't be exceeded by that of any other – and let him believe that the choice he will make will be effective. Suppose that he will not change his mind on any of this before he chooses. It follows from what he believes that some set S of his options is the set from which he will choose one, and that this option will come out. To simplify a bit, suppose that S contains only the single option o^x. The agent is now committed to believing that he will do x. He will also be bound to believe this at the point of choice. If he does believe it then, he won't have any issue of whether to do x or not, and so won't have any choice to make. Foreseeing his choice-situation precludes his having a choice.

We can generalise further. Shackle and Jeffrey both find a problem for rational-choice theory here. But the agent's assumption that he will choose rationally can give way to others. Let the agent believe that he will pursue choice-policy C – whether this is maximisation, or something else, whatever it is, as long as it allows him to predict his choice from the rest of his information. (This last clause means that policy C might be that of maximining or maximaxing or the like, but can't be that of flipping a coin or of consulting some oracle.) Given any suitable C and all the rest of what he knows, the agent is logically bound to believe that he will do x (or it may now be y). Again, this settles whether he'll do it, and so dissolves his issue. The problem remains: foresight undoes choosing.

Shackle is impressed by his version of this. He thinks it goes to the heart of our theory of how people choose, and that it shows we need a new theory. We must, for a start, deny that the agent is certain regarding his choice-situation. We shall get to Shackle's idea of uncertainty later.

Speaking of his own version, Jeffrey finds no difficulty. He holds that "... the alleged contradiction arises through conflation of two different preference rankings: Yours before deciding to make o^x true, and yours thereafter" (1977, p. 137). Your preferring o^x to T is part of your pre-choice ranking. Your being indifferent between them is part of your ranking *post*-choice. Jeffrey rejects the possibility that you might know you will do x and so be indifferent *before* you choose. (Asked to comment on Shackle's discussion, Jeffrey would say you don't have a choice only where you already have made it.)

We spoke of a person's foreseeing his issue and his preference ranking. Jeffrey denies this ever happens. He holds that in thinking out what to do we set up the options from which we will choose and clarify our preferences. We are never done with this until the point of choice is reached. Or perhaps the other way: where we wind up deliberating, there we choose on the spot, even where we can't yet act. So we are never in a position to predict how we will choose. As soon as we settle our minds, the choice is made and all is over.

Granted we often don't known till we get there where we will be. But things are not always this much in the dark. Sometimes we know very well, but hope that a new option will still come up – that in fact we won't face the issue we now anticipate facing – or even that we won't have the preferences we now expect to have. So too for the factors that Jeffrey doesn't mention. Sometimes the agent sees the possible sequels before he chooses. Sometimes he does not see them until the point of choice. Sometimes he sees his choice-policy; sometimes indeed not. Sometimes he sees his choice will be effective, and again, sometimes he doesn't. Where he does see all this beforehand, the problem is in force.

Let me lay out the problem once more, and bring out its basic structure. Suppose that an agent A is what we shall call *deductively thorough*, that he believes all the deductive consequences of whatever he believes. Also that he is *belief-retentive*, that between now and the point of choice he will not drop any belief that he holds. (For short, say simply that A is *thorough* and *retentive*.) Suppose also that A believes that

   1. A's issue at the point of choice will be made up of options $o_1, o_2,$ $\ldots, o_m$,

   2. The sequels (*conditional* sequels) of these options would be $s_{11},$ $s_{12}, \ldots, s_{1n}; s_{21}, s_{22}, \ldots s_{2n}; \ldots; s_{m1}, s_{m2}, \ldots s_{mn}$,

   3. A's preference ranking at the point of choice will be R,

   4. A will pursue choice-policy C, and

   5. A's choice will be effective.

We are assuming it follows from (1)–(4) that

   6. A will chose o^x.

It follows from (5) and (6) that

   7. It is true that o^x.

Note that (7) does not contradict (1)–(5). Notice in particular that it does not contradict (1): the fact that A will do x does not preclude his facing an issue of whether or not he will.

But let us recall A's total situation. We are supposing that

   8. A believes (1)–(5).

Since A is deductively thorough, it follows that

   9. A believes A will choose o^x

and also that

   10. A believes it is true that o^x.

Since A is belief-retentive, it follows that

   11. A will at the point of choice believe it is true that o^x.

Clearly, (8) alone is not troublesome either. But (1)–(5) plus (8) make for trouble, for this (given A's thoroughness and retentiveness) implies (11), and (11) undercuts (1). Since A *believes* that he will do x, he cannot ask himself whether he will. None of the possibilities mentioned in (1) remain (live) options for him. Where he sees what he will do, there is no issue left.

Now (8) itself refers to (1)–(5), so there is something wrong with these five propositions. They are not inconsistent. But if A believes them all, and is thorough and retentive, then at least one of them must be false. Hintikka (1962) has studied the logic of some closely related cases. In his terms, a set of propositions that could not all be true in a world in which A were thorough is *indefensible for A*. All sets of inconsistent propositions are indefensible for A, but the converse of this does not hold. The two propositions *A believes v* and *A does not believe w*

are not inconsistent, where v is not the same as w; but if v implies w, these two are conjointly indefensible for A. A set of propositions that could not all be true in a world in which A were thorough *and believed these propositions* is *doxastically* indefensible for A – for instance, the pair of propositions v and *A does not believe w*, where v and w are as above. Enlarging on this now, a set of propositions that couldn't all be true in a world in which A believed these propositions and was thorough *and also retentive through some period t* will be said to be doxastically t-indefensible for A, or for short, d. t-indefensible for him. The set of propositions (1)–(5) is d. t-indefensible for A, t here being the period from now until A chooses. There is no world of the sort described in which they could all be true. So A can't properly believe them all.

What does 'properly' mean here? Someone believes something *properly* where he can defend it without looking foolish. (This is very weak, but that is as it should be.) A person can't *properly* believe something if he can fend off a challenge only by denying that he is thorough or that he believes what he is defending or that he will continue to believe it. Thoroughness and self-awareness are part of being rational – I argued this in Schick (1966). Retentiveness through t is no part of rationality, but where t starts now and is to be short (take this for granted above), no one can disclaim it without sounding frivolous, or perhaps senile. A can't *properly* believe (1)–(5) because he cannot face up to the challenge that these propositions backfire. He can only defend these propositions at the cost of discrediting himself.

Again, what A believes is not inconsistent. But if we make some modest assumptions about what A is like, we can prove that what A believes is false. A can rebut only by contesting these assumptions, by insisting that he is not self-aware, or isn't thorough, or not retentive. He can hold on to (1)–(5) only by arguing that he is foolish (better perhaps: *shallow*). A sensible A would rather give up (1)–(5).

A is no better off if he does not believe (1)–(4) but does believe the derivative proposition (6) along with (5). For the set of propositions (5) and (6) is also d. t-indefensible for A, t being as above. (I am here supposing that a person can only choose what had been an option for him before.) It follows that the unit set of

12.   *A will effectively choose o^x.*

is d. t-indefensible for A. Again, this does not say that (12) is self-contradictory, or even that (12) is false. What it says is that (12) *plus A's* being thorough and retentive and believing (12) is self-contradictory: that *if A is thorough and retentive and believes (12), then (12) is false*. It says that A cannot properly believe (12). We might put this another way. Since knowledge is justified or *proper* belief, A can't *know* (12). A person can't ever know how he will (effectively) choose. (Thus the problem of forebelief comes to be one of foreknowledge after all.)

Jeffrey's form of the problem is also covered. A here believes that

13. *A* will prefer o^x to T,
14. *A* will prefer o^x to all his other options,
15. *A* will choose that option which he prefers to all the others, and
16. *A*'s choice will be effective.

These propositions are not contradictory. The problem is rather that our norms of self-awareness and the rest keep *A* from being able to defend them. Given the usual utility theory, the set of (13)–(16) is d. *t*-indefensible for *A*. Jeffrey's own intuitions come close: "In the light of your *awareness* of your options, your *awareness* of your preference for ... o^x over T reduces that preference to indifference" (1977, p. 136; emphasis added).

This has to do with *self*-knowledge only. The two-person case goes smoothly. (12) is not d. *t*-indefensible for *B*, but only for *A*. *B* can properly believe (12). He can know how *A* will choose. He can't, however, convey what he knows about *A* to *A* himself. This last reverses a familiar philosophical thesis on privacy. The usual thesis is that there is much about us that we can know but no one else can, unless we choose to tell them. The point here is that others can know things about us that we ourselves cannot know, not even if these others tell us. Privileged access turns out to go in both directions.

Time now to sum it up. We have seen that logic alone rules out our knowing the whole truth about ourselves. Where we will (effectively) choose during *t*, self-omniscience is out during *t*. This is the basic conclusion above. How disturbing is it? Not very, I think. No one has any commitment to full self-knowledge. We want to know ourselves better. Some of us only want (for protection) to know ourselves better than others do. This is not ruled out. All that must go is expendable.

Our conclusion is moreover not the first of its kind. We have been speaking of the foreknowledge of choices. Popper (1950) concluded the same about the foreknowledge of discoveries, or better: of realisations. *A* must be uneasy with the proposition that he will realise that *p*. In our terms, the unit set of this proposition is d. *t*-indefensible for him, *t* here being the period from now until his realisation of *p*. The proposition implies that *p*, for a person can't realise something not true. So if *A* believes this proposition, and is deductively thorough, he believes also that *p*. If he is retentive through *t*, he will still believe *p* when he realises it. But this contradicts itself, for no one can realise (or come to believe) what he already believes. It follows that *A* can't properly believe that he will realise *p*. He can't *know* he will realise this. (Popper's own argument is rather different. Also, he holds it disproves determinism, but this now not in Humean terms but in those of predictability.)

Back to our result about choices. Let me stress the weakness of this. It does not say that we can't known anything beforehand about how we will choose. *A* can certainly know he will choose as *B* would have chosen, or as *C* wants him to, or in a way that he will regret. He can know this if these aren't themselves his options (if he is not deliberating whether to choose as *B* would, or as *C* would, or as *D* would) and if he can't infer from this knowledge which of his options he will choose. What he cannot know is that he will choose o^x. He might properly predict his choices under many descriptions. All that is ruled out is his foreknowing his choices *in the terms of his options*.

Also, the above has little to say on the foreknowledge of *actions*. Some authors (for instance, Hampshire, 1959, chaps. 2 and 3) argue that we can't know beforehand what we will *do*, unless we already intend to do it and know we have this intention. Our result here does not support this, at least not in general. To the extent that what we will do depends on how we will choose, the limits on the foreknowledge of our choices restrict the foreknowledge of our actions. But we can act without having chosen. (I have just scratched my nose. I faced no issue and made no choice: it itched, so I scratched.) The fact that actions involve intentions (I *wanted* to scratch my nose) does not go against this. An action can be intentional and yet not derive from a choice. So the problem of foreknowing our choices does not (in each case) carry over to our actions.

Still, we can underplay it too far. Granted there is no paradox here. *A* may nonetheless be troubled, at least where a d. *t*-indefensible set is not single-membered. He cannot properly believe all the propositions (1)–(5), nor all of (13)–(16) nor both (5) and (6). Yet each of these propositions singly may be credible enough for him. Which of them should he reject? There is no general answer to this. *A* will have to find his way out of each quandary as it comes.

2. Shackle introduces his third problem as follows. Suppose the agent

... can discern no pattern of association between act and sequel, so that it appears to him that any sequel ... can follow any act. Then he will see no purpose to be served by choosing amongst acts. In such a case I shall say that decision is powerless (1966, p. 74).

What there is problematical here is harder to see than with the rest. The problem of *illusion* seemed to be that we have no choices whatever. The problem of *empty* choices is that we can't have full self-knowledge. We want to think we have choices, and it may unsettle our pride to think that full self-knowledge can't be had. But nothing at all is denied us here.

A familiar analysis moreover suggests itself in this context. If every option has the same set of possible sequels, each sequel still need not follow every option with the same probability. Suppose that the probabilities differ. If the agent is not indifferent as to which of the sequels comes out, the expected utilities of his options may differ, and choice might make sense for him after all.

Shackle rejects this reasoning. In the case that he describes, ". . . foresight would . . . be at the opposite extreme from perfect, . . . uncertainty would be unbounded" (1965, pp. 4–5). He sees all cases involving uncertainty in a way that rules out probabilities, and so also rules out expected utilities. This of course does leave him hanging. But how does he come to it?

The core of Shackle's position here is his concept of sequels. On his view, these are not the actual causal consequences of the options the agent has, nor their conditional consequences in this contingency or that. Rather, they are what the agent *thinks* would causally follow his options. The agent may suppose that there are n contingencies, so that o_1 would be followed either by s_{11} or by s_{12} or by . . . s_{1n}, that o_2 would be followed by s_{21} or by s_{22} or by . . . s_{2n}, *etc.* He may also suppose the number of contingencies to be infinite. Shackle assumes that an agent facing an issue sees things in neither of these ways – he calls this the assumption of *bounded uncertainty*. He holds that the agent thinks the contingencies are finite, but that he cannot assign any specific number to them.

This new assumption takes a stand on how a person sees things. How can we generalise *a priori* about what people think? In general, of course, we can't, but Shackle insists we must make an exception for what this assumption says. He points out that contingencies are not actual states or conditions of some sort. The physical eye can't see them. Contingencies are possibilities, and can only be imagined. This now connects with Shackle's own version of indeterminism: Shackle holds that a person's imaginings are outside the scope of causality. The agent is *creative* here. Suppose he is also self-aware, that he knows his thinking is free. However many contingencies he notes, he knows he could have brought out more. He knows that the number of possibilities – the number he *might* have noted – is finite but can't be fixed. The principle of bounded uncertainty says just this.

I have already had my say on determinism and freedom. We needn't reject determinism to grant that someone could have thought differently. But this is a side-issue here. The question is: suppose we accept Shackle's assumption – what difference does it make? Shackle argues that it rules out the probabilistic measurement of the contingencies. He notes two kinds of probability. Frequency probabilities are out of place, for all contingencies are unique, and we can only speak of the frequency of repeatable types of conditions. Logical probabilities are out too. These are ratios of possibilities, and call for the possibilities to be finite and all of them *equally* possible, or at least comparable as to their relative possibility. The assumption of bounded uncertainty now keeps us from knowing that this holds. (Strictly: only the part on uncertainty comes in here.) Where the agent knows that some (or all) of the contingencies he sees are unions of narrower possibilities, but can't say (since he can't number them) how many items are included in each union, he can't know that the contingencies he sees are all equally possible. Nor can he know that *this* one is twice or three times as possible as *that*. So he has no proper basis for any distribution of probabilities. Shackle concludes that probabilities are inapplicable altogether. This means of course that expected-utility analyses can't get started.

Shackle goes on to work out a logic based on a non-probabilistic measure of what he calls *potential surprise*. I will not follow him further. (Those who want to continue should see Levi (1966) and (1972) on surprise and also Watkins (1955) for the full theory.) The argument rejecting probabilities seems to me to be thin. Shackle ignores subjective probabilities (the Ramsey–deFinetti–Savage sort). I see no reason for doubting that a subjectivist analysis applies. If it does apply, expected utilities are back in the picture.

Still, two questions have been raised. Clearly we often choose in situations in which we have no determinate probabilities. The choice problems here are usually labelled problems *under uncertainty*. Where and how do these problems arise? And what sort of logic of choice is appropriate? I shall set out answers to these two questions below. The answer to the first appears in an extension of Ramsey's (1931) analysis, or rather in an extension of a revised form of it. The second answer derives from an analysis of Levi's (1974) [Chapter 15, this volume].

I start out with the concept of preference. Two initial definitions here. If a person neither prefers v to w nor prefers w to v, he is indifferent as to v and w. If he is indifferent as to v and w and moreover prefers v to every proposition to which he prefers w and also prefers to v every proposition that he prefers to w, and if the converses also hold, then he *equivalues* v and w. Equivaluation implies indifference, but not the other way.

The next part is more ambitious. I want to show that the concept of utilities can be spelled out in preference-terms, and that of probabilities in terms of utilities. Suppose that a person prefers some proposition to every other proposition, or that there is some set of propositions to none of which he prefers any other, none of them also being indifferent to any other to which some proposition is preferred. Call this topmost proposition, or any proposition in the topmost set, α. Suppose also that there is some proposition to which this person prefers all others, or some set of propositions none of which he prefers to any other, none of them also being indifferent to any other that is preferred to some proposition. Call this bottommost proposition, or any proposition in the bottommost set, ω. To rule out all these propositions being indifferent to each other, assume that α is preferred to ω. Suppose finally the existence of some proposition N such that the agent equivalues the propositions v if N, w if not and w if N, v if not, whatever v and w, and ranks these two equivalued propositions between v and w. (To pin this down, take N to be *the toss will be heads* or *the card drawn will be red*.)

One last preliminary. We shall speak of the α-*to*-ω *spectrum*. This is a set of propositions generated from α and ω in a certain way: α and ω are in the spectrum and also α if N, ω if not. Call this third proposition μ. The spectrum then takes in α if N, μ if not and μ if N, ω if not. Call these propositions η and σ. The spectrum has α if N, η if not and η if N, μ if not and μ if N, σ if not and σ if N, ω if not. On the same filling-in principle, we add eight further propositions, and so on.

We now assign utilities recursively. First, we assign utilities to α and to ω – any numbers will do, provided only that the number assigned to α is the higher. We go on to say that a proposition in the spectrum has a utility that is half the sum of the utilities of the two propositions generating it. The utility of α if N, ω if not is thus midway between the utilities of α and ω. This mid-ranked proposition is μ. The utility of α if N, μ if not – this proposition is η – is midway between the utilities of α and μ. The utility of η if N, μ if not is midway between the utilities of η and μ, and so on. This fixes utilities for all propositions in the spectrum. If for some proposition v that is *not* in the spectrum there is some proposition in the spectrum that the agent equivalues to v, the utility of this equivalued proposition is the utility of v. Or rather, it is the utility of v given our choice of the origin and the unit of the utility scale, this being what our initial valuations of α and ω provided. (An improved version of this reduction of utilities appears in Schick, 1984, pp. 20–22.)

Probabilities follow directly. Read $p(v,w)$ as the *probability* the agent *would* assign to v if he believed w, or the probability of v *conditional upon* w – for brevity's sake, the w-probability of v. Read $u(w)$ as the utility the agent assigns to w. Conditional probabilities can be defined as follows:

$$p(v, w) = \frac{u(w) - u(\text{not-}v \ \& \ w)}{u(v \ \& \ w) - u(\text{not-}v \ \& \ w)} \tag{1}$$

given only that $u(v \ \& \ w) \neq u(\text{not-}v \ \& \ w)$.

Consider the special case of $p(v, w)$ in which w is some logical truth (say, z-or-not-z). This gives us the agent's probabilities unqualified. Read $p(v)$ as the plain *probability* the agent assigns (unconditionally) to

v. The suggestion is that $p(v)$ is short for $p(v, T)$ where T is the truth involved. Putting T for w in (1) and simplifying, we get

$$p(v) = \frac{u(T) - u(\text{not-}v)}{u(v) - u(\text{not-}v)} \tag{2}$$

All logical truths are the same (vacuous, null) proposition. So the utility of all such truths is the same, and it makes no difference what we take as T.

To get some perspective here, consider the familiar principle:

$$u(w) = p(v, w)u(v \text{ \& } w) + p(\text{not-}v, w)u(\text{not-}v \text{ \& } w) \tag{3}$$

This says that the utility of w is the weighted average of the utilities of its coming out true along with v and along with not-v – the weights being the w-probabilities of v and not-v. It says that the utility of a proposition is a certain conditional-probability-weighted average utility. If we let $p(\text{not-}v, w) = 1 - p(v, w)$, (3) follows from (1), and also conversely where $u(v \text{ \& } w) \neq u(\text{not-}v \text{ \& } w)$. The probabilities of anyone of whom (3) is true are related to his utilities as in (1).

We see that where a preference structure establishes (origin-and-unit-relative) point-utilities for every proposition, it also establishes a (unique) conditional point-probability for every proposition relative to every (logically contingent) other. It also establishes a (unique) *non*conditional point-probability for every proposition. That is, where it establishes determinate utilities all around, it also establishes determinate probabilities all around. On the Ramseyan analysis, every proposition always has a determinate utility, and so each also always has determinate probabilities.

But establishing determinate utilities involves one special assumption: that for every proposition not in a given spectrum there is always some proposition in that spectrum equivalued to it. This assumption need not hold. Suppose that in fact it is false, that some proposition v that is not in the α-to-ω spectrum is not equivalued to any proposition in this spectrum. Suppose also that there are two propositions ϕ and ψ in the spectrum, ϕ preferred to ψ, such that the agent is indifferent both between v and ϕ and between v and ψ. Also that he prefers v to all those and only those propositions to which he prefers ψ and that he prefers to v all those and only those propositions that he prefers to ϕ. Here we can't speak of the point-utility of v, but can speak of its utility *range*: we can say that the utility of v ranges between the utilities assigned to ϕ and ψ. That is, the utility of v is here *in*determinate within these limits.

Likewise for indeterminate probabilities. Where a person's utilities are not all fully determinate, neither are his probabilities. We can say in general that a proposition has any probability established for it via our definitions by some selection of point-utilities from within the utility ranges of certain other propositions. That is, where either $u(w)$ or $u(v \ \& \ w)$ or $u(\text{not-}v \ \& \ w)$ are indeterminate, (1) does not establish a determinate $p(v, w)$, and so also for (2) and $p(v)$. But any selection of utilities from within the utility ranges gives us a corresponding probability. So there are now probability ranges: there are upper and lower bounds to the probabilities of v. The probabilities are indeterminate within these limits.

Ramsey works out point-utilities and point-probabilities only. We go further and work out ranges. We do this by allowing for a proposition's being indifferent to each of several non-indifferent others: v may be indifferent both to β and to γ even where β is preferred to γ. This means that we provide for a person's indifferences not being transitive. Ramsey's analysis assumes transitivity (1931, p. 179 [p. 34, this volume]). Our analysis does not. Of course, full transitivity is always possible. In that case, the ranges shrink into points. The situation that Ramsey considers is a special case of ours.

We now have an answer to our first question. Probabilistic uncertainty arises where there is utility-uncertainty, and this holds where the agent's indifferences are not always transitive. We might put it another way. Note that equivaluations must be transitive. (This by our definition of equivaluation.) If a person's indifferences are transitive, all his indifferences are equivaluations. Let us refer to any indifference that is not an equivaluation as a *mere* indifference. Where a person is *merely* indifferent in whatever regard, his position, as a whole, might be described as vague. So it is vagueness that makes for uncertainty.

What kind of a logic of choice applies in contexts of uncertainty? Levi's discussion shows how expected utilities still can be used. I define the *expected utility* of an option o as follows:

$$eu(o) = p(c_1, o)u(s_1) + p(c_2, o)u(s_2) + \ldots p(c_n, o)u(s_n) \qquad (4)$$

where the c's are the contingencies the agent sees and the s's are the sequels of o in these contingencies. Notice that (4) is not a generalisation of (3). The sequel s_1 is the upshot of the agent's taking option o in contingency c_1, all that he thinks will causally follow his taking o where c_1 holds. He need not equivalue s_1 and c_1-and-o. So $u(s_1)$ need not be the same as $u(c_1 \ \& \ o)$, and so on. Equation (3) is a corollary of our definition (1). Equation (4) defines a new concept.

We have seen that where a person does not assign a proposition some point-probability, he does assign it some probability *range*. It follows that though his position can't always be laid out in terms of some single probability function – some single comprehensive point-probability assignment – it can always be represented by some set P of such functions. A person's P contains all the probability functions compatible with his probability ranges, that is, all functions determinable by some conjointly possible constrictions of all these ranges to points. Suppose that a person's position is maximally specific, that he assigns some point-probability to each proposition conditionally on every other. His P here is single-membered. Suppose that his position is maximally vague, that he assigns to each proposition only the full probability range from zero to one. Here his P is the set of all possible probability functions. Most actual positions fall somewhere between these extremes.

Let the agent's utilities for the sequels of each of his options in each of the contingencies he notes be determinate. We can say that a rational person would choose some option that has the greatest expected utility relative to one of the probability functions (*any* one) in his P. Where there is no uncertainty, all the functions in P assign the same probabilities to the contingencies noted, and the general policy reduces to the rule for choice under risk. Notice that, in other cases, it may provide for keeping an issue open to some extent. In some cases, it may even give us *carte blanche*. Each option may come out best relative to some function in P, so any choice might do.

Now to generalise this: the agent's utilities may be of any degree of specificity. We proceed as we did with probabilities. A person's utility position can always be represented in terms of some (origin-and-unit-relative) set U of *utility* functions. His set U contains all the utility functions compatible with his utility ranges, all the functions determinable by some constriction of all these ranges to points. The special case of full determinacy is trivial – this is the case in which the utilities of all propositions are determinate, not just (as above) those of certain sequels. Here all the ranges are points to begin with, and U contains one function only. (In this case, P too is single-membered.) In more typical cases, U (and so also P) is a manifold set of functions. A rational person keeps to those options that have the greatest expected utility relative to some pair of functions, one of them in his U, the other the matched function in his P.

We can put it another way. Where the functions in the agent's U don't all assign the same utilities to the possible sequels of his options, these options don't have determinate expected utilities for him. But each matched pair of functions in his U and P establishes an expected utility for each option, the set of all the expected utilities provided for an option delimiting its expected-utility *range*. Label the expected utilities established for a set of options by the *same* matched functions in U and P *corresponding* expected utilities. Our choice policy says that a rational person will choose an option that has an establishable expected utility – an expected utility in the range fixed for it – at least as great as the corresponding expected utility of any of the others. He will choose an option whose expected-utility range is not wholly below that of any alternative option.

This is my answer to the question of the sort of logic of choice that applies. (Again, the answer is basically Levi's.) There is one further matter. Where there is uncertainty, several of the expected-utility ranges may overlap, and so it may be that several options would pass. This recalls Shackle's idea of powerless choices. Indeed it goes beyond Shackle's case. It brings out that people face powerlessness (better: *weakness* or inconclusiveness) in situations far less extreme than Shackle's.

The usual response to this is to adopt supplementary logics. The familiar principles meant for the purpose have been devised for cases in which the utilities of all the sequels are determinate. These principles could now be expanded. The agent might be assumed to focus on only the upper and lower limits of the sequels' utility ranges. Perhaps he always chooses one of the still-live options all of whose sequels have minimum utilities higher (or no lower) than that of any sequel of any other residual option. This new secondary policy would often close the issue. The trouble is that neither it nor its variants has any claim to being thought rational. (Samuelson has written somewhere that mini-maxing does not make for rationality, but for paranoia.)

What other line is possible? As I see it, the expected utility range-overlaps do bring out a weakness, but not a weakness in ourselves. The finger points to rationality: there is only so much that rationality can do. There are limits to how far it can take us. Where we are not content to consider every residual option acceptable, we must look to principles of some extra-rational sort. But this dark suggestion won't be pressed here.

15. On indeterminate probabilities

Isaac Levi

Some men disclaim certainty about anything. I am certain that they deceive themselves. Be that as it may, only the arrogant and foolish maintain that they are certain about everything. It is appropriate, therefore, to consider how judgments of uncertainty discriminate between hypotheses with respect to grades of uncertainty, probability, belief, or credence. Discriminations of this sort are relevant to the conduct of deliberations aimed at making choices between rival policies not only in the context of games of chance, but in moral, political, economic, or scientific decision making. If agent X wishes to promote some aim or system of values, he will (*ceteris paribus*) favor a policy that guarantees him against failure over a policy that does not. Where no guarantee is to be obtained, he will (or should) favor a policy that

Work on this essay was partially supported by N.S.F. grant GS 28992. Research was carried out while I was a Visiting Scholar at Leckhampton, Corpus Christi, Cambridge. I wish to thank the Fellows of Corpus Christi College and the Departments of Philosophy and History and Philosophy of Science, Cambridge University, for their kind hospitality. I am indebted to Howard Stein for his help in formulating and establishing some of the results reported here. Sidney Morgenbesser, Ernest Nagel, Teddy Seidenfeld, and Frederic Schick as well as Stein have made helpful suggestions.

This version of my 1974 paper includes the following revisions of the earlier paper: (i) A correction of the convexity condition; (ii) an improvement of the condition of confirmational conditionalization; (iii) an alteration of the substance of footnotes 3 and 4 to reflect more accurately the position I took in subsequent publications; and (iv) some relatively minor stylistic revisions of no substantive import.

In his interesting discussion of this paper, S. Spielman (1975) correctly points out that the view adopted in this paper represents a change from the position I took in (1970). In the earlier paper, I sought to reduce all revisions of probability judgment to revisions of knowledge in accordance with conditionalization. This paper marks a shift from that position.

Reprinted from *Journal of Philosophy*, **71** (1974), pp. 391–418, by kind permission of the author and the publisher.

reduces the probability of failure to the greatest degree feasible. At any rate, this is so when X is engaged in deliberate decision making (as opposed to habitual or routine choice).

Two problems suggest themselves, therefore, for philosophical consideration:

> *The problem of rational credence:* Suppose that an ideally rational agent X is committed at time t to adopting as certain a given system of sentences $K_{X,t}$ (in a suitably regimented L) and to assigning to sentences in L that are not in $K_{X,t}$ various degrees of (personal) probability, belief, or credence. The problem is to specify conditions that X's "corpus of knowledge" $K_{X,t}$ and his "credal state" $B_{X,t}$ (i.e., his system of judgments of probability or credence) should satisfy in order to be reasonable.
>
> *The problem of rational choice:* Given a corpus $K_{X,t}$ and a credal state $B_{X,t}$ at t, how should X make decisions between alternative policies from which he must choose one at t?

Consideration of these two problems should lead to examination of a third. A rational agent X is entitled to count as certain at t not only logical, mathematical, and set-theoretical truths supplemented by suitably produced testimony of the senses, but theories, laws, and statistical claims as well. At the same time, the revisability of X's corpus at t should be recognized not only by others but by X himself. Moreover, just as X's judgments of certainty are liable to revision, so too are his judgments of probability or credence. Indeed, the two types of modification are apparently interdependent, and this interdependence itself deserves examination. The third problem, therefore, is as follows:

> *The problem of revision:* Under what conditions should X modify his corpus $K_{X,t}$ or his credal state $B_{X,t}$, and, if he should do so, how should he choose between alternative ways of making revisions?

In this essay, I shall not attempt to solve the problem of revision. However, I shall indicate how a prima facie obstacle to offering anything other than a dogmatic or antirationalistic answer to the question can be eliminated.

The obstacle is a serious one; for it derives from a very attractive system of answers to the problem of rational credence and the problem of rational choice. I allude to what is called the "bayesian" view. Bayesians do not agree with one another in their answers to these questions in all respects. The views of Harold Jeffreys and the early views of Rudolf Carnap are not consonant in important ways with the ideas of Bruno de Finetti and Leonard J. Savage (or the later Carnap). Nonetheless, the answers these and a host of other authors offer to the first two questions share certain important ramifications for the problem of revision. One of these implications is the commitment to either dogmatism or antirationalism.

Of course, identifying an objectionable consequence of bayesianism, where the objection is grounded on a question of philosophical principle, is in itself unlikely to persuade devoted bayesians to abandon their position. Such authors will be tempted to modify philosophical principle so as to disarm the objection; and they will have good reasons for doing so. Bayesian doctrine does offer answers to the first two questions. These answers are derivable from a system of principles which are precise and simple. Even the disputes between bayesians can be formulated with considerable precision. Furthermore, the prescriptions bayesians recommend for making choices appear to conform to presystematic judgment at least in some contexts of decision. Rival attempts to answer the problems of rational credence and rational choice seem either eclectic or patently inadequate when compared with the bayesian approach.

Thus, it is not enough to complain of the defects of bayesianism. The serious challenge is to construct an alternative system of answers to the problems of rational credence and choice which preserves the virtues of bayesianism without its vices – in particular, the defects it exhibits relevant to the problem of revision.

In this paper, I shall outline just such a rival view.

1. X's corpus of knowledge $K_{X,t}$ at t identifies a set of options A_1, A_2, ..., A_n as the options from which he will choose (at t' identical with or later than t) at least and at most one. In addition, $K_{X,t}$ implies that at least and at most one of the hypotheses h_1, h_2, ..., h_m is true and that each of the h_j's is consistent with $K_{X,t}$. Finally, $K_{X,t}$ implies that, if X chooses A_i when h_j is true, the hypothesis o_{ij} asserting the occurrence of some "possible consequence" of A_i is true.

The problem of rational choice is to specify criteria for evaluating various choices of A_is from among those feasible for X according to what he knows at t. Such criteria may be construed as specifying conditions for "admissibility." Option A_i is admissible if and only if X is permitted as a rational agent to choose A_i from among the feasible options. If A_i is uniquely admissible, X is obliged, as a rational agent, to choose it. In general, however, unique admissibility cannot be guaranteed, and no theory of rational choice pretends to guarantee it.

Bayesians begin their answer to the problem of rational choice by assuming that X is an ideally rational agent in the following sense:

(i) X has a system of evaluations for the possible consequences (the o_{ij}s) representable by a real-valued "utility" function $u(o_{ij})$ unique up to a positive affine transformation (i.e., where utility assignments are nonarbitrary once a 0 point and a unit are chosen – as in the case of measuring temperature).

(ii) X has a system of assignments of degrees of credence to the o_{ij}s, given the choice of A_i representable by a real-valued function $Q(o_{ij}; A_i)$ conforming to the requirements of the calculus of probabilities. Often X will assign credence values to the "states of nature" h_1, h_2, ..., h_n so that the h_js are probabilistically independent of the option chosen. When this is so, $Q(o_{ij}; A_i)$ equals the unconditional credence (given $K_{X,t}$) $Q(h_j)$. In the sequel, I shall suppose that we are dealing with situations of this kind.

Given such a utility function $u(o_{ij})$ and Q-function $Q(h_j)$, let $E(A_i) = \sum_{j=1}^{m} u(o_{ij})Q(h_j)$. $E(A_i)$ is the expected utility of the option A_i.

Bayesians adopt as their fundamental principle of rational choice the principle that an option is admissible only if it bears maximum expected utility among all the feasible options.

Very few serious writers on the topic of rational choice object to the principle of maximizing expected utility in those cases where X's values and credal state can be represented by a utility function unique up to a positive affine transformation and a unique probability function. The doubts typically registered concern the applicability of this principle. That is to say, critics doubt that ordinary men have the ability under normal circumstances to satisfy the conditions of ideal rationality stipulated by strict bayesians even to a modest degree of approximation.

The bayesian riposte to doubts about applicability is to insist that rational men should meet the requirements for applying the principle of maximizing expected utility and that, appearances to the contrary notwithstanding, men are quite capable of meeting these requirements and often do so.

I am not concerned to speculate on our capacities for meeting strict bayesian requirements for credal (and value) rationality. But even if men have, at least to a good degree of approximation, the abilities bayesians attribute to them, there are many situations where, in my opinion, rational men *ought not* to have precise utility functions and precise probability judgments. This is to say, on some occasions, we should avoid satisfying the conditions for applying the principle of maximizing expected utility even if we have the ability to satisfy them.

In this essay, reference to the question of utility will be made from time to time. I shall not, however, attempt to explain why I think it is sometimes (indeed, often) irrational to evaluate consequences by means of a utility function unique up to a positive affine transformation. My chief concern is to argue that rational men should sometimes avoid adopting numerically precise probability judgments.

The bayesian answer to the problem of rational choice presupposes at least part of an answer to the problem of rational credence. For a strict bayesian, a rational agent has a credal state representable by a numerically precise function on sentences (or pairs of sentences when conditional probability is considered) obeying the dictates of the calculus of probabilities.

There are, to be sure, serious disputes among bayesians concerning credal rationality. In his early writings, Carnap (1962, pp. 219–241) believed that principles of "inductive logic" could be formulated so that, given X's corpus $K_{X,t}$, X's credal state at t would be required by the principles of inductive logic to be represented by a specific Q-function that would be the same for anyone having that corpus. Others (including the later Carnap, 1971a, p. 27) despair of identifying such strong principles. Nonetheless, bayesian critics of the early Carnap's program for inductive logic continue to insist that ideally rational agents should assign precise probabilities to hypotheses.

2. X's corpus of knowledge $K_{X,t}$ shall be construed to be the set of sentences (in L) to whose certain truth X is committed at t. I am not suggesting that X is explicitly or consciously certain of the truth of every sentence in $K_{X,t}$, but only that he is committed to being certain. X might be certain at t of the truth of h and, hence, be committed to being certain of $h \lor g$, without actually being certain. Should it be brought to X's attention, however, that $h \lor g$ is a deductive consequence of h, he would be obliged as a rational agent either to cease being certain of h or to take $h \lor g$ to be certain. The latter alternative amounts to retaining his commitment; the former to abandoning it.

In this sense, X's corpus of knowledge at t should be a deductively closed set of sentences. Insofar as we restrict our attention to changes in knowledge and credence which are changes in commitments, modifications of corpora of knowledge are shifts from deductively closed sets of sentences to other deductively closed sets of sentences. Such modifications come in three varieties:

1. *Expansions*, where X strengthens his corpus by adding new items. Some examples of expansion are acquiring new items via observation, from the testimony of others and through inductive or nondeductive inference leading to the "acceptance" of statistical claims, laws, or theories into the corpus.
2. *Contractions*, where X weakens his corpus by removing items. This can happen when X detects an inconsistency in his corpus due to his having added at some previous expansion step an observation report that contradicts assumptions already in his corpus, or when X finds himself in disagreement with Y (whose views he respects on the point at issue) and wishes to resolve the dispute without begging the question.

3. *Replacements*, where X shifts from a theory containing one assumption to another containing an assumption contradicting the first. This can happen when X substitutes one theory for another in his corpus.

No matter which kind of modification is made, I shall suppose that there is a "weakest" potential corpus UK (the "urcorpus") of sentences in L such that no rational agent should contract that corpus. UK is the deductively closed set of sentences in L such that every potential corpus in L is an expansion of UK (or is UK itself). I shall suppose that UK contains logical truths, set-theoretical truths, mathematical truths, and whatever else might be granted immunity from removal from the status of knowledge. (The items in UK are in this sense incorrigible.)

Replacement poses special problems for an account of the revision of knowledge. At t when X's corpus is $K_{X,t}$, why should he shift to a corpus K^* which is obtained by deleting items from $K_{X,t}$ and replacing them with other items inconsistent with the first? From X's point of view, at t, he is replacing a theory which he is certain is true by another which he is certain is false.

The puzzle can be avoided by regarding replacements for purposes of analysis as involving two steps: (a) contraction to a corpus relative to which no question is begged concerning the rival theories, and (b) subsequent expansion based on the information available in the contracted corpus, supplemented, perhaps, by the results of experiments conducted in the interim.

Those who insist on attempting to justify replacements without decomposing them into contractions followed by expansions confront the predicament that they cannot justify such shifts without begging questions. Such justification is no justification. The conclusion that beckons is that all replacements are forms of "conversion" to which men are subjected under revolutionary stress. This is the view which Thomas Kuhn has made so popular and which stands opposed to views that look to the formulation of objective criteria for the evaluation of proposed modifications of knowledge.

3. How does all this relate to bayesian views about the revision of credal states?

Consider X's corpus of knowledge $K_{X,t}$ at t. X's credal state $B_{X,t}$ at t is, according to strict bayesians, determined by $K_{X,t}$. Strict bayesians disagree among themselves concerning the appropriate way in which to formulate this determination. The following characterization captures the orthodox view in all its essentials.

Let K be any potential corpus (i.e., let it be UK or an expansion thereof). Let $C_{X,t}(K)$ be X's judgment at t as to what his credal state should be were he to adopt K as his corpus of knowledge. I shall suppose that X is committed to judgments of this sort for every feasible K in L. The resulting function from potential corpora of knowledge to potential credal states shall be called X's "confirmational commitment" at t.

According to strict bayesians, no matter what corpus K is (provided it is consistent), $C_{X,t}$ (K) is representable by a probability function where all sentences in K receive probability I. In particular, $C_{X,t}$ (UK) is representable by a function P $(x;y)$ – which I shall call a P-function, to contrast it with a Q-function representing $C_{X,t}(K)$ where K is an expansion of UK.

Strict bayesians adopt the following principle, which imposes restrictions upon confirmational commitments:

Confirmational conditionalization: If K is obtained from UK by adding e (consistent with UK) to UK and forming the deductive closure, $P(x;y)$ represents $C_{X,t}(UK)$ and $Q(x;y)$ represents $C_{X,t}(K)$, $Q(h;f) = P(h;f\&e)$.

In virtue of this principle, X's confirmational commitment is defined by specifying $C_{X,t}(UK) = C_{X,t}$ and employing confirmational conditionalization.[1] X's credal state at t, $B_{X,t}$, is then determined by $K_{X,t}$ and $C_{X,t}$ according to the following principle:

$$\text{Total Knowledge: } C_{X,t}(K_{X,t}) = B_{X,t}$$

Notice that the principle of confirmational conditionalization, even when taken together with the principle of total knowledge, does not prescribe how X should modify his credal state given a change in his corpus of knowledge.

To see this, suppose that at t_1 X's corpus is K_1 and that at t_2 his corpus K_2 is obtained from K_1 by adding e (consistent with K_1) and forming the deductive closure. From confirmational conditionalization and total knowledge, we can conclude that *if X does not alter his confirmational commitment in the interim from t_1 to t_2*, then, if Q_1 represents B_{X,t_1}, and Q_2 represents B_{X,t_2}, $Q(h;f) = Q_1(h;f\&e)$. Should X renege at t_2 on the confirmational commitment he adopted at t_1, the change in knowledge just described need not and will not, in general, lead to a modification of credal state of the sort indicated.

---

[1] Confirmational commitments built on the principle of confirmational conditionalization are called "credibilities" by Carnap (1971a, pp. 17–19). The analogy is not quite perfect. According to Carnap, a credibility function represents a permanent disposition of X to modify his credal states in the light of changes in his corpus of knowledge. When credibility is rational, it can be represented by a "confirmation function." Since I wish to allow for modifications of confirmational commitments as well as bodies of knowledge and credal states, I assign dates to confirmational commitments. Throughout I gloss over Carnap's distinction between credibility functions and confirmation functions (1971a, pp. 24–27).

Nonetheless, strict bayesians unanimously suppose that a rational agent will, save under unusual circumstances, modify his credal state in the fashion indicated. This mode of revising credal states is often called "conditionalization." To distinguish it from confirmational conditionalization and other types of conditionalization, I shall call it "intertemporal credal conditionalization." I contend that the strict bayesian endorsement of intertemporal credal conditionalization presupposes commitment to the following principle:

Confirmational tenacity: For every X, t, and t', $C_{X,t} = C_{X,t'}$

Thus, strict bayesians have an answer to the problem of revising credal states. X's confirmational commitment is to be held fixed over time. Given such a fixed commitment, the credal state he should adopt is determined for each possible modification of his corpus of knowledge which is a consistent expansion of UK. The problem of revising credal states reduces, therefore, to the problem of revising corpora of knowledge.

Is this answer to the problem of revision satisfactory? It would be, in my opinion, if the program for inductive logic envisaged by Carnap in his early writings on the subject could be realized. Inductive logic would then be strong enough to single out a standard P-function that all rational agents should adopt as their confirmational commitment. A fortiori, all such agents should hold that commitment fast at all times.

Few bayesians now think an inductive logic of the requisite power can be constructed. Their reasons (which, in my opinion, are sound) need not detain us. In response to this skepticism, most bayesians no longer require that all rational agents endorse a single standard confirmational commitment. They hold that rational X is perfectly free to pick any confirmational commitment consonant with the principles of inductive logic. Rational Y is quite free to pick a different commitment. However, bayesians tend to insist that, once X and Y have chosen their respective commitments, they should hold them fixed. To do this is to follow the probabilistic analogue of the method of tenacity so justly criticized by Peirce in "Fixation of Belief."

In the spirit of Peirce, it would have been far better to say that a rational X should not modify his confirmational commitment capriciously – i.e., without justification. To follow this approach, however, demands consideration of criteria for justified modifications of confirmational commitments. Bayesians not only fail to do this, but, as I shall now argue, they cannot do so without great difficulty. Given the bayesian answer to the problem of rational credence, no shift can be justified. If I am right, for bayesians, either tenacity should be obeyed, or, if not, justification is gratuitous. I think this implication of bayesian doctrine is to be deplored and should lead to scrutiny of other approaches.

4. Modifying a confirmational commitment is not quite the same as modifying a corpus of knowledge. Yet, shifting from a confirmational commitment represented by a precise probability function to another confirmational commitment represented by a different precise probability function seems analogous to replacement in the following sense: The shift from confirmational commitment C_1 to confirmational commitment C_2 involves a shift to a confirmational commitment conflicting with C_1 in the sense that the P-function X uses to determine his credal state relative to his corpus when C_1 is adopted yields different precise subjective probability or credence assignments for hypotheses from those which X would make were he to adopt C_2 (and keep his corpus constant).

From X's vantage point at t when he endorses C_1, C_2 is illegitimate. He cannot justify shifting to C_2. At least, he cannot justify a direct shift. Can he do so indirectly by first performing a shift analogous to contraction from C_1 to C_3, which begs no questions concerning the merits of C_1 and C_2? Not from a strict bayesian point of view; for C_3 would, like C_1 and C_2, have to be representable by a precise P-function. The shift from C_1 to C_3 would be as problematic as the shift from C_1 to C_2.

Thus, from a bayesian point of view, no shift from one confirmational commitment to another can be justified. A rational man should conform to confirmational tenacity so that no justification is needed or else hold that some shifts are permitted without justification. Carnap (1971b, 51–52) sometimes seems to recognize shifts in confirmational commitments as a result of conceptual change. Alternatively, one might allow shifts in confirmational commitment due to conversion under revolutionary stress. Except for the minimal requirement that the shift be to a commitment obeying requirements of inductive logic, no critical control is to be exercised. Bayesians are committed to being dogmatically tenacious or arbitrarily capricious.

The source of the difficulty should be apparent. Bayesians restrict the confirmational commitments a rational agent may adopt to those representable by numerically precise probability functions. This precludes shifting from a confirmational commitment C_1 to a confirmational commitment C_3 that begs no questions as to the merits of C_1 and another commitment C_2 that conflicts with C_1. My thesis is that not only are rational men allowed to make shifts to non-question-begging commitments but that on many occasions they ought to do so. That is to say, it is sometimes appropriate for a rational agent to adopt a confirmational commitment that is indeterminate in the sense that it cannot be represented by a numerically precise probability function. If we relax the stringent requirements imposed by bayesians on confirmational commitments and credal states so as to allow for such shifts, there is at least some hope that we can avoid endorsement of tenacity or capriciousness. Within the strict bayesian framework, we cannot expect to do so except by clinging desperately to Carnap's early program for constructing an inductive logic so strong as to single out a standard P-function to represent the uniquely rational confirmational commitment (for a given language).

I propose to explore one way of relaxing strict bayesian requirements. The basic idea is to represent a credal state (confirmational commitment) by a *set* of Q-functions (P-functions). When the set is single-membered, the credal state (confirmational commitment) will be indistinguishable in all relevant respects from a strict bayesian credal state (confirmational commitment).

On this view, if X starts at t with the precise (i.e., single-membered) confirmational commitment C_1, he can then shift to a confirmational commitment that has as members all the P-functions in C_1 as well as the P-functions in some other confirmational commitment C_2. (As the technical formulation will indicate, other P-functions will be members of C_3 as well.)

C_3 will be "weaker" than C_1 or C_2 in that it will allow more P-functions to be "permissible" than either of the other two confirmational commitments alone does. It will allow as permissible all P-functions recognized as such according to C_1 and according to C_2. In this sense, the shift to C_3 will beg no questions as to the permissibility of the P-functions in the other two confirmational commitments.

Of course, the notion of a permissible P-function (and the correlative notion of a permissible Q-function according to a credal state) require elucidation. I shall offer only an indirect clarification. The account of rational credence (and confirmational commitment) based on the new proposal will be supplemented by criteria for rational choice which indicate how permissibility determines the admissibility of options. By indicating the connections between permissibility and rational choice, permissibility will have been characterized indirectly.

5. To simplify the technical details, I shall restrict the discussion to characterizing credal states and confirmational commitments for sentences in a given language L which belong to a set M generated as follows: Let $h_1, h_2, \ldots, h_n$ be a finite set of sentences in L all consistent with the urcorpus UK for L and such that UK logically implies the truth of at least and at most one h_i. M is the set of sentences in L which are equivalent, given UK, to a disjunction of zero or more distinct h_{iS}. (A disjunction of zero h_{iS} is, as usual, a sentence inconsistent with UK.)

With this understanding, X's credal state at t will be a set $B_{X,t}$ of functions $Q(x;y)$ where the sentences substituted for "x" are in M and the sentences substituted for "y" are in M and are consistent with $K_{X,t}$. When the sentence substituted for "y" is a member of $K_{X,t}$, I shall write $Q(x) = Q(x;y)$.

The set $B_{X,t}$ must satisfy the following three conditions:

1. *Nonemptiness*: $B_{X,t}$ is nonempty.
2. *Convexity*: Let $B_{e,X,t}$ be the set of functions of the form $Q_e(x) = Q(x;e)$ for fixed e in M consistent with $K_{X,t}$ and Q in $B_{X,t}$. $B_{e,X,t}$ is a convex set for every e consistent with $K_{X,t}$ – i.e., every weighted average of finitely many Q_e in $B_{e,X,t}$ is also in $B_{e,X,t}$.
3. *Coherence*: Every Q-function in $B_{X,t}$ is a finitely additive probability measure on M where $Q(h;e) = 1$ if $K,e\ 1 - h$ and $Q(h';e') = Q(h;e)$ if $K\ 1 - e = e'$ and $h = h'$.

Every Q-function in $B_{X,t}$ is "permissible" according to $B_{X,t}$.

As before, X's confirmational commitment $C_{X,t}(K)$ is a function from feasible corpora of knowledge to potential credal states that X at t considers to be the credal states he should adopt were he to adopt K as his corpus of knowledge. The value of the function for given K, therefore, is a nonempty, convex set of Q-functions relative to K. $C_{X,t}(UK) = C_{X,t}$ is, therefore, a nonempty convex set of P-functions. The principle of confirmational conditionalization introduced previously must now be modified to conform to the new characterization of confirmational commitments and credal states:

Confirmational conditionalization: Let K be obtained from UK by adding e (consistent with UK) to UK and forming the deductive closure. If Q is a member of $C_{X,t}(K)$, there is a P in $C_{X,t}(UK)$ such that $Q(h;f) = P(h;f\&e)$ and for every P in $C_{X,t}(UK)$, there is a Q in $C_{X,t}(K)$ such that $Q(h;f) = P(h;f\&e)$.

$B_{X,t}$ can be determined, as before, as follows:

> *Total knowledge*: $B_{X,t} = C_{X,t}(K_{X,t})$

Thus, X's confirmational commitment is defined by specifying the value of $C_{X,t}(UK)$.

A strict bayesian confirmational commitment, of course, allows a single P-function to be uniquely permissible. However, confirmational commitments are possible which contain more than one P-function. In general, I shall say that one confirmational commitment is stronger than another if the set of its P-functions is a subset of the set of P-functions in the other commitment.

On this view, the weakest confirmational commitment possible is that which contains all the P-functions that meet the requirements of inductive logic. I shall continue to follow Carnap in understanding inductive logic to be a system of principles that impose constraints on probability functions eligible for membership in confirmational commitments.

In contrast the strongest confirmational commitment would be the empty one – which is inconsistent with our first requirement of nonemptiness. A strongest "consistent" confirmational commitment is single-membered.

We can, by the way, extend the notion of a confirmational commitment so as to define it for an inconsistent corpus. We can require that $C_{X,t}(K)$ where K is inconsistent, be empty. This means that our previous requirement that a credal state be nonempty is to be restricted to cases where K is consistent. Thus, X might adopt a consistent confirmational commitment (i.e., one that is nonempty). Yet, if he should, unfortunately, endorse an inconsistent K, his credal state should be empty.

As noted previously, strict bayesians have differed among themselves as to what constitutes a complete system of principles of inductive logic. These differences persist on the view I am now proposing. They may be viewed, however, in a new light. The disagreements over inductive logic turn out to be disagreements over what constitutes the "weakest" possible confirmational commitment – which I shall call "$CIL(UK)$."

"Coherentists" like de Finetti and Savage claim that the principle of coherence constitutes a complete inductive logic. On their view, $CIL(UK)$ is the set of all P-functions obeying the calculus of probabilities defined over M.

Some authors are prepared to add a further principle to the principle of coherence. This principle determines permissible Q-values for hypotheses about the outcome of a specific experiment on a chance device, given suitable knowledge about the experiment to be performed and the chances of possible outcomes of experiments of that type.

There is considerable controversy concerning the formulation of such a principle of "direct inference." In large measure, the controversy reflects disagreements over the interpretation of "chance" or "statistical probability," concerning the so-called "problem of the reference class" and random sampling. Indeed, the reason coherentists do not endorse a principle linking objective chance with credence is that they either deny the intelligibility of the notion of objective chance or argue in favor of dispensing with that notion.

Setting these controversies to one side, I shall call anyone who holds that a complete inductive logic consists of the coherence principle and an additional principle of direct inference from knowledge of chance to outcomes of random experiments an "objectivist."

There are many authors who are neither coherentists nor objectivists because they wish to supplement the principles of coherence and direct inference with additional principles. Some follow J. M. Keynes, Jeffreys, and Carnap in adding principles of symmetry of various kinds. Others, like I. Hacking (1965, p. 135), introduce principles of irrelevance or other criteria which attempt to utilize knowledge about chances in a manner different from that employed in direct inference. Approaches of this sort stem by and large from the work of R. A. Fisher. I lack a good tag for this somewhat heterogeneous group of viewpoints. They all agree, however, in denying that objectivist inductive logic is a complete inductive logic.

Attempting to classify the views of historically given authors concerning inductive logic is fraught with risk. I shall not undertake a tedious and thankless task of textual analysis in the vain hope of convincing the reader that many eminent authors have been committed to an inductive logic whether they have said so or not. Yet much critical insight into controversies concerning probability, induction, and statistical inference can be obtained by reading the parties to the discussion as if they were committed to some form of inductive logic. If I am right, far from being a dead issue, inductive logic remains very much alive and debated (at least implicitly) not only by bayesians of the Keynes-Jeffreys-Carnap persuasion but by objectivists to whose number I think belong J. Neyman, H. Reichenbach, and authors like H. Kyburg and I. Hacking (in his first book) who are associated in different ways with the tradition of R. A. Fisher.

Assuming, for the sake of the argument, that the debate concerning what constitutes a complete set of principles of inductive logic is settled (I, for one, would defend and will defend elsewhere adopting a variant of an objectivist inductive logic), there is yet another dimension to debates among students of probability, induction, and statistical inference.

Some authors seem to endorse the view that a rational agent should adopt the weakest confirmational commitment, *CIL*, consonant with inductive logic and hold it fast. They are, in effect, advocating confirmational tenacity. They do so, however, on the grounds that one should not venture to endorse a confirmational commitment stronger than the weakest allowed by inductive logic. (Their view is analogous to one that would require adopting the weakest corpus of knowledge *UK* and holding it fast.) I shall call advocates of such a view "necessitarians."

Again, classifying historically given authors is a risky business. However, Keynes, Jeffreys, and Carnap (in his early work) seem to be clear examples of necessitarians. What is more interesting is the implication that anyone is a necessitarian who insists that the only conditions under which a numerically precise probability can be assigned to a statement (other than a statement that is certainly true or false) are those derivable via direct inference from knowledge of chances. Such authors, on my view, are committed to saying that, when numerical probabilities are not assignable in this way, any numerical value is a permissible assignment provided that it is derived from Q-functions allowed by inductive logic.

To illustrate, suppose that X knows that a given coin has a .4 or a .6 chance of landing heads on a toss. Let h_1 be the first hypothesis that the chance is .4, and h_2 the second hypothesis. Let g be the hypothesis that the coin will land heads on the next toss. By direct inference, every permissible Q-function in X's credal state must be such that $Q(g;h_1) = .4$ and $Q(g;h_2) = .6$. By coherence, every Q-function in his credal state must be such that $Q(h_2) = 1 - Q(h_1)$, where $Q(h_1)$ is some real number between 0 and 1 and $Q(g) = Q(g;h_1)Q(h_1) + Q(g;h_2)Q(h_2) = .4Q(h_1) + .6(1 - Q(h_1))$.

According to the authors I have in mind, there is no unique numerical value that a rational X should adopt as uniquely permissible for $Q(h_1)$. As I am interpreting such authors as Kyburg, Neyman, Reichenbach, and Salmon, they mean to say that X's credal state should consist of all Q-functions meeting the conditions specified. The upshot is that the set of permissible Q-values for g should consist of all Q-values in the interval from .4 to .6. If I am reading them right, they endorse an objectivist logic and, at the same time, insist that X should adopt *CIL* as his confirmational commitment. They are "objectivist necessitarians."

The early Carnap, as noted previously, had hoped to identify an inductive logic that singled out a unique P-function as eligible for membership in confirmational commitments. Had his hope been realized, a rational agent would perforce have had to be a necessitarian. The weakest confirmational commitment would have been the strongest consistent one as well. Confirmational tenacity would have been necessitated by the principles of inductive logic.

But if Carnap's program is abandoned, necessitarianism is by no means the only response that one can make. Indeed, it seems to be of doubtful tenability, if for no other reason than that credal states formed on a necessitarian basis seem to be too weak for use in practical decision making or statistical inference. (Many objectivist necessitarians seem to deny this; but the matter is much too complicated to discuss here.)

Personalists, like de Finetti and Savage, abandon necessitarianism but continue to endorse confirmational tenacity – at least during normal periods free from revolutionary stress. It is this position that I contended earlier leads to dogmatism or capriciousness with respect to confirmational commitment.

The view I favor is *revisionism*. This view agrees with the personalist position in allowing rational men to adopt confirmational commitments stronger than *CIL*. It insists, however, that such commitments are open to revision. It sees as a fundamental epistemological problem the task of providing an account of the conditions under which such revision is appropriate and criteria for evaluating proposed changes in confirmational commitment on those occasions when such shifts are needed.

I shall not offer an account of the revision of confirmational commitments. The point I wish to emphasize here is that, once one abandons the strict bayesian approach to credal rationality and allows credal states to contain more than one permissible Q-function in the manner I am suggesting, the revisionist position can be seriously entertained. The strict bayesian view precludes it and leaves us with the dubious alternatives of necessitarianism and personalism. By relaxing the strict bayesian requirements on credal rationality, we can at least ask a question about revision which could not be asked before.

6. According to the approach I am proposing, X's credal state at t is characterized by a set of Q-functions defined over sentences in a set M. Such a representation describes X's credal state globally. Nothing has been said thus far as to how individual sentences in M are to be assigned grades of credence or how the degrees of credence assigned to two or more sentences are to be compared with one another. The following definitions seem to qualify for this purpose:

Definition 1: $Cr_{X,t}(h;e)$ is the set of real numbers r such that there is a Q-function in $B_{X,t}$ according to which $Q(h;e) = r$.

Definition 2: $c_{X,t}(h;e)$ is the set of real numbers r such that there is a P-function in $C_{X,t}$ according to which $P(h;e) = r$.

In virtue of the convexity requirement, both the credence function $Cr_{X,t}(h;e)$ and the confirmation function $c_{X,t}(h;e)$ will take sets of values that are subintervals of the unit line – i.e., the interval from 0 to 1. The lower and upper bounds of such intervals have properties which have been investigated by I. J. Good (1962), C. A. B. Smith (1961), and A. P. Dempster (1967).

A partial ordering with respect to comparative credence or with respect to comparative confirmation can be defined as follows:

Definition 3: $(h;e) \overset{Cr_{X,t}}{\leq} (h';e')$ if and only if, for every Q-function in $B_{X,t}$, $Q(h;e) \leq Q(h';e')$.

Definition 4: $(h;e) \overset{C_{X,t}}{\leq} (h';e')$ if and only if, for every P-function in $C_{X,t}$, $P(h;e) \leq P(h';e')$.

The partial orderings induced by credal states and confirmational commitments conform to the requirements of B. O. Koopman's (1940) axioms for comparative probability. Koopman pioneered in efforts to relax the stringent requirements imposed by bayesians on rational credence. Within the framework of his system, he was able not only to specify conditions of rational comparative probability judgment but to identify ways of generating interval-valued credence functions.

According to Koopman's approach, however, any two credal states (confirmational commitments) represented by the same partial ordering of the elements of M are indistinguishable. My proposal allows for important differences. Several distinct convex sets of probability distributions over the elements of M can induce the same partial ordering on the elements of M according to definitions 3 and 4.

Dempster (1967), Good (1962), Kyburg (1961), Smith (1961), and F. Schick (1958) have all proposed modifying bayesian doctrine by allowing credal states and confirmational commitments to be represented by interval-valued probability functions. Good, Smith, and Dempster have also explored the representation of credal states defined by interval-valued credence functions by means of sets of probability measures. Smith and Dempster explicitly consider convex sets of measures. Nonetheless, all these authors, including Dempster and Smith, seem to regard credal states (and confirmational commitments) represented by the same interval-valued function as indistinguishable. In contrast, my proposal recognizes credal states as different even though they generate the identical interval-valued function – provided they are different convex sets of Q-functions.[2]

Thus, the chief difference between my proposal and other efforts to come to grips with "indeterminate" probability judgments is that my proposal recognizes significant differences between credal states (confirmational commitments) where other proposals recognize none. Is this a virtue, or are the fine distinctions allowed by my proposal so much excess conceptual baggage?

I think that the distinctions between credal states recognized by the proposals introduced here are significant. Agents X and Y, who confront the same set of feasible options and evaluate the possible consequences in the same way may, nonetheless, be obliged as rational agents to choose different options if their credal states are different, even though their credal states define the same interval-valued credence function. That is to say, according to the decision theory that supplements the account of rational credence just introduced, differences in credal states recognized by my theory but not by Dempster's or Smith's, do warrant different choices in otherwise similar contexts of choice.

---

[2] The difference between my approach and Smith's was drawn to my attention by Howard Stein. To all intents and purposes, both Dempster and Smith represent credal states by the largest convex sets that generate the interval-valued functions characterizing those credal states. Dempster (332/3) is actually more restrictive than Smith. Dempster, by the way, wrongly attributes to Smith the position I adopt. To my knowledge, Dempster is the first to consider this position in print – even if only to misattribute it to Smith.

To explain this claim, we must turn to a consideration of rational choice. We would have to do so anyhow. One of the demands that can fairly be made of those who propose theories rival to bayesianism is that they furnish answers not only to the problems of rational credence and revision but to the questions about rational choice. Furthermore, the motivation for requiring credal states to be non-empty, convex sets of probability measures and the explanation of the notion of a permissible Q-function are best understood within the context of an account of rational choice. For all these reasons, therefore, it is time to discuss rational choice.

7. Consider, once more, a situation where X faces a decision problem of the type described in section 1. No longer, however, will it be supposed that X's credal state for the "states of nature" $h_1, h_2, \ldots, h_n$ and for the possible consequences $o_{i1}, o_{i2}, \ldots, o_{im}$ conditional on X choosing A_i are representable by a single Q-function. Instead, the credal state will be required only to be a nonempty convex set of Q-functions.[3]

Although I have not focused attention here on the dubiety of requiring X's evaluations of the o_{ij}s to be representable by a utility function unique up to a positive affine transformation, I do believe that rational men can have indeterminate preferences and will, for the sake of generality, relax the bayesian requirement as follows: X's system of evaluations of the possible consequences of the feasible options is to be represented by a set G of "permissible" u-functions defined over the o_{ij}s which is (a) nonempty, (b) convex, and such that all positive affine transformations of u-functions in G are also in G. A bayesian G is, in effect, such that all u-functions in it are positive affine transformations of one another. It is this latter requirement that I am abandoning.

In those situations where X satisfies strict bayesian conditions so that his credal state contains only a single Q-function and G contains all and only those u-functions which are positive affine transformations of some specific u-function u_1, an admissible option A_i is, according to the principle of maximizing expected utility, an option that bears maximum

---

[3] I am assuming here that the option chosen is confirmationally irrelevant to the states of nature. Let B_{A_i} be the credal state B (= $B_{X,t}$), restricted to the permissible functions of the form $Q(x;A_i$ is chosen) and where the values of x are boolean combinations of hypotheses specifying the states of nature of the form hj and B_t be the credal state B restricted to the permissible functions of the form $Q(x;t) = Q(x)$ where t is entailed by $K_{X,t}$. The choosing of A_i is confirmationally irrelevant in the weak sense to the hj's if and only if the set $B_{A_i} = B_t$. This notion of irrelevance does not imply that $Q(hj;A_i) = Q(h_j)$ for every permissible Q-function in B. I call this latter condition "strong confirmational irrelevance" whereas the former is "weak confirmational irrelevance." When strong confirmational irrelevance is satisfied, the E-admissible options can be determined using the unconditional Q-functions in B_t to compute expected utilities. For further discussion of confirmational irrelevance, see I. Levi (1980, pp. 225–233, and 1978, pp. 263–273). This note represents a revision of footnote 12 in the original version of this paper.

expected utility $E(A_i) = \sum_{i=1}^{m} Q(h_i)u_1(o_{ij})$. Notice that, if any positive

affine transformation of u_1 is substituted for u_1 in the computation of expected utility, the ranking of options with respect to expected utility remains unaltered. Hence we can say that, according to strict bayesians, an option is admissible if it bears maximum expected utility relative to the uniquely permissible Q-function and to any of the permissible ~~Q-function and to any of the permissible~~ u-functions in G (all of which are positive affine transformations of u_1).

There is an obvious generalization of this idea applicable to situations where $B_{X,t}$ contains more than one permissible Q-function and G contains u-functions that are not positive affine transformations of one another. I shall say that A_i is *E-admissible* if and only if there is at least one Q-function in $B_{X,t}$ and one u-function in G such that $E(A_i)$ defined relative to that Q-function and u-function is a maximum among all the feasible options. The generalization I propose is the following:

E-admissibility: All admissible options are E-admissible.

The principle of E-admissibility is by no means novel. I. J. Good (1952, p. 114), for example, endorsed it at one time. Indeed, Good went further than this. He endorsed the converse principle that all E-admissible options are admissible as well.

I disagree with Good's view on this. When X's credal state and goals select more than one option as E-admissible, there may be and sometimes are other considerations than E-admissibility which X, as a rational agent, should employ in choosing between them.

There are occasions where X identifies two or more options as E-admissible and where, in addition, he has the opportunity to defer decision between them. If that opportunity is itself E-admissible, he should as a rational agent "keep his options open." Notice that in making this claim I am not saying that the option of deferring choice between the other E-admissible options is "better" than the other E-admissible options relative to X's credence and values and the assessments of expected utility based thereon. In general, E-admissible options will not be comparable with respect to expected utility (although sometimes they will be). The injunction to keep one's options open is a criterion of choice that is based not on appraisals of expected utility but on the "option-preserving" features of options. Deferring choice is better than the other E-admissible options in this respect, but not with respect to expected utility.

Thus, a *P-admissible* option is an option that is (a) E-admissible and (b) "best" with respect to E-admissible option preservation among all E-admissible options. I shall not attempt to provide an adequate explication of clause (b) here. In the subsequent discussion, I shall consider situations where there are no opportunities to defer choice. Nonetheless, it is important to notice that, given a suitably formulated surrogate for (b), the following principle holds:

P-admissibility: All admissible options are P-admissible.

My disagreement with Good goes still further than this; for I reject not only the converse of E-admissibility but that of P-admissibility as well.

To illustrate, consider a situation that satisfies strict bayesian requirements. X knows that a coin with a .5 chance of landing heads is to be tossed once. g is the hypothesis that the coin will land heads. Under the circumstances, we might say that X's credal state is such that all permissible Q-functions assign g the value $Q(g) = .5$. Suppose that X is offered a gamble on g where X gains a dollar if g is true and loses one if g is false. (I shall assume that X has neither a taste for nor an aversion to gambling and that, for such small sums, money is linear with utility.) He has two options: to accept the gamble and to reject it. If he rejects it, he neither gains nor loses.

Under the circumstances described, the principle of maximizing expected utility may be invoked. It indicates that both options are optimal and, hence, in my terms E-admissible. Since there are no opportunities for delaying choice, both options (on a suitably formulated version of P-admissibility) become P-admissible.

Bayesians – and Good would agree with this – tend to hold that rational X is free to choose either way. Not only are both options E-admissible. They are both admissible. Yet, in my opinion, rational X should refuse the gamble. The reason is not that refusal is better in the sense that it has higher expected utility than accepting the gamble. The options come out equal on this kind of appraisal. Refusing the gamble is "better," however, with respect to the security against loss it furnishes X. If X refuses the gamble, he loses nothing. If he accepts the gamble, he might lose something. This appeal to security does not carry weight, in my opinion, when accepting the gamble bears higher expected utility than refusing it. However, in that absurdly hypothetical situation where they bear precisely the same expected utility, the question of security does become critical.

These considerations can be brought to bear on the more general situation where two or more options are E-admissible (even though they are not equal with respect to expected utility) and where the principle of P-admissibility does not weed out any options.

An S-admissible option (i.e., option admissible with respect to security) is an option that is P-admissible and such that there is a permissible u-function in G relative to which the minimum u-value assigned a possible consequence o_{ij} of option A_i is a maximum among all P-admissible options.[4]

S-admissibility: All admissible options are S-admissible.

I cannot think of additional criteria for admissibility which seem adequate. (But then I have no precise conditions of adequacy.) I think, perhaps, we should keep an open mind on this matter. Nonetheless, for the present, I shall tentatively assume that the converse of S-admissibility holds. This assumption will not alter the main course of the subsequent argument.

Even without detailed exploration of the ramifications of this decision theory, some of its main features are immediately apparent. It conforms to the strict bayesian injunction to maximize expected utility in those situations where X has a precise credal state and G contains u-functions that are all linear transformations of one another. In this sense, bayesian decision theory is a special case of mine.

Similarly, the proposed decision theory identifies situations where the well-known maximin criterion is applied legitimately. Customarily maximin is used to select that option from among all the *feasible* options which maximizes the minimum gain. This recommendation is legitimate, according to my theory, provided (1) G contains all and only u-functions that are positive affine transformations of one another, and (2) all feasible options are P-admissible. But even if condition (1) is satisfied, it can be the case that the maximin solution from among all

---

[4] Maximin principles in particular and criteria of S-admissibility in general suffer from a serious ambiguity. For example, the set of possible consequences of a "mixed act" constructed by choosing between "pure options" A_i and A_j with the aid of a chance device with known chance probability of selecting one or the other option may be construed as the set of possible consequences of either A_i or A_j. According to this conception of possible consequences of the mixed act, the security level for a given u-function is the lowest of the security levels belonging to A_i and A_j. Hence, one cannot raise the security level by taking a mixture of two options. Wald, von Neumann, and Morgenstern proceed differently. They take a possible consequence of the mixed option to be uniquely determined by the "state of nature." Given the state of nature, the mixed option is equivalent to a lottery whose value is equal to its expected utility given the state of nature. Such a lottery is then the "consequence" of the mixed option given the state of nature. The set of consequences so determined by the possible states of nature are the possible consequences of the mixed option for the purposes of determining security. Security levels for mixed options may then be determined differently and can, indeed, be higher than the security levels of the pure options involved in the mixture. In an earlier version of this footnote, I favored construing S-admissibility using the first conception of a security level. In later publications, I abandoned this view. I now think that the way in which a decision maker individuates possible consequences of options for the purpose of identifying security levels is up to the agent and reflects a feature of the agent's value commitments which ought not to be dictated to the agent by principles of "thin" rationality. For further discussion of this point, see I. Levi (1980, pp. 156–163). In any case, as was pointed out in the original version of this note, mixtures of E-admissible options are not always E-admissible. In this paper, I leave mixed options out of account.

the feasible options is not itself E-admissible and so cannot be considered to be S-admissible.

Finally, my proposal is able to discriminate between and cover a wider variety of situations where neither maximizing expected utility nor maximining can be invoked with much plausibility. Moreover, it does so with the aid of a unified system of criteria of rational credence and rational choice. Thus, it does offer answers to just those questions which Bayesian theory purports to solve. Moreover, it escapes the bayesian commitment to the dubious doctrines of necessitarianism or personalism.

8. Some elementary properties of credal states as nonempty convex sets will be illustrated and explained by applying the decision theory just outlined to simple gambling situations. Suppose X knows that a coin is to be tossed and has either a .4 or .6 chance of landing heads. g is the hypothesis that the coin will land heads. I shall suppose that X has neither a taste for nor an aversion to gambling and that X's values are such that G is a set of u-functions that are linear transformations of the monetary payoffs of the gambles to be considered.

Case 1: X is offered a gamble on a take-it-or-leave-it basis where he receives $S-P$ dollars if g is true and loses P dollars if g is false. (Both S and P are positive.)

Case 2: X is offered a gamble on a take-it-or-leave-it basis where he loses P dollars if g is true and receives $S-P$ dollars if g is false. (S and P have the same values as in case 1.)

h_1 is the hypothesis that the chance of heads is .4, and h_2 is the hypothesis that the chance of heads is .6. By the reasoning of page 300, every permissible Q-function in X's credal state should be such that $Q(g) = .4Q(h_1) + .6[1 - Q(h_1)]$.

According to strict bayesians, X should, therefore, adopt a credal state that selects a single such Q-function as permissible. This can be done by selecting a single value for $Q(h_1)$. If that value is r, $Q(g) = .4r + .6(1 - r) = .6 - .2r$.

Hence, the bayesian will find that accepting the case 1 gamble is uniquely admissible if and only if $Q(g) > P/S$, and will find accepting the case 2 gamble uniquely admissible if and only if $Q(\sim g) > P/S$. (Otherwise rejecting the gamble for the appropriate case is uniquely admissible, assuming that ties in expected utility are settled in favor of rejection.) Hence, if P/S is less than .5, a bayesian must preclude the possibility of accepting the gamble being inadmissible both in case 1 and in case 2.

Suppose, however, that $Cr_{X,t}(h_1)$ takes a nondegenerate interval as a value. For simplicity, let that interval be [0,1]. The set of permissible

Q-values for g must be all values of $.6 - .2r$ where r takes any value from 0 to 1. Hence, $Cr_{X,t}(g) = [.4, .6]$.

Under these conditions, my proposal holds that, when P/S falls in the interval from $.4$ to $.6$, both options are E-admissible (and P-admissible) in case 1. The same is true in case 2. But in both case 1 and case 2, rejecting the gamble is uniquely S-admissible. Hence, in both cases, X should reject the gamble. *This is true even when P/S is less than* $.5$. In this case, my proposal allows a rational agent a system of choices that a strict bayesian would forbid. In adopting this position, I am following the analysis advocated by C. A. B. Smith for handling pairwise choices between accepting and rejecting gambles. Smith's procedure, in brief, is to characterize X's degree of credence for g by a pair of numbers (the "lower pignic probability" and the "upper pignic probability" for g) as follows: The lower pignic probability $\underline{s}$ represents the least upper bound of betting quotients P/S for which X is prepared to accept gambles on g for positive S. The upper pignic probability $\bar{s}$ for g is $1 - t$, where t is the least upper bound of betting quotients P/S for which X is prepared to accept gambles on $\sim g$ for positive S. Smith requires that $\underline{s} \leq \bar{s}$, but does not insist on equality as bayesians do. Given Smith's definitions of upper and lower pignic probabilities, it should be fairly clear that, in case 1 and case 2 where $Cr_{X,t}(g) = [.4, .6]$, Smith's analysis and mine coincide.[5]

Before leaving cases 1 and 2, it should be noted that, if X's credal state were empty, no option in case 1 would be admissible and no option in case 2 would be admissible either. If X is confronted with a case 1 predicament and an empty credal state, he would be constrained to act and yet as a rational agent enjoined not to act. The untenability of this result is to be blamed on adopting an empty credal state. Only when X's corpus is inconsistent, should a rational agent have an empty credal state. But, of course, if X finds his corpus inconsistent, he should contract to a consistent one.

---

[5] Smith (1961, pp. 3–5, 6–7). The agreement applies only to pairwise choices where one option is a gamble in which there are two possible payoffs and the other is refusing to gamble with 0 gain and 0 loss. In this kind of situation, it is clear that Smith endorses the principle of E-admissibility, but not its converse. However, in the later sections of his paper where Smith considers decision problems with three or more options or where the possible consequences of an option to be considered are greater than 2, Smith seems (but I am not clear about this) to endorse the converse of the principle of E-admissibility – counter to the analysis on the basis of which he defines lower and upper pignic probabilities. Thus, it seems to me that either Smith has contradicted himself or (as is more likely) he simply does not have a general theory of rational choice. The latter sections of the paper may then be read as interesting explorations of technical matters pertaining to the construction of such a theory, but not as actually advocating the converse of E-admissibility. At any rate, since it is the theory Smith propounds in the first part of his seminal essay which interests me, I shall interpret him in the subsequent discussion as having no general theory of rational choice beyond that governing the simple gambling situations just described.

Case 3: A_1 is accepting both the case 1 and the case 2 gamble jointly with a net payoff if g is true or false of $S - 2P$.

This is an example of decision making under certainty. Everyone agrees that if P is greater than $2S$ the gamble should be rejected; for it leads to certain loss. If P is less than $2S$ X should accept the gamble; for it leads to a certain gain. These results, by the way, are implied by the criteria proposed here as well as by the strict bayesian view.

Strict bayesians often defend requiring that Q-functions conform to the requirements of the calculus of probabilities by an appeal to the fact that, when credal states contain but a single Q-function, a necessary and sufficient condition for having credal states that do not license sure losses (dutch books) is having a Q-function obeying the calculus of probabilities. The arguments also support the conclusion that, even when more than one Q-function is permissible according to a credal state, if all permissible Q-functions obey the coherence principle, no dutch book can become E-admissible and, hence, admissible.

Case 4: B_1 is accepting the case 1 gamble, B_2 is accepting the case 2 gamble, and B_3 is rejecting both gambles.

Let the credal state be such that all values between 0 and 1 are permissible Q-values for h_1 and, hence, all values between .4 and .6 are permissible for g.

If P/S is greater than .6, B_3 is uniquely E-admissible and, hence, admissible. If P/S is less than .4, B_3 is E-inadmissible. The other two options are E-admissible and admissible.

If P/S is greater than or equal to .4 and less than .5, B_3 remains inadmissible and the other two admissible.

If P/S is greater than or equal to .5 and less than .6, all three options are E-admissible; but B_3 is uniquely S-admissible. Hence, B_3 should be chosen when P/S is greater than or equal to .5.

Three comments are worth making about these results.

1. I am not sure what analysis Smith would propose of situations like case 4. At any rate, his theory does not seem to cover it (but see footnote 5).
2. When P/S is between .4 and .5, my theory recommends rejecting the gamble in case 1, rejecting the gamble in case 2, and yet recommends accepting one or the other of these gambles in case 4. This violates the so-called "principle of independence of irrelevant alternatives."[6]

[6] See Luce and Raiffa (1958, pp. 288–289, pp. 61–63, this volume). Because the analysis offered by Smith and me for cases 1 and 2 seems perfectly appropriate and the analysis for case 4 also appears impeccable, I conclude that there is something wrong with the principle of independence of irrelevant alternatives.

 A hint as to the source of the trouble can be obtained by noting that if "E-

3. If the convexity requirement for credal states were violated by removing as permissible values for g all values from $(S - P)/S$ to P/S, where P/S is greater than .5 and less than .6, but leaving all other values from .4 to .6, then – counter to the analysis given previously, B_3 would not be E-admissible in case 4. The peculiarity of that result is that B_1 is E-admissible because, for permissible Q-values from .6 down to P/S, it bears maximum expected utility, with B_3 a close second. B_2 is E-admissible because, for Q-values from .4 to $(S - P)/S$, B_2 is optimal, with B_3 again a close second. If the values between $(S - P)/S$ and P/S are also permissible, B_3 is E-admissible because it is optimal for those values. To eliminate such intermediate values and allow the surrounding values to retain their permissibility seems objectionable. Convexity guarantees against this.

Case 5: X is offered a gamble on a take-it-or-leave-it basis in which he wins 15 cents if f_1 is true, loses 30 cents if f_2 is true, and wins 40 cents if f_3 is true.

Suppose X's corpus of knowledge contains the following information:

Situation a: X knows that the ratios of red, white, and blue balls in the urn are either (i) 1/8, 3/8, 4/8 respectively; (ii) 1/8, 4/8, 3/8; (iii) 2/8, 4/8, 2/8; or (iv) 4/8, 3/8, 1/8.

X's credal state for the f_is is determined by his credal state for the four hypotheses about the contents of the urn according to a more complex variant of the arguments used to obtain credence values for g in the first four cases. If we allow all Q-functions compatible with inductive logic or an objectivist kind to be permissible, X's credal state for the f_is is the convex set of all weighted averages of the four triples of ratios. $Cr_{X,t}(f_1) = (1/8, 4/8)$, $Cr_{X,t}(f_2) = (3/8, 4/8)$, and $Cr_{X,t}(f_3) = (1/8, 4/8)$. Both accepting and rejecting the gamble are E-admissible. Rejecting the gamble, however, is uniquely S-admissible. X should reject the gamble.

Situation b: X knows that the ratios of red, white, and blue balls is correctly described by (i), (ii), or (iv), but not by (iii). Calculation reveals that the interval-valued credence function is the same as in situation *a*. Yet it can be shown that accepting the gamble is uniquely E-admissible and, hence, admissible. X should accept the gamble.

admissible" is substituted for "optimal" in the various formulations of the principle cited by Luce and Raiffa, p. 289 [p. 62, this volume], the principle of independence of irrelevant alternatives stands. The principle fails because S-admissibility is used to supplement E-admissibility in weeding out options from the admissible set.

Mention should be made in passing that even when "E-admissible" is substituted for "optimal" in Axiom 9 of Luce and Raiffa, p. 292 [p. 65, this volume], the axiom is falsified. Thus, when $.5 \le P/S \le .6$ in case 4, all three options are E-admissible, yet some mixtures of B_1 and B_2 will not be.

Now we can imagine situations that are related as a and b are to one another except that the credal states do not reflect differences in statistical knowledge. Then, from the point of view of Dempster and Smith, the credal states would be indistinguishable. Because the set of permissible Q-distributions over the f_is would remain different for situations a and b, my view would recognize differences and recommend different choices. If the answer to the problem of rational choice proposed here is acceptable, the capacity of the account of credal rationality to make fine distinctions is a virtue rather than a gratuitous piece of pedantry.

The point has its ramifications for an account of the improvement of confirmational commitments; the variety of discriminations that can be made between confirmational commitments generates a variety of potential shifts in confirmational commitments subject to critical review. For intervalists, a shift from situation a to b is no shift at all. On the view proposed here, it is significant.

The examples used in this section may be used to illustrate one final point. The objective or statistical or chance probability distributions figuring in chance statements can be viewed as assumptions or hypotheses. Probabilities in this sense can be unknown. We can talk of a set of simple or precise chance distributions among which X suspends judgment. Such *possible* probability distributions represent hypotheses which are possibly true and which are themselves objects of appraisal with respect to credal probability. *Permissible* probability distributions which, in our examples, are defined over such *possible* probability distributions (like the hypotheses h_1 and h_2 of cases 1, 2, 3 and 4) are not themselves possibly true hypotheses. No probability distributions of a still higher *type* can be defined over them.[7]

I have scratched the surface of some of the questions raised by the proposals made in this essay. Much more needs to be done. I do believe, however, that these proposals offer fertile soil for cultivation

---

[7] I mention this because I. J. Good, whose seminal ideas have been an important influence on the proposals made in this essay, confuses permissible with possible probabilities. As a consequence, he introduces a hierarchy of types of probability (Good, 1962, p. 327). For criticism of such views, see Savage (1954, p. 58). In fairness to Good, it should be mentioned that his possible credal probabilities are interpreted not as possibly true statistical hypotheses but as hypotheses entertained by X about his own unknown strictly bayesian credal state. Good is concerned with situations where strict bayesian agents having precise probability judgments cannot identify their credal states before decision and must make choices on the basis of partial information about themselves. [P. C. Fishburn (1964) devotes himself to the same question.] My proposals do not deal with this problem. I reject Good's and Fishburn's view that every rational agent is at bottom a strict bayesian limited only by his lack of self-knowledge, computational facility, and memory. To the contrary, I claim that, even without such limitations, rational agents should not have precise bayesian credal states. The difference in problem under consideration and presuppositions about rational agents has substantial technical ramifications which cannot be developed here.

not only by statisticians and decision theorists but by philosophers interested in what in my opinion, ought to be the main problem for epistemology – to wit, the improvement (and, hence, revision) of human knowledge and belief.

16. Unreliable probabilities, risk taking, and decision making

Peter Gärdenfors and Nils-Eric Sahlin

1. The limitations of strict Bayesianism

A central part of Bayesianism is the doctrine that the decision maker's knowledge in a given situation can be represented by a subjective probability measure defined over the possible states of the world. This measure can be used to determine the expected utility for the agent of the various alternatives open to him. The basic decision rule is then that the alternative which has the maximal expected utility should be chosen.

A fundamental assumption for this strict form of Bayesianism is that the decision maker's knowledge can be represented by a *unique* probability measure. The adherents of this assumption have produced a variety of arguments in favor of it, the most famous being the so-called Dutch book arguments. A consequence of the assumption, in connection with the rule of maximizing expected utility, is that in two decision situations which are identical with respect to the probabilities assigned to the relevant states and the utilities of the various outcomes the decisions should be the same.

It seems to us, however, that it is possible to find decision situations which are identical in all the respects relevant to the strict Bayesian, but which nevertheless motivate different decisions. As an example to illustrate this point, consider Miss Julie who is invited to bet on the outcome of three different tennis matches.[1] As regards match A, she is

The order of the authors' names is based only upon age and (or?) wisdom. The authors wish to thank Robert Goldsmith, Sören Halldén, Bengt Hansson, and Isaac Levi for helpful criticism and comments.

[1] This example was chosen in order to simplify the exposition. We believe, however, that similar examples can be found within many areas of decision making, e.g. medical diagnosis and portfolio selection.

very well-informed about the two players – she knows everything about the results of their earlier matches, she has watched them play several times, she is familiar with their present physical condition and the setting of the match, etc. Given all this information, Miss Julie predicts that it will be a very even match and that a mere chance will determine the winner. In match *B*, she knows nothing whatsoever about the relative strength of the contestants (she has not even heard their names before) and she has no other information that is relevant for predicting the winner of the match. Match *C* is similar to match *B* except that Miss Julie has happened to hear that one of the contestants is an excellent tennis player, although she does not know anything about which player it is, and that the second player is indeed an amateur so that everybody considers the outcome of the match a foregone conclusion.

If pressed to evaluate the probabilities of the various possible outcomes of the matches, Miss Julie would say that in all three matches, given the information she has, each of the players has a 50 percent chance of winning. In this situation a strict Bayesian would say that Miss Julie should be willing to bet at equal odds on one of the players winning in one of the matches if and only if she is willing to place a similar bet in the two other matches. It seems, however, perfectly rational if Miss Julie decides to bet on match *A*, but not on *B* or *C*, for the reason that a bet on match *A* is more *reliable* than a bet on the others. Furthermore she would be very suspicious of anyone offering her a bet at equal odds on match *C*, even if she could decide for herself which player to back.

The main point of this example is to show that the amount and quality of information which the decision maker has concerning the possible states and outcomes of the decision situation in many cases is an important factor when making the decision. In order to describe this aspect of the decision situation, we will say that the information available concerning the possible states and outcomes of a decision situation has different degrees of *epistemic reliability*. This concept will be further explicated later. We believe that the epistemic reliability of a decision situation is *one* important factor when assessing the *risk* of the decision. In our opinion, the major drawback of strict Bayesianism is that it does not account for the variations of the epistemic reliability in different decision situations.

The concept of epistemic reliability is useful also in other contexts than direct decision making. In the next section, after presenting the models of decision situations, we will apply this concept in a discussion of Popper's "paradox of ideal evidence."

In order to determine whether empirical support could be obtained for the thesis that the epistemic reliability of the decision situation affects the decision, Goldsmith and Sahlin (1982) performed a series of experiments. In one of these, test subjects were first presented with descriptions of a number of events and were asked to estimate for each event the probability of its occurrence. Some events were of the well-known parlor game type, e.g., that the next card drawn from an ordinary deck of cards will be a spade; while other events were ones about which the subjects presumably had very limited information, e.g., that there will be a bus strike in Verona, Italy next week. Directly after estimating the probability of an event, subjects were asked to show, on a scale from 0 to 1, the perceived reliability of their probability estimate. The experiment was constructed so that for each subject several sets of events were formed, such that all the events in a set had received the same probability estimate but the assessed reliability of the various estimates differed. For each set, the subject was then asked to choose between lottery tickets involving the same events, where a ticket was to be conceived as yielding a win of 100 SwKr if the event occurred but no monetary loss if it did not occur. One hypothesis that obtained support in this experiment was that for probabilities other than fairly low ones, lottery tickets involving more reliable probability estimates tend to be preferred. This, together with the results of similar experiments, suggested the reliability of probability estimates to be an important factor in decision making.

The aim of the present paper is to outline a decision theory which is essentially Bayesian in its approach but which takes epistemic reliability of decision situations into consideration. We first present models of the knowledge relevant in a decision situation. One deviation from strict Bayesianism is that we use a class of probability measures instead of only one to represent the knowledge of an agent in a given decision situation. Another deviation is that we add a new measure which ascribes to each of these probability measures a degree of epistemic reliability. The first step in a decision, according to the decision theory to be presented here, is to select a class of probability measures with acceptable degrees of reliability on which a decision is to be based. Relative to this class, one can then, for each decision alternative, compute the minimal expected utility of the alternative. In the second step the alternative with the largest minimal expected utility is chosen.[2] This decision theory is then compared to some other generalized Bayesian decision theories, in particular Levi's theory as presented in (1974).

2. Models of decision situations

Our description of a decision situation will have many components in common with the traditional Bayesian way of describing decision prob-

---

[2] This decision rule is a generalization of the rule suggested by Gärdenfors in (1979), p. 169.

lems. A decision is a choice of one of the *alternatives* available in a given situation. For simplicity, we will assume that in any decision situation there is a finite set $\mathscr{A} = \{a_1, a_2, \ldots, a_n\}$ of alternatives.

Though the decision maker presumably has some control over the factors which determine the outcome of the decision, he does not, in general, have complete control. The uncertainty as to what the outcome of a chosen alternative will be is described by referring to different *states of nature* (or just *states*, for brevity). We will assume that, in any given decision situation, only a finite number of states are relevant to the decision. These states will be denoted $s_1, s_2, \ldots, s_m$.

The result or *outcome* of choosing the alternative a_i, if the true state of nature is s_j will be denoted o_{ij}. An important factor when making a decision is that of the *values* the decision maker attaches to outcomes. We will make the standard assumption that this valuation can be represented by a utility measure u.[3] The utility of the outcome o_{ij} will be denoted u_{ij}. It is assumed that all information on how the decision maker values the outcomes is summarized by the utility measure.

A final factor in describing a decision situation is that of the beliefs the decision maker has concerning which of the possible states of nature is the true state. Within strict Bayesianism it is assumed that these beliefs can be represented by a single probability measure defined over the states of nature. This assumption is very strong since it amounts to the agent having *complete* information in the sense that he is *certain* of the probabilities of the possible states of nature. The assumption is unrealistic, since it is almost only in mathematical games with coins and dice that the agent has such complete information, while in most cases of practical interest the agent has only partial information about the states of nature.

In the strict form of Bayesianism which is advocated by de Finetti (1937) and Savage (1954) among others, it is assumed that the agent's subjective probability of a state of nature can be determined by his inclination to accept bets concerning the state.[4] The so-called Dutch book theorem states that if it is not possible to construct a bet where the agent will lose money independently of which state turns out to be the actual one, then the agent's degrees of beliefs satisfy Kolmogoroff's axioms, i.e., there is a unique probability measure that describes these degrees of belief.

---

[3] This measure is assumed to be unique up to a positive linear transformation. In (1974) [Chapter 15, this volume] and (1980), Levi has generalized another dimension of the traditional Bayesian decision theory by allowing sets of utility functions which are not linear transformations of each other. We believe that this generalization is beneficiary in some contexts, but we will not discuss it further in the present paper. Cf. note 17.

[4] It is interesting to note that de Finetti (1980, p. 62, note [a]) has recognized some problem in such an operational definition as a way of representing the agent's beliefs.

However, a presupposition of this theorem is that the agent be willing to take either side of a bet, i.e., if the agent is not willing to bet on the state s_j at odds of $a{:}b$, then he should be willing to bet on not-s_j at odds of $b{:}a$. But this assumption makes too heavy demands on people's willingness to make bets. One is often not willing to accept either of the two bets.[5] In our opinion, this is explained by the fact that the estimated probability of the different states of nature are unreliable and one is not willing to take the risk connected with this uncertainty.[6] This criticism is directed against the assumptions behind the Dutch book theorem, but similar criticism can be constructed against other arguments in favor of the assumption of representing beliefs by a unique probability measure.

In this paper we will relax this assumption and, as a first step in the description of the beliefs which are relevant in a decision situation, we instead assume that the beliefs about the states of nature can be represented by a set $\mathcal{P}$ of probability measures. The intended interpretation of the set $\mathcal{P}$ is that it consists of all epistemically possible probability measures over the states of nature, where we conceive of a probability measure as epistemically possible if it does not contradict the decision maker's knowledge in the given decision situation.[7] In this way, we associate with each state s_j a set of probability values $P(s_j)$, where $P \in \mathcal{P}$. The values may be called the epistemically possible probabilities of the state s_j. For simplicity, we will assume that the probabilities of the outcomes o_{ij} are independent of which alternative is chosen, so that $P(o_{ij}) = P(s_j)$, for all $P \in \mathcal{P}$ and all alternatives a_i. Since this assumption can be relaxed, the decision theory to be presented can be extended to the more general case.[8]

The idea of representing a state of belief by a class of probability

---

[5] The central axiom is the so-called coherence criterion which assumes that if the agent is willing to bet on state s_j at the least odds of $a{:}b$, then he should be willing to bet on not-s_j at odds of $b{:}a$. The first of these betting ratios will thus be equal to one minus the second betting ratio, i.e. $a/(a + b) = 1 - b/(a + b)$. Smith (1965), among others, points out that this assumption need not be satisfied. An agent may very well be willing to bet on least odds of $a{:}b$ for s_j, but at the same time bet on least odds of $c{:}d$ against s_j, where $a/(a + b) \neq 1 - c/(c + d)$, which contradicts the coherence criterion.

[6] This aspect of risk taking will be further discussed in the next section.

[7] In the present paper we do not aim at an elaborate analysis of the concept of knowledge, but we take this as a primitive notion.

[8] In this paper we use a decision theory similar to Savage's (1954/1972). We thus deliberately exclude problems connected with conditional probabilities and probabilities of conditionals. One reason for this is that it is rather straightforward to generalize a decision theory based on such probabilities in the same way as we have generalized Savage's theory. The second reason is that Luce and Krantz (1971) have shown that in decision situations with only finitely many states and outcomes it is possible to translate a decision situation containing conditional probabilities into a Savage type situation (and vice versa). For a discussion of this result, cf. Jeffrey (1977). As is easily seen, this result also holds for sets $\mathcal{P}$ of probability measures.

measures is not new but has been suggested by various authors (see, e.g., Dempster, 1967; Good, 1962; Smith, 1961, 1965). It has been most extensively discussed by Levi (1974 [Chapter 15, this volume], 1980), but, as will be seen in the sequel, he does not use the class of probability measures in the same way as we do.

Levi also assumes that the set of probability measures is *convex*, i.e., that if P and P' are two measures in the set, then the measure $\alpha \cdot P + (1 - \alpha) \cdot P'$ is also in the set, for any α between 0 and 1.[9] The motivation for this assumption is that if P and P' both are possible probability distributions over the states, then any mixture of these distributions is also possible. We will discuss the requirement of convexity in section 5.

If $\mathscr{P}$ is assumed to be convex, then the set of epistemically possible probabilities associated with a state s_j by the elements of $\mathscr{P}$ will form an *interval* from the lowest probability assigned to s_j to the highest. Some authors have taken such intervals as basic when describing beliefs about the states of nature – to each state is assigned a probability interval and this assignment is governed by some consistency restrictions (see, e.g., Dempster, 1967; Edman, 1973; Gärdenfors, 1979; Good, 1962; Halldén, 1973; Sahlin, 1986; Smith, 1961, 1965). The representation by a convex set of probability measures is, however, more general, since from such a set one can always compute a unique set of associated intervals, but starting from an assignment of consistent probability intervals, there will in general be a large number of convex sets of probability measures that will generate the intervals.[10]

We believe that not all of an agent's beliefs about the states of nature relevant to a decision situation can be captured by a set $\mathscr{P}$ of probability measures. As a second element in describing the beliefs relevant to a decision situation, we introduce a (real-valued) measure ρ of the *epistemic reliability* of the probability measures in $\mathscr{P}$. Even if several probability distributions are epistemically possible, some distributions are more reliable – they are backed up by more information than other distributions.

---

[9] Levi (1980), p. 402, requires that the set of "permissible" probability measures be convex. The interpretation of "permissible" is discussed in section 4. In this connection it is interesting to note that Savage [1972, p. 58, note (+)] mentions that "one tempting representation of the unsure is to replace the person's single probability measure P by a set of such measures, especially a convex set".

[10] We say that a set of probability intervals associated with the states of a decision situation is *consistent* if and only if, for any state s_i and for any number x within the interval associated with s_i, there is a combination of numbers, which lie within the intervals associated with the remaining states, such that the sum of x and these numbers equal 1. Levi (1974, pp. 416–417, [pp. 308–309, this volume]) gives an example which shows that there may be two decision situations with the same alternatives, states and outcomes, but with different sets of "permissible" probability measures, which give different decisions when his decision theory is used, although the intervals that can be associated with the states are identical.

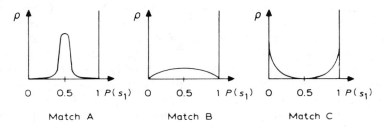

Figure 16.1

The measure ρ is intended to represent these different degrees of-reliability. In the introductory examples, Miss Julie ascribes a much greater epistemic reliability to the probability distribution where each player has an equal chance of winning in match *A* where she knows a lot about the players than in match *B* where she knows nothing relevant about the players. In match *C*, where she knows that one player is superior to the other, but not which, the epistemically most reliable distributions are the two distributions where one player is certain to win. Since there are only two relevant states of nature in these examples, viz., the first player wins (s_1) and the second player wins (s_2), a probability distribution can be described simply by the probability of one of the states. We can then illustrate the epistemic reliability of the various distributions in the three matches by diagrams as in Figure 16.1.

Even if examples such as these illustrate the use of the measure ρ, its properties should be specified in greater detail. Technically, the only property of ρ that will be needed in this paper is that the probability distributions in $\mathscr{P}$ can be *ordered* with respect to their degrees of epistemic reliability. However, it seems natural to postulate that ρ has an upper bound, representing the case when the agent has complete information about a probability distribution, and a lower bound, representing the case when the agent has no information at all about these distributions. However, we will not attempt a full description of the properties of the measure ρ, since we believe that this can be done only in a more comprehensive decision theory in which the relations between different decision situations are exploited.[11]

---

[11] An interesting possibility is to take ρ to be a second order probability measure, i.e., to let ρ be a probability measure defined over the set $\mathscr{P}$ of epistemically possible probability measures. If $\mathscr{P}$ is finite there seem to be no problems connected with such a measure. But if $\mathscr{P}$ is taken to be a convex set of probability measures and we at the same time want all measures in $\mathscr{P}$ to have a non-zero second order probability, we run into problems. However, nothing that we have said excludes the possibility of taking ρ to be a non-standard probability measure. For a discussion of such measures see Bernstein and Wattenberg (1969).

A fundamental feature of the epistemic reliability of the probability distributions possible in a decision situation, as we conceive of the measure, is that the less relevant information the agent has about the states of nature, the less epistemic reliability will be ascribed to the distributions in $\mathscr{P}$. Where little information is available, therefore, all distributions will, consequently, have about the same degree of epistemic reliability. Conversely, in a decision situation where the agent is well-informed about the possible states of nature, some distributions will tend to have a considerably higher degree of epistemic reliability than others.

A problem which strongly supports our thesis that the measure of epistemic reliability is a necessary ingredient in the description of a decision situation is Popper's paradox of ideal evidence. Popper asks us to consider the following example (1974, pp. 407–408):

Let z be a certain penny, and let a be the statement "the nth (as yet unobserved) toss of z will yield heads". Within the subjective theory, it may be assumed that the absolute (or prior) probability of the statement a is equal to $\frac{1}{2}$, that is to say,

$$P(a) = \frac{1}{2} \tag{1}$$

Now let e be some *statistical evidence*; that is to say, a *statistical report*, based upon the observation of thousands or perhaps millions of tosses of z; and let this evidence e be *ideally favourable* to the hypothesis that z is strictly symmetrical.... We then have no other option concerning $P(a, e)$ than to assume that

$$P(a, e) = \frac{1}{2} \tag{2}$$

This means that the probability of tossing heads remains unchanged in the light of the evidence e; for we now have

$$P(a) = P(a, e). \tag{3}$$

But, according to the subjective theory, (3) means that e is, on the whole (absolutely) *irrelevant* information with respect to a.

Now this is a little startling; for it means, more explicitly, that our so-called *"degree of rational belief" in the hypothesis, a, ought to be completely unaffected by the accumulated evidential knowledge, e*; that the absence of any statistical evidence concerning z justifies precisely the same "degree of rational belief" as the weighty evidence of millions of observations which, *prima facie*, support or confirm or strengthen our belief.

The "subjective theory" which Popper is referring to in this example is what we have here called strict Bayesianism.

Now, with the aid of the models of decision situations presented above, we can describe Popper's example in the following way. There is a set $\mathscr{P}$ of possible probability measures concerning the states of nature described by a and not-a. If one is forced to state the probability of a, before the evidence e is obtained, the most reliable answer seems to be ½. The degree of epistemic reliability of this estimate is, however, low, and there are many other answers which seem almost as reliable. After the evidence e is obtained, the most reasonable probability assessment concerning a is still ½, but now the distribution associated with this answer has a much higher degree of epistemic reliability than before, and the other distributions in $\mathcal{P}$ have correspondingly lower degrees of reliability. It should be noted that this distinction between the two cases cannot be formulated with the aid of the set $\mathcal{P}$ only, but the measure ρ of epistemic reliability is also necessary.[12]

We will conclude this section by briefly mentioning some related attempts to extend models of belief by some measure of "reliability".[13] An interesting concept was introduced by Keynes in (1921, p. 7):

> As the relevant evidence at our disposal increases, the magnitude of the probability of the argument may either decrease or increase, according as the new knowledge strengthens the unfavourable or favourable evidence; but *something* seems to have increased in either case – we have a more substantial basis upon which to rest our conclusion. I express this by saying that an accession of new evidence increases the *weight* of an argument. New evidence will sometimes decrease the probability of an argument, but it will always increase its "weight".

Here Keynes writes about the probability of an argument, while we are concerned with probability distributions over states of nature. Even if the intuitions behind Keynes's "weight of evidence" and our "epistemic reliability" are related, it is difficult to say how far this parallel can be drawn.

Carnap (1950, pp. 554–555) discusses "the problem of the reliability of a value of degree of confirmation" which obviously is the same as Keynes's problem. Carnap remarks that Keynes's concept of "the weight of evidence" was forestalled by Peirce who mentioned it in (1932, p. 421), in the following way:

> to express the proper state of belief, not *one* number but *two* are requisite, the first depending on the inferred probability, the second on the amount of knowledge on which that probability is based.

[12] We can compare this example with the difference between match A and match B in the introductory example. The reliability measure connected with match A which is depicted in Figure 16.1, can be seen as corresponding to the reliability measure after the evidence e has been obtained, and the measure connected with match B as corresponding to the reliability measure before e is obtained. Ideas similar to those presented here have been discussed by Bar-Hillel (1979) and Rosenkrantz (1973).

[13] Models of belief similar to ours have been presented in terms of fuzzy set theory by Watson, et al. (1979) and Freeling (1980). However, we will not consider this theory in the present paper.

The models of the agent's beliefs about the states of nature in a decision situation which have been presented here contain the two components $\mathscr{P}$ and ρ, i.e., the set of epistemically possible probability distributions, and the measure of epistemic reliability. These two components can be seen as an explication of the two numbers required by Peirce.[14]

3. A decision theory

The models of decision situations which were outlined in the previous section will be used now as a basis for a theory of decision. This theory can be seen as a generalization of the Bayesian rule of maximizing expected utility.

A decision, i.e., a choice of one of the alternatives in a decision situation, will be arrived at in two steps. The first step consists in restricting the set $\mathscr{P}$ to a set of probability measures with a "satisfactory" degree of epistemic reliability. The intuition here is that in a given decision situation certain probability distributions over the states of nature, albeit epistemically possible, are not considered as serious possibilities. For example, people do not usually check whether there is too little brake fluid in the car or whether the wheels are loose before starting to drive, although, for all they know, such events are not impossible, and they realize that if any such event occurred they would be in danger.

Now examples of this kind seem to show that what the agent does is to disregard certain states of nature rather than probability distributions over such states. But if a certain state of nature is not considered as a serious possibility, then this means that all probability distributions which assign this state a positive probability are left out of consideration. And there may be cases when some probability distributions are left out of account even if all relevant states of nature are considered to be serious possibilities. So, restricting the set $\mathscr{P}$ is a more general way of modelling the process than restricting the set of states.

Deciding to consider some distributions in $\mathscr{P}$ as not being serious

---

[14] In the quotation above it is obvious that Popper uses the traditional definition of relevance, i.e., e is *relevant* to a if and only if $P(a) \neq P(a, e)$. We believe that this definition is too narrow. Instead we propose the definition that e is relevant to a iff $P(a) \neq P(a, e)$ or the evidence e changes the degree of epistemic reliability of P. Keynes is also dissatisfied with the traditional definition of "relevance". He wants to treat "weight of evidence" and "relevance" as correlative terms so that "to say that a new piece of evidence is 'relevant' is the same thing as to say that it increases the 'weight' of the argument" (1921, p. 72). For a proof that Keynes's definition of "relevance" leads to a trivialization result, and for a discussion of some general requirements on a definition of "relevance" the reader is referred to Gärdenfors (1978).

possibilities means that *one takes a risk*. The less inclined one is to take risks, the greater the number of distributions in $\mathscr{P}$ will be that are taken into account when making the decision.

A fundamental question is how the agent determines which probability distributions in $\mathscr{P}$ are "satisfactorily reliable" and which are not. In our view, the answer is that the measure ρ of epistemic reliability should be used when selecting the appropriate subset $\mathscr{P}/\rho_0$ of $\mathscr{P}$. The agent selects a *desired level ρ_0 of epistemic reliability* and only those probability distributions in $\mathscr{P}$ which pass this ρ-level are included in $\mathscr{P}/\rho_0$, but not the others.[15] $\mathscr{P}/\rho_0$ can be regarded as the set of probability distributions that the agent takes into consideration in making the decision. An obvious requirement on the chosen level of reliability is, of course, that there be *some* distribution in $\mathscr{P}$ which passes the level.

Which "desired level of epistemic reliability" the agent will choose depends on how large the risks are he is willing to take. The more risk aversive the agent is, the lower the chosen level of epistemic reliability will be. It is important to note that two agents in identical epistemic situations, here identified by a set $\mathscr{P}$ and a measure ρ, may indeed choose different values of ρ_0 depending on their different risk taking tendencies. This is the reason why it is assumed that ρ yields an ordering of $\mathscr{P}$ and not merely a dichotomy.

If the agent is willing to take a maximal risk as regards which probability distribution to consider a "satisfactory" it may happen that there will be only one distribution which passes the desired level of reliability. After such a choice his relevant information about the states of nature will be of the same type as for the strict Bayesian, i.e. a unique probability measure, but this situation will arise for quite different reasons.[16]

The second step in the decision procedure starts with the restricted

---

[15] The "desired level of epistemic reliability" can be interpreted in terms of levels of aspiration so that the ρ_0 chosen by the decision maker is his level of aspiration as regards epistemic reliability.

[16] We have mentioned that an agent is taking a risk by not taking all epistemically possible measures under consideration. "Risk" is a notion of great complexity and the literature is flooded with papers trying to capture all aspects of risk in *one* measure. However, we do not believe that this is possible and we will give a brief explanation why. Let us, as an example, return to Miss Julie. It seems reasonable that she perceives a greater risk in betting on match *B* (and an even greater one in match *C*) than in match *A*. She may very well, if forced to do so, estimate that each player has a 50 percent chance of winning in both match *A* and *B*, but still regard match *B* as riskier. This is due to the fact that her state of knowledge is very different in the two decision situations and it shows that the degree of epistemic reliability is an important factor when determining the risk involved in a decision situation. This aspect of risk taking has not been considered in the traditional theories of risk.

This dimension of risk taking is represented in our theory by the selection of a subset $\mathscr{P}/\rho_0$ of $\mathscr{P}$. An agent who takes all epistemically possible measures into consideration takes no "epistemic" risk at all. If it is assumed that ρ is a second order set $\mathscr{P}/\rho_0$ of probability distributions. For each alternative a_i and each probability distribution P in $\mathscr{P}/\rho_0$ the expected utility e_{ik} is computed in the ordinary way. The *minimal expected utility* of an alternative a_i, relative to a set $\mathscr{P}/\rho_0$, is then determined, this being defined as the lowest of these expected utilities e_{ik}. Finally the decision is made according to the following rule (cf. Gärdenfors, 1979, p. 169):

The maximin criterion for expected utilities (MMEU): The alternative with the largest minimal expected utility ought to be chosen.

In order to illustrate how this decision procedure works, we will return to the introductory examples. For simplicity, let us, for all three tennis matches, denote by s_1 the event that the first player to serve wins the match, and by s_2 the event that the other player wins. Now assume that Miss Julie is offered the following bet for each of the three matches: She wins \$30 if s_1 occurs and loses \$20 if s_2 occurs. For each match, she must choose between the alternative a_1 of accepting the bet and the alternative a_2 of declining it. Let us furthermore assume that Miss Julie's utilities are mirrored by the monetary values of the outcomes.

In match A, where Miss Julie is very well informed about the players, she considers that the only probability distribution that she needs to take into consideration is the distribution P_1, where $P_1(s_1) = P_1(s_2) = 0.5$. She is willing to take the (small) risk of letting $\mathscr{P}/\rho_0$ consist of this distribution only. The only, and hence minimal, expected utility to compute is then $0.5 \cdot 30 + 0.5 \cdot (-20)$ for a_1 and $0.5 \cdot 0 + 0.5 \cdot 0$ for a_2. Hence, according to MMEU, she should accept the bet in match A.

In match B, where she has no relevant information at all, the epistemic reliability of the epistemically possible distributions is more evenly spread out. Consequently, Miss Julie is not so willing to leave distributions out of account when forming the subset $\mathscr{P}/\rho_0$ as in the previous case. For simplicity, let us assume that $\mathscr{P}/\rho_0 = \{P_1, P_2, P_3\}$, where P_1 is as before, P_2 is defined by $P_2(s_1) = 0.25$ and $P_2(s_2) = 0.75$, and P_3 is defined by $P_3(s_1) = 0.75$ and $P_3(s_2) = 0.25$. The expected utilities for the alternative a_1 of accepting the bet, determined by these three distributions, are 5, -7.5 and 17.5 respectively; while the expected utilities for the alternative a_2 of not accepting the bet are 0 in all three cases. Since the minimal expected utility of a_1 is less than that of a_2, the MMEU

---

probability distribution [cf. note 11 and Sahlin (1980, 1983)], we suggest the following measure of the epistemic risk R taken by an agent in a decision situation:

$$R(\mathscr{P}/\rho_0) = 1 - \rho(\mathscr{P}/\rho_0)/\rho(\mathscr{P}),$$

where $\rho(\mathscr{P})$ is equal to $\Sigma_{P \in \mathscr{P}} \rho(P)$ and similar for $\rho(\mathscr{P}/\rho_0)$. As is easily seen, Miss Julie will take a rather great epistemic risk if she acts as a strict Bayesian in match B. But whether she will do so or not is dependent on her risk preferences.

For a criticism of the traditional risk concept and for a discussion of other solutions see Sahlin (1984) and Hansson (1981, and Chapter 8, this volume).

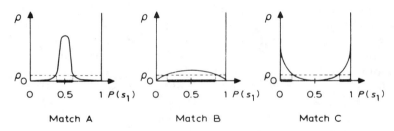

Figure 16.2

criterion demands that a_2 be the chosen alternative, i.e., Miss Julie should decline the bet offered.

In match C, it is reasonable to assume that $\mathscr{P}/\rho_0$ contains some probability distribution which assigns s_1 a very high probability and some distribution which assigns it a very low probability. A similar analysis as above then shows that Miss Julie should not accept the bet in this case either.

This example can be heuristically illustrated as in Figure 16.2. In this figure the broken horizontal line indicates the desired level of epistemic reliability.

To give a further illustration of the decision theory, it should be noted that the hypothesis from the Goldsmith-Sahlin experiments, mentioned in the introduction, is well explained by the MMEU criterion. When an agent is asked to choose between tickets in two lotteries which are estimated to have the same primary probability of winning, he should choose the ticket from the lottery with the epistemically most reliable probability estimate, since this alternative will have the highest minimal expected utility.[17] Still other applications of the decision theory will be presented in the next two sections.

A limiting case of information about the states of nature in a decision situation is to have *no* information at all. In the decision models presented here, this would mean that all probability distributions over the states are epistemically possible and that they have equal epistemic reliability. In such a case, the minimal expected utility of an alternative

[17] The results of the majority of subjects of the Goldsmith–Sahlin experiments support the thesis that the degree of epistemic reliability of the probability estimates is an important factor when choosing lottery tickets, but, it should be admitted that not all of these results can be explained by the MMEU criterion. The main reason for this is, in our opinion, that the agent's *values* are not completely described by a utility measure. In this paper we have concentrated on the epistemic aspects of decision making and used the traditional way, i.e., utility measures, to represent the values of the decision maker. We believe that this part of the traditional Bayesian decision theory should be modified as well, perhaps by including a "level of aspiration", but such an extension lies beyond the scope of this paper.

is obtained from the distribution which assigns the probability 1 to the worst outcome of the alternative. This is, however, just another way of formulating the classical maximin rule, which has been applied in what traditionally has been called "decision making under uncertainty" (a more appropriate name would be "decision making under ignorance"). Hence, the classical maximin turns out to be a special case of the MMEU criterion.

At the other extreme, having *full* information about the states of nature implies that only one probability distribution is epistemically possible.[18] In this case the MMEU criterion collapses into the ordinary rule within strict Bayesianism, i.e., the rule of maximizing expected utility, which has been applied to what traditionally, but somewhat misleadingly, has been called "decision making under risk".

The decision theory which has been presented here thus covers the area between the traditional theories of "decision making under uncertainty" and "decision making under risk" and it has these theories as limiting cases.

4. Relation to earlier theories

Several authors have proposed decision theories which are based on more general ways of representing the decision maker's beliefs about the states than what is allowed by strict Bayesianism. The most detailed among these is Levi's theory (1974 [Chapter 15, this volume], 1980) which will be discussed in a separate section. In this section we will compare the present theory with some earlier statistical decision theories.

In (1950), Wald formulates a theory of "statistical decision functions" where he considers a set Ω of probability measures and a "risk" function. He says that "the class Ω is to be regarded as a datum of the decision problem" (p. 1). He also notes that the class Ω will generally vary with the decision problem at hand and that in most cases "will be a proper subset of the class of all possible distribution functions" (p. 1). In the examples, Ω is often taken to be a parametric family of known functional form. A risk function is a function which determines the "cost" of a wrong decision. Such a function can be seen as an inverted utility function restricted to negative outcomes. On this interpretation it is easier to compare Wald's theory to the present decision theory.

Wald suggests two alternative decision rules. The first is the tradi-

---

[18] Such a distribution may, in many cases, assign the probability 1 to one of the states, but we do not assume that it always will. We interpret "full information" in a pragmatic way, meaning something like "having as much information as is practically possible", so, even if the world is deterministic, having full information does not entail that one knows which is the true state of nature.

tional Bayesian method when an "a priori" probability measure in Ω can be selected, or, as Wald puts it, when "it exists and is known to the experimenter" (p. 16). The second case is when the entire set Ω is employed to determine the decision. Wald suggests that in such cases one should "minimize the maximum risk". In our terminology, using utility functions instead of risk functions, this is the same as maximizing the minimal expected utility with respect of the set Ω.

If Wald's theory is interpreted as above, the difference between his and our theory is mainly of epistemological character. Since Wald does not say anything about how Ω is to be determined it is difficult to tell whether it corresponds to our set $\mathscr{P}$ of epistemically possible probability functions or to the set $\mathscr{P}/\rho_0$ of "reliable" functions. In particular, he does not introduce any factor corresponding to the measure ρ of epistemic reliability, nor does he associate the choice of Ω with any form of risk taking.

Hurwicz (1951) apparently interprets Wald's set Ω as corresponding to our set $\mathscr{P}$. He notes that sometimes some of the distributions in Ω are more "likely" than others (p. 343). For example, assume that Ω consists of all normal distributions with mean zero and standard deviation σ. In a particular decision situation evidence at hand may support the assumption that σ is considerably small. It thus seems reasonable to select a proper subset Ω_0 of Ω which is restricted to those distributions with standard deviation less than or equal to some value σ_0.

Hurwicz assumes that such a subset Ω_0 ($\Xi_{\mathscr{F}}^{(0)}$ in his terminology) of Ω can be selected in any decision situation. He then suggests a "generalized Bayes-minimax principle" which amounts to using Ω_0 as the base when maximizing the minimal expected utility (minimizing the maximal risk). Obviously, the set Ω_0 corresponds closely to our set $\mathscr{P}/\rho_0$. The main difference between Hurwicz's theory and the present one is that he does not give any account of how the set Ω_0 is to be determined. In particular he, as Wald, does not introduce any factor corresponding to the measure ρ.

In (1952), Hodges and Lehmann suggest an alternative to Wald's minimax solution which they call a "restricted Bayes solution". It is of interest here since it is adopted by Ellsberg (1961, Chapter 13, this volume) as a solution to his "paradox". Let us start by considering Ellsberg's problem (Fellner, 1961, considers problems similar to Ellsberg's).

Ellsberg (1961, pp. 653–654 [pp. 254–255, this volume]) asks us to consider the following decision problem. Imagine an urn known to contain 30 red balls and 60 black and yellow balls, the latter in unknown proportion. One ball is to be drawn at random from the urn. In the first situation you are asked to choose between two alternatives a_1 and a_2. If you choose a_1 you will receive $100 if a red ball is drawn and nothing if a black or yellow ball is drawn. If you choose a_2 you will receive $100 if a black ball is drawn, otherwise nothing. In the second situation you are asked to choose, under the same circumstances, between the two alternatives a_3 and a_4. If you choose a_3 you will receive $100 if a red or a yellow ball is drawn, otherwise nothing and if you choose a_4 you will receive $100 if a black or yellow ball is drawn, otherwise nothing. This decision problem is shown in the following decision matrix.

	Red	Black	Yellow
a_1	$100	$0	$0
a_2	$0	$100	$0
a_3	$100	$0	$100
a_4	$0	$100	$100

The most frequent pattern of response to these two decision situations is that a_1 is preferred to a_2 and a_4 is preferred to a_3. As Ellsberg notes, this preference pattern violates Savage's "sure thing principle" (postulate $P2$ in 1972, [see p. 82, this volume]), which requires that the preference ordering between a_1 and a_2 be the same as the ordering between a_3 and a_4.

When applying the present decision theory to this problem, the main step is to determine the set $\mathscr{P}/\rho_0$. The set $\mathscr{P}$ should be the same in the two decision situations, since they do not differ with respect to the information about the states. Now, unless the decision maker believes that he is being cheated about the content of the urn, $\mathscr{P}$ is most naturally taken as the class of distributions ($\frac{1}{3}$, x, $\frac{2}{3} - x$), where x varies from 0 to $\frac{2}{3}$.

If the decision maker chooses a low ρ_0, $\mathscr{P}/\rho_0$ will contain most of the distributions in this class. For simplicity, let us for the moment assume that $\mathscr{P}/\rho_0 = \mathscr{P}$. With this choice, the minimal expected utilities of the alternatives are $\frac{1}{3} \cdot u(\$100)$, $1 \cdot u(\$0)$, $\frac{1}{3} \cdot u(\$100)$ and $\frac{2}{3} \cdot u(\$100)$, for a_1, a_2, a_3 and a_4, respectively. Assuming that $\frac{1}{3} \cdot u(\$100)$ is greater than $u(\$0)$, the MMEU criterion requires that a_1 be preferred to a_2 and a_4 to a_3, which accords with Ellsberg's findings. Intuitively, a_2 and a_3 involve greater "epistemic risks" than a_1 and a_4 – thus a_2 and a_3 are avoided by most subjects. This feature is well captured by the present decision theory.

If the decision maker is willing to take an epistemic risk and chooses a higher ρ_0, fewer distributions will be included in $\mathscr{P}/\rho_0$. If the decision maker has no further information about the distribution of black and yellow balls, then, because of symmetry, it is likely that he judges the distribution ($\frac{1}{3}$, $\frac{1}{3}$, $\frac{1}{3}$) as being the highest in the ρ-ordering. If this is the only distribution included in $\mathcal{P}/\rho_0$, he should be indifferent between a_1 and a_2 and between a_3 and a_4 according to the MMEU criterion. In this case Savage's sure thing principle is not violated.[19]

In order to explain the "paradox" that decision makers do not act according to Savage's sure thing principle, Ellsberg (1961, p. 661, p. 262, this volume) first introduces for a decision maker in a given situation a set Y^0 of probability distributions

that still seem "reasonable", reflecting judgements that he "might almost as well" have made, or that his information – perceived as scanty, unreliable, *ambiguous* – does not permit him confidently to rule out.

Ellsberg also considers a particular probability distribution y^0, the "estimated" probability distribution, which can be viewed as the distribution a strict Bayesian decision maker would have adopted in the decision situation. Ellsberg also ascribes to y^0 a degree ρ_e (*e* for Ellsberg) of the decision maker's "confidence" in the estimate.

The decision rule suggested by Ellsberg, which is the restricted Bayesian solution developed by Hodges and Lehmann, can now be described as follows: Compute for each action a_i the expected utility according to the distribution y^0 and the minimal expected utility relative to the set Y^0. Associate with each action an idex based on a weighed average of these two factors, where ρ_e is the weight ascribed to the first factor and $1 - \rho_e$ is the weight ascribed to the latter factor. Finally, choose that act with the highest index.

When comparing this decision theory with the theory presented in this paper, one notes that there are differences both of epistemological and formal character. Firstly, since Ellsberg, like his predecessors, does not say anything about how the set Y^0 is to be determined, and it is therefore difficult to say whether it corresponds to our set $\mathcal{P}/\rho_0$. Secondly, even if we identify Y^0 with $\mathcal{P}/\rho_0$, Ellsberg exploits the degree ρ_e of "confidence", which is defined for only one distribution y^0, in a way that differs considerably from our use of the measure ρ, which is assumed to be defined for all distributions in $\mathcal{P}$. In particular we need not assume that ρ gives a numerical value, only that it orders the distributions in $\mathcal{P}$. Thirdly, since the decision rules are different, the theories will recommend different decisions in many situations. The most important disagreement here is that we reject the need of an

---

[19] However, if the agent has some information that he judges relevant for the distribution of the black and yellow balls, then the epistemic reliability of the distributions in $\mathcal{P}$ may be quite different. He may, for example, believe that the distributions in $\mathcal{P}/\rho_0$ cluster around e.g. ($\frac{1}{3}$, $\frac{1}{2}$, $\frac{1}{6}$). Then the MMEU criterion will recommend that a_2 be preferred to a_1 and that a_4 be preferred to a_3. This recommendation does not conflict with Savage's sure-thing principle either (cf. Gärdenfors & Sahlin, 1983).

estimated distribution y^0. We believe that once the set $\mathscr{P}/\rho_0$ has been selected, the distribution with the highest degree of reliability (corresponding to Ellsberg's y^0) does not play any outstanding role in the decision making.

This difference between the two theories is in principle testable, assuming that ρ_e is not always close to zero. The experiment performed by Becker and Brownson (1964) is relevant here. They offered subjects ambiguous bets differing in the range of possible probabilities but of equal expected value and found that subjects were willing to pay money for obtaining bets with narrower ranges of probability. This finding seems to highlight the importance of a measure of reliability. But they did not, however, obtain support for Ellsberg's hypothesis that the distribution y^0 is relevant for the decision making.

5. A comparison with Levi's theory

In this section we compare our theory with Levi's theory which is presented in (1974, Chapter 15, this volume) and elaborated on in (1980). Levi starts out from a description of the decision maker X's information at the time t about the states of nature. This information is contained in a convex set $B_{X,t}$ of probability distributions. The distributions in $B_{X,t}$ are, according to Levi, the "permissible" distributions. As to the meaning of "permissible", he offers only indirect clarification by indicating the connections between permissibility and rational choice. In order to compare the theories, we will here assume that the set $B_{X,t}$ corresponds to the set $\mathscr{P}/\rho_0$ (or its convex hull) as presented in section 3.[20]

Levi also generalizes the traditional way of representing the utilities of the outcomes by introducing a class G of "permissible" utility measures, such that not all of these measures need be linear transformations of one another.

An alternative a_i is said to be *E-admissible* if and only if there is some probability distribution P in $B_{X,t}$ and some utility function u in G such that the expected utility of a_i relative to P and u is maximal among all the available alternatives. A first requirement on the alternative to be chosen in a given decision situation is then that it should be E-admissible.

The second step in Levi's decision procedure concerns the opportu-

---

[20] It should be noted that even if $\mathscr{P}$, the set of all epistemically possible probability measures, is convex, the set $\mathscr{P}/\rho_0$ selected with the aid of the desired level of epistemic reliability need not be convex. For example, in the representation of the epistemic reliability connected with match C as depicted in Figure 16.1, the corresponding set $\mathscr{P}/\rho_0$ will consist of two disconnected intervals at the end points 0 and 1. If Levi's $B_{X,t}$ is identified with $\mathscr{P}/\rho_0$, this example shows that his requirement of convexity is not always realistic.

nity to defer decision between two or more E-admissible alternatives. He argues that a rational agent should "keep his options open" whenever possible. An alternative is said to be *P-admissible* if it is E-admissible and it is "best" with respect to E-admissible option preservation. Levi does not, however, explicate what he means by "best", and we will ignore the effects of this requirement here, since we have not imposed any structure on the set of alternatives. (Further information on the notion of *P*-admissibility is to be found in chap. 6 of Levi, 1980.)

Let us say that a *P*-admissible alternative a_i is *security optimal relative to a utility function u* if and only if the minimum *u*-value assigned to some possible outcome o_{ij} of a_i is at least as great as the minimal *u*-value assigned to any other *P*-admissible alternative. Levi then, finally, calls an alternative *S-admissible* if it is *P*-admissible and security optimal relative to *some* utility function in *G*.

Levi states (1974, p. 412, p. 304, this volume) that he "cannot think of additional criteria for admissibility which seem adequate" so he, tentatively, assumes that all *S*-admissible alternatives are 'admissible' for the final choice, which then, supposedly, is determined by some random device.

In order to illustrate the differences between the decision theory presented in the previous section and Levi's theory, we will consider the following example which contains two states and three alternatives:

	s_1	s_2
a_1	−10	12
a_2	11	−9
a_3	0	0

In this matrix the numbers denote the utilities of the outcomes. Assume that the set $\mathscr{P}/\rho_0$, which here is identified with Levi's set $B_{X,t}$, consists of the two probability distributions P, defined by $P(s_1) = 0.4$ and $P(s_2) = 0.6$, and P', defined by $P'(s_1) = 0.6$ and $P'(s_2) = 0.4$, together with all convex combinations of P and P'.

The minimal expected utility of a_1 is −1.2 and the minimal expected utility of a_2 is −1.0. The minimal expected utility of a_3 is of course 0, so MMEU requires that a_3 be chosen.

In contrast to this, only a_1 and a_2 are E-admissible. P-admissibility has no effect here, but a_2 is security optimal relative to the utility measure given in the matrix, so a_2 is the only S-admissible alternative. Thus, according to Levi's theory, a_2 should be chosen.

Kyburg (1983*a*) presents a related decision theory based on probability intervals. The central decision rule, which he calls Principle III, says that the decision maker ought to reject any choice a_i for which there is an a_j whose minimum expected utility exceeds the maximum expected utility of a_i. The relation between this rule, Levi's theory, and the MMEU principle is discussed in Sahlin (1985). It can, for example, be shown that in a decision problem similar to the one above these three theories recommend different decisions.

When the uncertainty about the states of nature in this decision situation, represented by $\mathcal{P}/\rho_0$, is considered, we believe that a_3 is intuitively the best alternative. Against this it may be argued that using a maximin principle is unnecessarily risk aversive. It should be remembered, however, that when restricting $\mathcal{P}$ to $\mathcal{P}/\rho_0$ the agent is already taking a risk and his choice of $\mathcal{P}/\rho_0$ indicates that he is not willing to take any further epistemic risks. On the other hand, Levi's requirement of E-admissibility has the consequence that, in many cases, the choices made by his theory seem unrealistically optimistic.

A strange feature of Levi's theory is that if the previous decision situation is restricted to a choice between a_2 and a_3, then his theory recommends choosing a_3 instead of a_2! In their chapter on individual decision making under uncertainty, Luce and Raiffa (1957, pp. 288–290, pp. 61–63, this volume) introduces the condition of *independence of irrelevant alternatives* which in its simplest form demands that if an alternative is not optimal in a decision situation it cannot be made optimal by adding new alternatives to the situation. The example presented here shows that Levi's theory does not satisfy this condition since in the decision situation where a_2 and a_3 are the only available alternatives and where a_3 is optimal according to Levi's theory, a_2 can be made optimal by adding a_1. It is easy to show, however, that the MMEU criterion which has been presented here satisfies the condition of independence of irrelevant alternatives.[21]

---

[21] It is also easy to see that according to Levi's theory there may be a change in the set of optimal alternatives if two (or more) states are conjoined. This is the case if it is assumed that if s_i and s_j are the two states to be conjoined, then, for any probability measure, P, the probability of the conjoined state is equal to $P(s_i) + P(s_j)$ and the utility of the outcome in the new state, if alternative a_k is chosen, is equal to $(P(s_i) \cdot u_{ki} + P(s_j) \cdot u_{kj})/(P(s_i) + P(s_j))$. Such a contraction of the decision problem is reasonable, if, for example, it is realized that the partitioning adopted initially was unnecessarily refined. It is easy to verify that the set of alternatives which are optimal according to MMEU is not altered by such a conjoining of states.

Levi discusses these matters (1980, pp. 161–162). He notes that one can obtain different classes of S-admissible alternatives by using different partitions of the states, but he contends that the adoption of a method for fixing security levels in determining S-admissibility is a moral or political value judgement distinct from a principle of rational choice. We do not have any such problems, since we do not obtain different results when conjoining states as above.

Levi's theory also seems to have problems in explaining some of the experimental results considered in this paper. If Levi's theory is applied to Ellsberg's two decision situations as presented earlier, it gives the result that both alternatives in the two situations are S-admissible, and hence that a_1 is equally good as a_2 and a_3 is equally good as a_4. This contrasts with Ellsberg's findings which are in accordance with the recommendations of the present theory. Similar considerations apply to the experiments presented in Becker and Brownson (1964).

Decision theories of the kind presented in this paper are based on several idealizations and they will unavoidably be exposed to some refractory empirical material. We believe, however, that the considerations of this section show that the decision theory presented in section 3 is a more realistic theory than Levi's.

6. Conclusion

The starting-point of this paper is that in many decision situations the assumption, made in the strict form of Bayesianism, that the beliefs of an agent can be represented by a single probability distribution is unrealistic. We have here presented models of belief which contain firstly, a class of probability distributions, and, secondly, a measure of the epistemic reliability of these probability distributions. Several authors before us have suggested that a class of probability distributions should be exploited when describing the beliefs of the agent. A main thesis of this paper is that this is not sufficient, but an assessment of the information on which the class of probability distributions is based is also necessary. We have here tried to capture this assessment by the measure ρ of epistemic reliability. With the aid of this measure we can account for one form of risk taking in decision situations.

On the basis of the models of the beliefs which are relevant in a decision situation we have formulated a decision theory. We have argued that this theory has more desirable properties and is better supported than other decision theories which also generalize the traditional Bayesian theory.

The MMEU criterion, which has been suggested here as the main rule of the decision theory, is generally applicable to decision situations where the possible outcomes are non-negative from the point of view of the decision maker. However, there are situations where the MMEU criterion seems to be too cautious. For example, the "reflection effect" and the "isolation effect" suggested by Kahneman and Tversky (1979, Chapter 11, this volume) cannot be explained directly with the decision theory of this paper. We believe that in order to cover these phenomena a more general and comprehensive decision theory is needed which includes references to the decision maker's "levels of aspiration". A special case of the effects of levels of aspiration would be the "shifts of reference point" discussed by Kahneman and Tversky. Introducing "levels of aspiration" means that the part of traditional Bayesian theory which refers to utilities has to be considerably extended and modified.

Part V
Causal decision theory

Ingenious problems which cannot be properly handled by the traditional decision theory form an obvious basis for developments of the theory. Earlier in the volume we have seen how the "paradoxes" introduced by Allais and Ellsberg have led to lively discussions and fruitful investigations. In the present part a new line of development will be presented, namely causal decision theory. This theory derives from the following decision problem:

Newcomb's problem. In front of you on the table are two boxes. One box is transparent and you can see that it contains $1000. The other box is opaque, but you are told that it contains $1,000,000 ($1M) or nothing ($0). You have two choice alternatives: Take the opaque box only or take both boxes. The hook is that what is in the opaque box is determined by a gifted person who is a very reliable predictor of your behavior: If he predicts that you will take only the opaque box, he will put a million dollars in it; however, if he predicts that you will take both boxes, then he puts nothing in the opaque box. The person makes his prediction prior to your choice, i.e., when you decide the content of the opaque box is already determined. But you also know that the person is almost a perfect predictor of your behavior and has proved very successful in predicting other people's behavior in similar situations. Would you choose one or two boxes?

Perhaps a decision matrix will help you reach a decision:

State

Act	p_1	p_2
	Predicts one box	Predicts two boxes
t_1: Take one box	\$1,000,000	\$0
t_2: Take two boxes	\$1,001,000	\$1,000

This problem, invented by William Newcomb and first presented and discussed by Robert Nozick (1969), has divided the decision theorists into two camps. (For a selection of articles on this problem, see Campbell and Sowden, 1985.) First we have those who think it is rational to take but one box. The core of the one-boxer's argument is as follows. The conditional probability $P(p_1/t_1)$ that the person predicts that I take one box given that I in fact take one box is close to one. Let us for the sake of the argument, assume that this probability equals 0.99 (thus $P(p_2/t_1) = 0.01$). Similarly, the probability $P(p_2/t_2)$ that the person predicts that I will take two boxes given that I in fact take two, is close to one, say 0.99 in this case, too. Given these values it is easy to calculate the conditional expected values of the two options. We find that $CEU(t_1) = 0.99.u(\$1M) + 0.01.u(\$0)$ and that $CEU(t_2) = 0.01.u(\$1M + \$1,000) + 0.99.(\$1,000)$. Assuming that dollars and utilities are roughly exchangeable, we note that $CEU(t_1) > CEU(t_2)$ since \$990,000 is much larger than \$11,000.

The one-boxer thus appeals to the principle MCEU of maximizing conditional expected utility that was presented in the introduction (see pp. 9–10).

This view of the problem is challenged by the two-boxers, i.e., those who argue that the only rational thing to do is to take both boxes. There is no need to be cautious and take only one box, since the person has already made his prediction and determined the content of the opaque box. And since there is no such thing as backwards causation (we cannot change the past even if we would sometimes like to have this power), our choice will have no effect on the content of the box: When we make our choice there is either a million or nothing in the opaque box. Our present choice will not change this fact. But, why shouldn't we pick the dominating act? If we take both boxes we know that we will come out \$1000 ahead, whatever the prediction was.

The two-boxer appeals to the following decision principle:

Dominance: If, for every state of the world, the outcome of alternative *a* is better than the outcome of alternative *b*, then it is not rational to choose *b*.

Newcomb's problem shows that causal independence may occur without probabilistic independence. In the problem, our actual choice will not influence the prediction, indicating causal independence, but the conditional probabilities differ depending on which act is chosen. In a situation like this where we have causal independence but probabilistic dependence the dominance principle and the principle of maximizing conditional expected utility sometimes conflict with each other.

It has been argued (*locus classicus* is the article by Gibbard and Harper, Chapter 17, this volume) that Bayesian decision theory, for example in the form given by Jeffrey, cannot account for this type of problems since it gives the wrong recommendation. The theory should therefore be abandoned and replaced by a new and better theory which is sensitive to the relevant *causal* factors of the problem. The main feature of the causal decision theories that have been proposed for this aim is that we should calculate expected utilities in terms of causal dependencies rather than in terms of conditional probabilities. The most common way of analysing the relevant type of causal dependency is by exploiting *counterfactual* propositions.

Newcomb's problem is "solved" by the causal decision theories by the claim that if no causal dependencies can be found, we should act as if we had probabilistic independence. But in these cases there is no longer any conflict between maximizing (conditional) expected utility and choosing in accordance with the dominance principle. The causal theories therefore recommend that we take both boxes.

Naturally, the causal decision theories have been criticized in a number of different ways by the defenders of the traditional theory (most persistently by Ellery Eells, 1981, 1982, 1984*a*, *b*, 1985*a*, *b*). One conceptual problem concerns the *semantics* of counterfactual sentences. It takes a good deal of metaphysics to determine the truth value of a counterfactual proposition, let alone its probability. This is, however, not the place to discuss the morass of theories in this area (see, e.g., Lewis, 1973*a*).

Another problem is that it is difficult to see how a causal decision theory could be *tested*. Even if the traditional theory has not fared well in empirical testing, it is at least testable. The reason why causal decision theory is not falsifiable is that by using counterfactual propositions, which have undetermined truth values, as a central element in the theory, we open the door for as many interpretations of experimental results as we like. How can we know that the decision makers tested evaluate the relevant counterfactuals in the same way as we do? Employing counterfactuals means that we are fishing in muddy waters. However, in defense of the causal theories, it can be said that psychological reports on people's choice behaviour suggest that they base their arguments for the choices on counterfactual reasoning.

Influenced by the Newcomb problem and similar "causal" decision situations, Jeffrey (1981, 1965/1983, second edition) has abandoned his old theory. In the new edition of *The Logic of Decision* he suggests a new decision rule which asks us to make *ratifiable* decisions. An act a is said to be ratifiable if it has maximal expected utility on the assumption that the decision maker has *already decided* to perform a. In Jeffrey's epitome: "To put it romantically: 'Choose for the person you expect to be when you have chosen'" (1965/1983, p. 16).

In order to illustrate how this rule functions, let us work through Newcomb's problem again. First assume that you have decided to take only one box, i.e. to perform t_1, but have not yet taken the box. This decision (let us denote it d_1) affects the probabilities in the decision matrix above in the sense that it "screens off" the influence of the act itself on the probabilities of the states. In the original story $P(p_1/t_1)$ was close to one and $P(p_1/t_2)$ was close to zero. But the new probabilities $P(p_1/d_1 \& t_1)$ and $P(p_1/d_1 \& t_2)$ will be *identical* (and very high) if it is assumed that you have already decided to take only one box since the probability of p_1 is very high when you have decided to take one box and, according to the story, this probability will not be affected by what you actually do.

Similarly, if we assume that you have already decided to take both boxes (let us denote this decision d_2), the probabilities $P(p_2/d_2 \& t_2)$ and $P(p_2/d_2 \& t_1)$ will be identical and of the same magnitude as $P(p_1/d_1)$. Due to the extra $1000, the expected value of taking two boxes, given that you have already decided to take two boxes, will be higher than the expected utility of taking one box only, given that you have already decided to take one box. Thus taking two boxes is the only ratifiable decision. Recall that Jeffrey's older principle MCEU recommended taking one box only.

One thing that will strike the psychologically oriented reader is that this argument relies on the assumption that the decision to act is *as reliable* as a clue to what the predictor does as the act itself. To see the problem with this assumption, consider the smoker who on New Year's Eve solemnly decides to quit smoking, only to find himself starting New Year's Day by lighting a cigarette. This example indicates that sometimes it is not by your decisions but only through your actions that you can learn what kind of person you are.

A problematic assumption for Jeffrey's theory of ratifiable decisions is that the decision maker is assumed to make decisions on *hypothetical* beliefs. Some of these hypothetical beliefs may *contradict* his present beliefs, and, even worse, they may contradict the beliefs he has after making the decision. Another problem for Jeffrey's theory is that there exist decision problems where there is no ratifiable act (see Rabinowicz, 1985, and the article in this volume, Chapter 19). In such cases the theory leaves us without guidance.

Except for the theories presented by Gibbard and Harper and by Jeffrey, there are several other variants of causal decision theories [e.g., Sobel (1978), Skyrms (1980a), and Lewis (1981), Chapter 18, this volume]. Lewis compares his theory to the versions offered by several other authors, and suggests that the versions have more in common than meets the eye.

17. Counterfactuals and two kinds of expected utility

Allan Gibbard and William L. Harper

1. Introduction

This article develops a proposal made by Robert Stalnaker, in response to a theorem by David Lewis.* We begin with a rough theory of rational decision-making. In the first place, rational decision-making involves conditional propositions: when a person weighs a major decision, it is rational for him to ask, for each act he considers, what would happen if he performed that act. It is rational, then, for him to consider propositions of the form "If I were to do a, then c would happen". Such a proposition we shall call a *counterfactual*, and we shall form counterfactuals with a connective "$\square\!\!\rightarrow$" on this pattern: "If I were to do a, then c would happen" is to be written "I do a $\square\!\!\rightarrow$ c happens".

Now ordinarily, of course, a person does not know everything that would happen if he performed a given act. He must resort to probabilities: he must ascribe a probability to each pertinent counterfactual "I do a $\square\!\!\rightarrow$ c happens". He can then use these probabilities, along with the desirabilities he ascribes to the various things that might happen if he

* Lewis's theorem and Stalnaker's proposal were presented in a symposium with Harper at the meetings of the Canadian Philosophical Association in 1972. See Lewis (1976), and part of a 1972 Stalnaker letter to Lewis reprinted as Stalnaker (1978).

An earlier draft of this paper was circulated in January 1976. A much shorter version was presented to the 5th International Congress of Logic, Methodology, and Philosophy of Science, London, Ontario, August 1975. There, and at the earlier University of Western Ontario research colloquium on Foundations and Applications of Decision Theory we benefited from discussions with many people; in particular we should mention Richard Jeffrey, Isaac Levi, Barry O'Neill and Howard Sobel.

Czeslaw Porebski pointed out to us a number of errors in previously published versions.

Reprinted from Foundations and Applications of Decision Theory, vol. 1, Reidel Publishing Company, Dordrecht, 1978, pp. 125–162, by kind permission of the authors and the publisher.

did a given act, to reckon the expected utility of a. If a has possible outcomes $o_1, \ldots, o_n$, the expected utility of a is the weighted sum

$$\Sigma_i \; prob \; (\text{I do } a \; \square\!\!\rightarrow o_i \text{ obtains)} \, \mathcal{D} \, o_i,$$

where $\mathcal{D} \, o_i$ is the desirability of o_i. On the view we are sketching, then, the probabilities to be used in calculating expected utility are the probabilities of certain counterfactuals.

That is not the story told in familiar Bayesian accounts of rational decision; those accounts make no overt mention of counterfactuals. We shall discuss later how Savage's account (1954/1972) does without counterfactuals; consider first an account given by Jeffrey (1965, pp. 5–6).

A formal Bayesian decision problem is specified by two rectangular arrays (matrices) of numbers which represent probability and desirability assignments to the act-condition pairs. The columns represent a set of incompatible conditions, an unknown one of which actually obtains. Each row of the desirability matrix,

$$d_1 \, d_2 \, \ldots \, d_n$$

represents the desirabilities that the agent attributes to the n conditions described by the column headings, on the assumption that he is about to perform the act described by the row heading; and the corresponding row of the probability matrix,

$$p_1 \, p_2 \, \ldots \, p_n$$

represents the probabilities that the agent attributes to the same n conditions, still on the assumption that he is about to perform the act described by the row heading. To compute the expected desirability of the act, multiply the corresponding probabilities and desirabilities, and add:

$$p_1 \, d_1 + p_2 \, d_2 + \cdots + p_n \, d_n.$$

On the Bayesian model as presented by Jeffrey, then, the probabilities to be used in calculating "expected desirability" are "probabilities that the agent attributes" to certain conditions "on the assumption that he is about to perform" a given act. These, then, are conditional probabilities; they take the form $prob(S/A)$, where A is the proposition that the agent is about to perform a given act and S is the proposition that a given condition holds.

On the account Jeffrey gives, then, the probabilities to be used in decision problems are not the unconditional probabilities of certain counterfactuals, but are instead certain conditional probabilities. They take the form $prob(S/A)$, whereas on the view we sketched at the outset, they should take the form $prob(A \; \square\!\!\rightarrow S)$. Now perhaps, for all we have said so far, the difference between these accounts is merely one of presentation. Perhaps for every appropriate A and S, we have

$$prob(A \;\Box\!\!\rightarrow\; S) = prob(S/A); \tag{1}$$

the probability of a counterfactual $A \;\Box\!\!\rightarrow\; S$ always equals the corresponding conditional probability. That would be so if (1) is a logical truth. David Lewis, however, has shown (1976) that on certain very weak and plausible assumptions, (1) is not a logical truth: it does not hold in general for arbitrary propositions A and S.[1] That leaves the possibility that (1) holds at least in all decision contexts: that it holds whenever A is an act an agent can perform and *prob* gives that agent's probability ascriptions at the time.

In Section 3, we shall state a condition that guarantees the truth of (1) in decision contexts. We shall argue, however, that there are decision contexts in which this condition is violated. The context we shall use as an example is patterned after one given by Stalnaker. We shall follow Stalnaker in arguing that in such contexts, (1) indeed fails, and it is probabilities of counterfactuals rather than conditional probabilities that should be used in calculations of expected utility. The rest of the paper takes up the ramifications for decision theory of the two ways of calculating expected utility. In particular, the two opposing answers to Newcomb's problem (Nozick, 1969) are supported respectively by the two kinds of expected utility maximization we are discussing.

We are working in this paper within the Bayesian tradition in decision theory, in that the probabilities we are using are subjective probabilities, and we suppose an agent to ascribe values to all probabilities needed in calculations of expected utilities. It is not our purpose here to defend this general tradition but rather to work within it, and to consider two divergent ways of developing it.

2. Counterfactuals

What we shall be saying requires little in the way of an elaborate theory of counterfactuals. We do suppose that counterfactuals are genuine propositions. For a proposition to be a counterfactual, we do not require that its antecedent be false: on the view we are considering, a rational agent entertains counterfactuals of the form "I do $a \;\Box\!\!\rightarrow\; S$" both for the act he will turn out to perform and for acts he will turn out not to perform. To say $A \;\Box\!\!\rightarrow\; S$ is not to say that A's holding would bring about S's holding: $A \;\Box\!\!\rightarrow\; S$ is indeed true if A's holding would bring about S's holding, but $A \;\Box\!\!\rightarrow\; S$ is true also if S would hold regardless of whether A held.

These comments by no means constitute a full theory of counterfactuals. In what follows, we shall appeal not to a theory of counterfac-

---

[1] Lewis first presented this result at the June 1972 meeting of the Canadian Philosophical Association.

tuals, but to the reader's intuitions about them – asking the reader to bear clearly in mind that "I do $a \;\Box\!\!\rightarrow S$" is to be read "If I were to do a, then S would hold".

It may nevertheless be useful to sketch a theory that would support what we shall be saying; the theory we sketch here is somewhat like that of Stalnaker and Thomason (Stalnaker, 1968; Stalnaker and Thomason, 1970). Let a be an act which I might decide at time t to perform. An a-world will be a possible world which is like the actual world before t, in which I decide to do a at t and do it, and which obeys physical laws from time t on. Let W_a be the a-world which, at t, is most like the actual world at t. Thus W_a is a possible world which unfolds after t in accordance with physical law, and whose initial conditions at time t are minimally different from conditions in the actual world at t in such a way that "I do a" is true in W_a. The differences in initial conditions should be entirely within the agent's decision-making apparatus. Then "I do $a \;\Box\!\!\rightarrow S$" is true iff S is true in W_a.[2]

Two axioms that hold on this theory will be useful in later arguments. Our first axiom is just a principle of modus ponens for the counterfactual.

Axiom 1. $(A \;\&\; (A \;\Box\!\!\rightarrow S)) \supset S$.

Our second axiom is a Stalnaker-like principle.

Axiom 2. $(A \;\Box\!\!\rightarrow \bar{S}) \equiv \overline{(A \;\Box\!\!\rightarrow S)}$.

The rationale for this is that "I do $a \;\Box\!\!\rightarrow S$" is true iff S holds in W_a and "I do $a \;\Box\!\!\rightarrow \bar{S}$" is true iff $\bar{S}$ holds in W_a. We shall also appeal to a consequence of these axioms.

Consequence 1. $A \supset [(A \;\Box\!\!\rightarrow S) \equiv S]$.

We do not regard Axiom 2 and Consequence 1 as self-evident. Our reason for casting the rough theory in a form which gives these principles is that circumstances where these can fail involve complications

---

[2] Although the rough treatment of counterfactuals we propose is similar in many respects to the theories developed by Stalnaker and Lewis, it differs from them in some important respects. Stalnaker and Lewis each base their accounts on comparisons of overall similarity of worlds. On our account, what matters is comparative similarity of worlds at the instant of decision. Whether a given a-world is selected as W_a depends not at all on how similar the future in that world is to the actual future; whatever similarities the future in W_a may have to the actual future will be a semantical consequence of laws of nature, conditions in W_a at the instant of decision, and actual conditions at that instant. (Roughly, then, they will be consequences of laws of nature and the similarity of W_a to the actual world at the instant of decision.) We consider only worlds in which the past is exactly like the actual past, for since the agent cannot now alter the past, those are the only worlds relevant to his decision. Lewis (1973b, p. 566 and in conversation) suggests that a proper treatment of overall similarity will yield as a deep consequence of general facts about the world the conditions we are imposing by fiat.

which it would be best to ignore in preliminary work.[3] Our appeals to these Axioms will be rare and explicit. For the most part in treating counterfactuals we shall simply depend on a normal understanding of the way counterfactuals apply to the situations we discuss.

3. Two kinds of expected utility

We have spoken on the one hand of expected utility calculated from the probabilities of counterfactuals, and on the other hand of expected utility calculated from conditional probabilities. In what follows, we shall not distinguish between an act an agent can perform and the proposition that says that he is about to perform it; acts will be expressed by capital letters early in the alphabet. An act will ordinarily have a number of alternative outcomes, where an *outcome* of an act is a single proposition which, for all the agent knows, expresses *all* the consequences of that act which he cares about. An outcome, then, is a specification of what might eventuate which is complete in the sense that any further specification of detail is irrelevant to the agent's concerns, and it specifies something that, for all the agent knows, might really happen if he performed the act. The agent, we shall assume, ascribes a magnitude $\mathscr{D}O$ to each outcome O. He knows that if he performed the act, one and only one of its outcomes would obtain, although he does not ordinarily know which of its outcomes that would be.

Let $O_1, \ldots, O_m$ be the outcomes of act A. The *expected utility* of A *calculated from probabilities of counterfactuals* we shall call $\mathscr{U}(A)$; it is given by the formula

$$\mathscr{U}(A) = \Sigma_j \, prob(A \; \Box\!\!\rightarrow O_j) \, \mathscr{D}O_j.$$

The *expected utility of A calculated from conditional probabilities* we shall call $\mathscr{V}(A)$; it is given by the formula

$$\mathscr{V}(A) = \Sigma_j \, prob(O_j/A) \, \mathscr{D}O_j.$$

Perhaps the best mnemonic for distinguishing $\mathscr{U}$ from $\mathscr{V}$ is this: we shall be advocating the use of counterfactuals in calculating expected utility, and we shall claim that $\mathscr{U}(A)$ is the genuine expected utility of A. $\mathscr{V}(A)$, we shall claim, measures instead the welcomeness of the news that one is about to perform A. Remember $\mathscr{V}(A)$, then, as the *value of A*

---

[3] In characterizing our conditional we have imposed the Stalnaker-like constraint that there is a unique world W_a which would eventuate from performing a at t. Our rationale for Axiom 2 depends on this assumption and on the assumption that if a is actually performed then W_a is the actual world itself. Consequence 1 is weaker than Axiom 2, and only depends on the second part of this assumption. In circumstances where these assumptions break down, it would seem to us that using conditionals to compute expected utility is inappropriate. A more general approach is needed to handle such cases.

as news, and remember $\mathcal{U}(A)$ as what the authors regard as the genuine expected utility of A.

Now clearly $\mathcal{U}(A)$ and $\mathcal{V}(A)$ will be the same if

$$prob(A \;\square\!\!\!\rightarrow O_j) = prob(O_j/A) \tag{2}$$

for each outcome O_j. Unless (2) holds for every O_j such that $\mathcal{U} O_j \neq O$, $\mathcal{U}(A)$ and $\mathcal{V}(A)$ will be the same only by coincidence. We know from Lewis's work (1976) that (2) does not hold for all propositions A and O_j; can we expect that (2) will hold for the appropriate propositions?

One assumption, together with the logical truth of Consequence 1, will guarantee that (2) holds for an act and its outcomes. Here and throughout, we suppose that the function *prob* gives the probability ascriptions of an agent who can immediately perform the act in question, and that *prob* $\phi = 1$ for any logical truth ϕ.

Condition 1 on act A and outcome O_i. The counterfactual $A \;\square\!\!\!\rightarrow O_i$ is stochastically independent of the act A. That is to say,

$$prob(A \;\square\!\!\!\rightarrow O_i/A) = prob(A \;\square\!\!\!\rightarrow O_i).$$

(Read $prob(A \;\square\!\!\!\rightarrow O_i/A)$ as the conditional probability of $A \;\square\!\!\!\rightarrow O_i$ on A.)

Assertion 1. Suppose Consequence 1 is a logical truth. If A and O_i satisfy Condition 1, and $prob(A) > 0$, then

$$prob(A \;\square\!\!\!\rightarrow O_i) = prob(O_i/A).[4]$$

Proof. Since Consequence 1 is a logical truth, for any propositions P and Q,

$$prob(P \supset [P \;\square\!\!\!\rightarrow Q) \equiv Q]) = 1.$$

Hence if *prob* $P > 0$, then

$$prob([P \;\square\!\!\!\rightarrow Q) \equiv Q]/P) = 1;$$

$$\therefore prob(P \;\square\!\!\!\rightarrow Q/P) = prob(Q/P).$$

From this general truth we have

$$prob(A \;\square\!\!\!\rightarrow O_i/A) = prob(O_i/A),$$

and from this and Condition 1, it follows that

$$prob(A \;\square\!\!\!\rightarrow O_i) = prob(O_i/A).$$

That proves the Assertion.

Condition 1 is that the counterfactuals relevant to decision be

---

[4] This is stated by Lewis (1976, note 10).

stochastically independent of the acts contemplated. Stochastic independence is the same as epistemic independence. For $prob(A \mathrel{\Box\!\!\rightarrow} O_i/A)$ is the probability it would be rational for the agent to ascribe to the counterfactual $A \mathrel{\Box\!\!\rightarrow} O_i$ on learning A and nothing else – on learning that he was about to perform that act. Thus to say that $prob(A \mathrel{\Box\!\!\rightarrow} O_i/A) = prob(A \mathrel{\Box\!\!\rightarrow} O_i)$ is to say that learning that one was about to perform the act would not change the probability one ascribes to the proposition that if one were to perform the act, outcome O_i would obtain. We shall use the terms "stochastic independence" and "epistemic independence" interchangeably.

The two kinds of expected utility $\mathcal{U}$ and $\mathcal{V}$ can also be characterized in a way suggested by Jeffrey's account of the Bayesian model. Let acts $A_1, \ldots, A_m$ be open to the agent. Let states $S_1, \ldots, S_n$ partition the possibilities in the following sense. For any propositions $S_1, \ldots, S_n$, the truth-function $aut\ (S_1, \ldots, S_n)$ will be their exclusive disjunction: $aut\ (S_1, \ldots, S_n)$ holds in and only in circumstances where exactly one of $S_1, \ldots, S_n$ is true. Let the agent know $aut\ (S_1, \ldots, S_n)$. For each act A_i and state S_j, let him know that if he did A_i and S_j obtained, the outcome would be O_{ij}. Let him ascribe each outcome O_{ij} as desirability $\mathcal{D}\,O_{ij}$. This will be a *matrix formulation* of a decision problem; its defining features are that the agent knows that $S_1, \ldots, S_n$ partition the possibilities, and in each of these states $S_1, \ldots, S_n$, each act open to the agent has a unique outcome. A set $\{S_1, \ldots, S_n\}$ of states which satisfy these conditions will be called *the states of a matrix formulation* of the decision problem in question.

Both $\mathcal{U}$ and $\mathcal{V}$ can be characterized in terms of a matrix formulation:

$$\mathcal{U}(A_i) = \Sigma_j\ prob(A_i \mathrel{\Box\!\!\rightarrow} S_j)\,\mathcal{D}\,O_{ij};$$

$$\mathcal{V}(A_i) = \Sigma_j\ prob(S_j/A_i)\,\mathcal{D}\,O_{ij}.$$

If $\mathcal{D}\,O_{ij}$ can be regarded as the desirability the agent attributes to S_j "on the assumption that" he will do A_i, then $\mathcal{V}(A_i)$ is the desirability of A_i as characterized in the account we quoted from Jeffrey.

On the basis of these matrix characterizations of $\mathcal{U}$ and $\mathcal{V}$, we can state another sufficient condition for the $\mathcal{U}$-utility and $\mathcal{V}$-utility of an act to be the same.

Condition 2 on act A_i, states $S_1, \ldots, S_n$, and the function prob. For each A_i and S_j,

$$prob(A_i \mathrel{\Box\!\!\rightarrow} S_j/A_i) = prob(A_i \mathrel{\Box\!\!\rightarrow} S_j).$$

Assertion 2. Suppose Consequence 1 is a logical truth. If a decision problem satisfies Condition 2 for act A_i, then $\mathcal{U}(A_i) = \mathcal{V}(A_i)$. The proof is like that of Assertion 1.

4. Act-dependent states in the Savage framework

Savage's representation of decision problems (1954) is roughly the matrix formulation just discussed. Ignorance is represented as ignorance about which of a number of states of the world obtains. These states are mutually exclusive, and as specific as the problem requires (p. 15). The agent ascribes desirability to "consequences", or what we are calling *outcomes*. For each act open to the agent, he knows what outcome obtains for each state of the world; if he does not, the problem must be reformulated so that he does. Savage indeed defines an act as a function from states to outcomes (Savage, 1954, p. 14).

It is a consequence of the axioms Savage gives that a rational agent is disposed to choose as if he ascribed a numerical desirability to each outcome and a numerical probability to each state, and then acted to maximize expected utility, where the expected utility of an act A is

$$\Sigma_S \, prob(S) \mathscr{D} O(A, S). \tag{3}$$

(Here $O(A, S)$ is the outcome of act A in state S.) Another consequence of Savage's axioms is the principle of dominance: If for every state S, the outcome of act A in S is more desirable than the outcome of B in S, then A is preferable to B.

Consider this misuse of the Savage apparatus; it is of a kind discussed by Jeffrey (1965, pp. 8–10)

Case 1. David wants Bathsheba, but since she is the wife of Uriah, he fears that summoning her to him would provoke a revolt. He reasons to himself as follows: "There are two possibilities: R, that there will be a revolt, and $\bar{R}$, that there won't be. The outcomes and their desirabilities are given in Matrix 1, where B is that I take Bathsheba and A is that I abstain from her. Whether or not there is a revolt, I prefer having Bathsheba to not having her, and so taking Bathsheba dominates over abstaining from her.

	R	$\bar{R}$
A	$R\bar{B}(0)$	$\bar{R}\bar{B}(9)$
B	$RB(1)$	$\bar{R}B(10)$

Matrix 1

This argument is of course fallacious: dominance requires that the states in question be independent of the acts contemplated, whereas taking Bathsheba may provoke revolt. To apply the Savage framework to a decision problem, one must find states of the world which are in some sense act-independent.

We now pursue a suggestion by Jeffrey on how to deal with states that are act-dependent. Construct four new conditionalized[5] states:

S_{00}: There would be no revolt whatever I did.
S_{01}: A would not elicit revolt, whereas B would.
S_{10}: A would elicit revolt, whereas B would not.
S_{11}: There would be a revolt whatever I did.

If these states hold independently of A and B, we can now work from Matrix 2 without fallacy. Since in Matrix 2 neither row dominates, the decision must be made on the basis of probabilities ascribed to the states $S_{00}, \ldots, S_{11}$.

	S_{00}	S_{01}	S_{10}	S_{11}
A	$\bar{R}\bar{B}(9)$	$\bar{R}\bar{B}(9)$	$R\bar{B}(0)$	$R\bar{B}(0)$
B	$\bar{R}B(10)$	$RB(1)$	$\bar{R}B(10)$	$RB(1)$

Matrix 2

What should the probabilities of these states be? One possible answer would be this: Each of the four states $S_{00}, \ldots, S_{11}$ can be expressed as a conjunction of counterfactuals. S_{01}, for instance, is the proposition $(A \mathbin{\square\!\!\rightarrow} \bar{R})$ & $(B \mathbin{\square\!\!\rightarrow} R)$. The probability of S_{01}, then, is simply the probability of this proposition, $prob([A \mathbin{\square\!\!\rightarrow} \bar{R}]$ & $[B \mathbin{\square\!\!\rightarrow} R])$.

The expected utility of an act can now be calculated in the standard way given by (3). The expected utility of A, for instance, will be

$$\Sigma_S \, prob(S)\mathcal{D}O(A, S). \tag{4}$$

where the summation is over the new states S, and $O(A, S)$ is the outcome of A in state S.

Does this procedure give the correct expected utility for the act? What it gives as the expected utility of A, we can show, is $\mathcal{U}(A)$ – at least that is what it gives if Axiom 2 is part of the logic of counterfactuals. For (4) expands to

$$prob(S_{00}) \, \mathcal{D}O(A, S_{00}) + prob(S_{01}) \, \mathcal{D}O(A, S_{01}) +$$
$$prob(S_{10}) \, \mathcal{D}O(A, S_{10}) + prob(S_{11}) \, \mathcal{D}O(A, S_{11}).$$

[5] This is our understanding of a proposal made by Jeffrey at the colloquium on Foundations and Applications of Decision Theory, University of Western Ontario, 1975. J. H. Sobel shows (in an unpublished manuscript) that, for all we have said, these new, conditionalized states may not themselves be act-independent. This section is slightly changed in light of Sobel's result.

We have

$$O(A, S_{00}) = O(A, S_{01}) = \bar{R}\bar{B};$$
$$O(A, S_{10}) = O(A, S_{11}) = R\bar{B}.$$

Thus since S_{00} and S_{01} are mutually exclusive, as are S_{10} and S_{11}, (4) becomes

$$prob(S_{00} \vee S_{01})\mathscr{D}\bar{R}\bar{B} + prob(S_{10} \vee S_{11})\mathscr{D}R\bar{B}.$$

Now $S_{00} \vee S_{01}$ is $([A \; \square\!\!\rightarrow \bar{R}] \; \& \; [B \; \square\!\!\rightarrow \bar{R}]) \vee [A \; \square\!\!\rightarrow \bar{R}] \; \& \; [B \; \square\!\!\rightarrow R])$, and in virtue of the logical truth of Axiom 2, this is $A \; \square\!\!\rightarrow \bar{R}$. Similarly, $S_{10} \vee S_{11}$ is $A \; \square\!\!\rightarrow R$. Thus (4) becomes

$$prob(A \; \square\!\!\rightarrow \bar{R})\mathscr{D}\bar{R}\bar{B} + prob(A \; \square\!\!\rightarrow R)\mathscr{D}R\bar{B},$$

which is $\mathscr{U}(A)$. This proof can of course be generalized.

We have considered one way to construct conditionalized states from act-dependent states; it is a way that makes use of counterfactuals. Suppose, though, we want to avoid the use of counterfactuals and rely instead on conditional probabilities. Jeffrey, as we understand him, suggests the following: ascribe to each new, conditionalized state the product of the pertinent conditional probabilities. We shall call this probability $prob^*$; thus, for instance,

$$prob^*(S_{01}) = prob\,(\bar{R}/A)\,prob(R/B),$$

and corresponding formulas hold for the other new states S_{00}, S_{10}, and S_{11}.

Using $prob^*$, we can again calculate expected utility in the standard way given by (3). The expected utility of A, for instance, will be

$$\Sigma_S\,prob^*\,(S)\mathscr{D}O(A, S). \tag{5}$$

where again the summation is over the new states S_{00}, S_{01}, S_{10}, and S_{11}. Now (5), it can be shown, has the value $\mathscr{V}(A)$. For (5) is the sum of terms

$$prob^*\,(S_{00})\mathscr{D}O(A, S_{00}) = prob\,(\bar{R}/A)\,prob\,(\bar{R}/B)\mathscr{D}\bar{R}\bar{B},$$
$$prob^*\,(S_{01})\mathscr{D}O(A, S_{01}) = prob\,(\bar{R}/A)\,prob\,(R/B)\mathscr{D}\bar{R}\bar{B},$$
$$prob^*\,(S_{10})\mathscr{D}O(A, S_{10}) = prob\,(R/A)\,prob\,(\bar{R}/B)\mathscr{D}R\bar{B},$$
$$prob^*\,(S_{11})\mathscr{D}O(A, S_{11}) = prob\,(R/A)\,prob\,(R/B)\mathscr{D}R\bar{B}.$$

Thus (5) equals

$$[prob(\bar{R}/B) + prob(R/B)]\,prob(\bar{R}/A)\mathscr{D}\bar{R}\bar{B}$$
$$+ [prob(\bar{R}/B) + prob(R/B)]\,prob(R/A)\mathscr{D}R\bar{B}$$
$$= prob(\bar{R}/A)\mathscr{D}\bar{R}\bar{B} + prob(R/A)\mathscr{D}R\bar{B},$$

and this is $\mathscr{V}(A)$.

Where, then, a decision problem is misformulated in the Savage framework with act-dependent states, we now have two ways of reformulating the problem with conditionalized states. The first way is to express each conditionalized state as a conjunction of counterfactuals. If the expected utility of an act A is then calculated in the standard manner and Axiom 2 holds, the result is $\mathcal{U}(A)$. The second way to reformulate the problem is to ascribe to each new conditionalized state the product of the pertinent conditional probabilities. If the expected utility of an act A is then calculated in the standard manner, the result is $\mathcal{V}(A)$. If Axiom 2 holds, then, the two reformulations yield respectively the two kinds of expected utility we have been discussing.

So far we have given the two reformulations only for an example. Here is the way the two methods of reformulation work in general. Let acts $A_1, \ldots, A_m$ be open to the agent, let states $S_1, \ldots, S_n$ not all be act-independent, and for each A_i and S_j, let the outcome of act A_i in S_j be O_{ij}. For each possible sequence $T_1, \ldots, T_m$ consisting of states in $\{S_1, \ldots, S_n\}$, there will be a new, conditionalized state $S(T_1, \ldots, T_m)$. The outcome of an act A_i in the new state $S(T_1, \ldots, T_m)$ will simply be the outcome of A_i in the old state T_i. What has been said so far applies to both methods. Now, according to the first method of reformulation, this new state $S(T_1, \ldots, T_m)$ will be

$$(A_1 \mathbin{\Box\!\!\rightarrow} T_1) \, \& \ldots \& \, (A_m \mathbin{\Box\!\!\rightarrow} T_m),$$

and hence, of course, its probability will be the probability of this proposition. According to the second method of reformulation, the probability of new state $S(T_1, \ldots, T_m)$ will be

$$prob \, (T_1/A_1) \times \cdots \times prob \, (T_m/A_m).$$

Once the problem is reformulated, expected utility is to be calculated in the standard way by formula (3).

Are these two ways of reformulating a decision problem equivalent or distinct? They are, of course, equivalent if Axiom 2 holds and $\mathcal{U}(A_i) = \mathcal{V}(A_i)$ for each act A_i, since the first method yields $\mathcal{U}(A_i)$ if Axiom 2 holds and the second method yields $\mathcal{V}(A_i)$. We already know that Condition 2 and the logical truth of Consequence 1 guarantee that $\mathcal{U}(A_i) = \mathcal{V}(A_i)$. Therefore, we may conclude that if Condition 2 holds and Axioms 1 and 2 are logical truths, the two reformulations are equivalent. Condition 2, recall, is that the counterfactuals $A_i \mathbin{\Box\!\!\rightarrow} S_j$ are epistemically act-independent: that for each of the old, act-dependent states in terms of which the problem is formulated, learning that one is about to perform a given act will not change the probability one ascribes to the proposition that if one were to perform that act, that state would obtain.

The upshot of the discussion is this. For the Savage apparatus to apply to a decision problem, the states of the decision matrix must be independent of the acts. We have considered two ways of dealing with a problem stated in terms of act-dependent states; both ways involve reformulating the problem in terms of new states which are act-independent. Given the logical truth of Axioms 1 and 2, a sufficient condition for the equivalence of the two reformulations is that the counterfactuals $A_i \;\square\!\!\rightarrow S_j$ be epistemically act-independent.

5. Act-dependent counterfactuals

Should we expect Condition 2 to hold? In the case of David, it seems that we should. Suppose David somehow learned that he was about to send for Bathsheba; that would give him no reason to change the probability he ascribes to the proposition "If I were to send for Bathsheba, there would be a revolt". Similarly, if David learned that he was about to abstain from Bathsheba, that would give him no reason to change the probability he ascribes to the proposition "If I were to abstain from Bathsheba, there would be a revolt". In the case of David, it seems, the pertinent counterfactuals are epistemically act-independent, and hence for each act he can perform, the $\mathcal{U}$-utility and the $\mathcal{V}$-utility are the same.

When, however, a common factor is believed to affect both behaviour and outcome, Condition 2 may fail, and $\mathcal{U}$-utility may diverge from $\mathcal{V}$-utility. The following case is patterned after an example used by Stalnaker to make the same point.[6]

Case 2. Solomon faces a situation like David's, but he, unlike David, has studied works on psychology and political science which teach him the following: Kings have two basic personality types, charismatic and uncharismatic. A king's degree of charisma depends on his genetic make-up and early childhood experiences, and cannot be changed in adulthood. Now charismatic kings tend to act justly and uncharismatic kings unjustly. Successful revolts against charismatic kings are rare, whereas successful revolts against uncharismatic kings are frequent. Unjust acts themselves, though, do not cause successful revolts; the reason that uncharismatic kings are prone to successful revolts is that they have a sneaky, ignoble bearing. Solomon does not know whether or not he is charismatic; he does know that it is unjust to send for another man's wife.

Now in this case, Condition 2 fails for states R and $\bar{R}$. The counterfactual $B \;\square\!\!\rightarrow R$ is not epistemically independent of B: we have

$$prob(B \;\square\!\!\rightarrow R/B) > prob(B \;\square\!\!\rightarrow R).$$

---

[6] Meeting of the Canadian Philosophical Association, 1972. Nozick gives a similar example (1969, p. 125).

For the conditional probability of anything on B is the probability Solomon would rationally ascribe to it if he learned that B. Since he knows that B's holding would in no way tend to bring about R's holding, he always ascribes the same probability to $B \;\Box\!\!\rightarrow R$ as to R. Hence both $prob(B \;\Box\!\!\rightarrow R) = prob(R)$ and $prob(B \;\Box\!\!\rightarrow R/B) = prob\,(R/B)$. Now if Solomon learned that B, he would have reason to think that he was uncharismatic, and thus revolt-prone. Hence $prob(R/B) > prob(R)$, and therefore

$$prob(B \;\Box\!\!\rightarrow R/B) = prob(R/B) > prob(R) = prob(B \;\Box\!\!\rightarrow R). \qquad (6)$$

Here, then, the counterfactual is not epistemically act-independent.

Equation (6) states also that $prob(B \;\Box\!\!\rightarrow R) < prob(R/B)$, so that in this case, the probability of the counterfactual does not equal the corresponding conditional probability. By similar argument we could show that $prob(A \;\Box\!\!\rightarrow R) > prob(R/A)$. Indeed in this case a $\mathscr{U}$-maximizer will choose to send for his neighbour's wife whereas a $\mathscr{Y}$-maximizer will choose to abstain from her – although we shall need to stipulate the case in more detail to prove the latter.

Consider first $\mathscr{U}$-maximization. We have that

$$\mathscr{U}(B) = prob(B \;\Box\!\!\rightarrow \bar{R})\,\mathscr{D}\bar{R}B + prob(B \;\Box\!\!\rightarrow R)\,\mathscr{D}RB;$$
$$\mathscr{U}(A) = prob(A \;\Box\!\!\rightarrow \bar{R})\,\mathscr{D}\bar{R}\bar{B} + prob(A \;\Box\!\!\rightarrow R)\,\mathscr{D}R\bar{B}.$$

We have argued that $prob(B \;\Box\!\!\rightarrow R) = prob(R)$. Similarly, $prob(A \;\Box\!\!\rightarrow R) = prob(R)$, and so $prob(A \;\Box\!\!\rightarrow R) = prob(B \;\Box\!\!\rightarrow R)$. Likewise $prob(A \;\Box\!\!\rightarrow \bar{R}) = prob(B \;\Box\!\!\rightarrow \bar{R})$. We know that $\mathscr{D}\bar{R}B > \mathscr{D}\bar{R}\bar{B}$ and $\mathscr{D}RB > \mathscr{D}R\bar{B}$. Therefore $\mathscr{U}(B) > \mathscr{U}(A)$. This is in effect an argument from dominance, as we shall discuss in Section 8.

Now consider $\mathscr{Y}$-maximization. Learning that A would give Solomon reason to think he was charismatic and thus not-revolt prone, whereas learning that B would give him reason to think that he was uncharismatic and revolt-prone. Thus $prob(R/B) > prob(R/A)$. Suppose the difference between these probabilities is greater than $\frac{1}{9}$, so that where $prob(R/A) = \alpha$ and $prob(R/B) = \alpha + \varepsilon$, we have $\varepsilon > \frac{1}{9}$. From Matrix 1, we have

$$\mathscr{Y}(A) = prob(\bar{R}/A)\,\mathscr{D}\bar{R}\bar{B} + prob(R/A)\,\mathscr{D}R\bar{B}$$
$$= 9(1 - \alpha) + 0.$$
$$\mathscr{Y}(B) = prob(\bar{R}/B)\,\mathscr{D}\bar{R}B + prob(R/B)\,\mathscr{D}RB$$
$$= 10(1 - \alpha - \varepsilon) + 1(\alpha + \varepsilon).$$

Therefore $\mathscr{Y}(A) - \mathscr{Y}(B) = 9\varepsilon - 1$, and since $\varepsilon > \frac{1}{9}$, this is positive. We have shown that if $\varepsilon > \frac{1}{9}$, then although $\mathscr{U}(B) > \mathscr{U}(A)$, we have $\mathscr{Y}(A) > \mathscr{Y}(B)$. Thus $\mathscr{U}$-maximization and $\mathscr{Y}$-maximization in this case yield conflicting prescriptions.

Which of these prescriptions is the rational one? It seems clear that in this case it is rational to perform the $\mathcal{U}$-maximizing act: unjustly to send for the wife of his neighbor. For Solomon cares only about getting the woman and avoiding revolt. He knows that sending for the woman would not cause a revolt. To be sure, sending for her would be an indication that Solomon lacked charisma, and hence an indication that he will face a revolt. To abstain from the woman for this reason, though, would be knowingly to bring about an indication of a desired outcome without in any way bringing about the desired outcome itself. That seems clearly irrational.

For those who find Solomon too distant in time and place or who mistrust charisma, we offer the case of Robert Jones, rising young executive of International Energy Conglomerate Incorporated. Jones and several other young executives have been competing for a very lucrative promotion. The company brass found the candidates so evenly matched that they employed a psychologist to break the tie by testing for personality qualities that lead to long run successful performance in the corporate world. The test was administered to the candidates on Thursday. The promotion decision is made on the basis of the test and will be announced on Monday. It is now Friday. Jones learns, through a reliable company grapevine, that all the candidates have scored equally well on all factors except ruthlessness and that the promotion will go to whichever of them has scored highest on this factor, but he cannot find out which of them this is.

On Friday afternoon Jones is faced with a new problem. He must decide whether or not to fire poor old John Smith, who failed to meet his sales quota this month because of the death of his wife. Jones believes that Smith will come up to snuff after he gets over his loss provided that he is treated leniently, and that he can convince the brass that leniency to Smith will benefit the company. Moreover, he believes that this would favorably impress the brass with his astuteness. Unfortunately, Jones has no way to get in touch with them until after they announce the promotion on Monday.

Jones knows that the ruthlessness factor of the personality test he has taken accurately predicts his behaviour in just the sort of decision he now faces. Firing Smith is good evidence that he has passed the test and will get the promotion, while leniency is good evidence that he has failed the test and will not get the promotion. We suppose that the utilities and probabilities correspond to those facing Solomon. $\mathcal{V}$-maximizing recommends firing Smith, while $\mathcal{U}$-maximizing recommends leniency. Firing Smith would produce evidence that Jones will get his desired promotion. It seems clear, however, that to fire Smith for this reason despite the fact that to do so would in no way help to bring about the promotion and would itself be harmful, is irrational.

6. The significance of $\mathcal{U}$ and $\mathcal{V}$

From the Solomon example, it should be apparent that the $\mathcal{V}$-utility of an act is a measure of the welcomeness of the news that one is about to perform that act. Such news may tend to be welcome because the act is likely to have desirable consequences, or tend to be unwelcome because the act is likely to have disagreeable consequences. Those, however, are not the only reasons an act may be welcome or unwelcome: an act may be welcome because its being performed is an indication that the world is in a desired state. Solomon, for instance, would welcome the news that he was about to abstain from his neighbor's wife, but he would welcome it not because he thought just acts any more likely to have desirable consequences than unjust acts, but because he takes just acts to be a sign of charisma, and he thinks that charisma may bring about a desired outcome.

$\mathcal{U}$-utility, in contrast, is a measure of the expected efficacy of an act in bringing about states of affairs the agent desires; it measures the expected value of the consequences of an act. That can be seen in the case of Solomon. The $\mathcal{U}$-utility of sending for his neighbor's wife is greater than that of abstaining, and that is because he knows that sending for her will bring about a consequence he desires – having the woman – and he knows that it will not bring about any consequences he wishes to avoid: in particular, he knows that it will not bring about a revolt.

What is it for an act to bring about a consequence? Here are two possible answers, both formulated in terms of counterfactuals.

In the first place, roughly following Sobel (1970, p. 400) we may say that act A brings about state S if $A \;\square\!\!\rightarrow S$ holds, and for some alternative A^* to A, $A^* \;\square\!\!\rightarrow S$ does not hold.[7] (An *alternative* to A is another act open to the agent on the same occasion). Now on this analysis, the $\mathcal{U}$-utility of an act as we have defined it is the sum of the expected value of its consequences plus a term which is the same for all acts open to the agent on the occasion in question; this latter term is the expected value of unavoidable outcomes. A state S is *unavoidable* iff for every act A^* open to the agent, $A^* \;\square\!\!\rightarrow S$ holds. Thus $A \;\square\!\!\rightarrow S$ holds iff S is a consequence of A or S is unavoidable. Hence in particular, for any outcome O,

$$prob(A \;\square\!\!\rightarrow O) = prob(O \text{ is a consequence of } A)$$
$$+ \; prob(O \text{ is unavoidable}),$$

---

[7] Sobel actually uses "$A^* \;\square\!\!\rightarrow \bar{S}$ does hold" where we use "$A^* \;\square\!\!\rightarrow S$ does not hold". With Axiom 2, these are equivalent.

and so we have

$$\mathscr{U}(A) = \Sigma_O \, prob(A \;\Box\!\!\to\; O)\,\mathscr{D}O$$
$$= \Sigma_O \, prob(O \text{ is a consequence of } A)\,\mathscr{D}O$$
$$+ \Sigma_O \, prob(O \text{ is unavoidable})\,\mathscr{D}O.$$

The first term is the expected value of the consequences of A, and the second term is the same for all acts open to the agent. Therefore on this analysis of the term "consequence", $\mathscr{U}$-utility is maximal for the act or acts whose consequences have maximal expected value.

Here is a second possible way of analyzing what it is to be a consequence. When an agent chooses between two acts A and B, what he really needs to know is not what the consequences of A are and what the consequences of B are, but rather what the consequences are of A as opposed to B and *vice versa*. Thus for purposes of decision-making, we can do without an analysis of the clause "S is a consequence of A", and analyze instead the clause "S is a consequence of A as opposed to B". This we can analyze as

$$(A \;\Box\!\!\to\; S) \;\&\; \sim (B \;\Box\!\!\to\; S).$$

Now on this analysis, $\mathscr{U}(A) > \mathscr{U}(B)$ iff the expected value of the consequences of A as opposed to B exceeds the expected value of the consequences of B as opposed to A. For any state S, $A \;\Box\!\!\to\; S$ holds iff either S is a consequence of A as opposed to B or $(A \;\Box\!\!\to\; S) \;\&\; (B \;\Box\!\!\to\; S)$ holds.

Thus $\quad \mathscr{U}(A) = \Sigma_O \, prob(A \;\Box\!\!\to\; O)\,\mathscr{D}O$
$$= \Sigma_O \, prob(O \text{ is a consequence of } A \text{ as opposed to } B)\,\mathscr{D}O$$
$$+ \Sigma_O \, prob([A \;\Box\!\!\to\; O] \;\&\; [B \;\Box\!\!\to\; O])\,\mathscr{D}O$$
$$\mathscr{U}(B) = \Sigma_O \, prob(O \text{ is a consequence of } B \text{ as opposed to } A)\,\mathscr{D}O$$
$$+ \Sigma_O \, prob([A \;\Box\!\!\to\; O] \;\&\; [B \;\Box\!\!\to\; O])\,\mathscr{D}O.$$

The second term is the same in both cases, and so $\mathscr{U}(A) > \mathscr{U}(B)$ iff

$$\Sigma_O \, prob \, (O \text{ is a consequence of } A \text{ as opposed to } B)\,\mathscr{D}O >$$
$$\Sigma_O \, prob(O \text{ is a consequence of } B \text{ as opposed to } A)\,\mathscr{D}O.$$

The left side is the expected value of the consequences of A as opposed to B; the right side is the expected value of the consequences of B as opposed to A. Thus for any pair of alternatives, to prefer the one with the higher $\mathscr{U}$-utility is to prefer the one the consequences of which as opposed to the other have the greater expected value.

We can now ask whether $\mathscr{U}$ or $\mathscr{V}$ is more properly called the "utility" of an act. The answer seems clearly to be $\mathscr{U}$. The "utility" of an act should be its expected genuine efficacy in bringing about states of affairs the agent wants, not the degree to which news of the act ought to cheer the agent. Since $\mathscr{U}$-utility is a matter of what the act can be expected to bring about whereas $\mathcal{V}$-utility is a matter of the welcomeness of news, $\mathcal{U}$-utility seems best to capture the notion of utility.

Jeffrey (1965, pp. 73–74) writes, "If the agent is deliberating about performing act A or act B, and if AB is impossible, there is no effective difference between asking whether he prefers A to B as a news item or as an act, for he makes the news". It should now be clear why it may sometimes be rational for an agent to choose an act B instead of an act A, even though he would welcome the news A more than that of B. The news of an act may furnish evidence of a state of the world which the act itself is known not to produce. In that case, though the agent indeed makes the news of his act, he does not make all the news his act bespeaks.

7. Two sure-thing principles

Case 3. Upon his accession to the throne, Reoboam wonders whether to announce that he will reign severely or to announce that he will reign leniently. He will be bound by what he announces. He slightly prefers a short severe reign to a short lenient reign, and he slightly prefers a long severe reign to a long lenient reign. He strongly prefers a long reign of any kind to a short reign of any kind. Where L is that he is lenient and D, that he is deposed early, his utilities are as in the Matrix 3.

	D	$\bar{D}$
L	0	80
$\bar{L}$	5	100

Matrix 3

The wise men of the kingdom give him these findings of behavioural science: There is no correlation between a king's severity and the length of his reign. Severity, nevertheless, often causes early deposition. The reason for the lack of correlation between severity and early deposition is that on the one hand, charismatic kings tend to be severe, and on the other hand, lack of charisma tends to elicit revolts. A king's degree of charisma cannot be changed in adulthood. There is at present no indication of whether Reoboam is charismatic or not.

These findings were based on a sample of 100 kings, 48 of whom had their reigns cut short by revolt. On post mortem examination of the pineal gland, 50 were found to have been charismatic and 50 uncharismatic. Eighty percent of the charismatic kings had been severe and

Table 17.1

	Charismatic	Uncharismatic	Total
Severe	16 deposed (40%) 24 long-reigned	8 deposed (80%) 2 long-reigned	24 deposed (48%) 26 long reigned
Lenient	2 deposed (20%) 8 long-reigned	22 deposed (55%) 18 long-reigned	24 deposed (48%) 26 long reigned

80 percent of the uncharismatic kings had been lenient. Of the charismatic kings, 40 percent of those who were severe were deposed whereas only 20 percent of those who were lenient were deposed. Of the uncharismatic kings, 80 percent of those who were severe were deposed whereas only 55 percent of those who were lenient were deposed. The totals were as in Table 17.1. This is Reoboam's total evidence on the subject.[8]

Reoboam's older advisors argue from a sure-thing principle. There are two possibilities, they say: that Reoboam is charismatic and that he is uncharismatic; what he does now will not affect his degree of charisma. On the assumption that he is charismatic, it is rational to prefer lenience. For since 40 percent of severe charismatic kings are deposed, the expected utility of severity in that case would be

$$0.4 \mathscr{D}SD + 0.6 \mathscr{D}S\bar{D} = 0.4 \times 5 + 0.6 \times 80 = 62,$$

whereas since only 20 percent of lenient charismatic kings are deposed, the expected utility of lenience in that case would be

$$0.2 \mathscr{D}LD + 0.8 \mathscr{D}L\bar{D} = 0.2 \times 0 + 0.8 \times 80 = 64.$$

On the assumption that he is uncharismatic, it is again rational to prefer lenience. For since 80 percent of severe uncharismatic kings are deposed, the expected utility of severity in this case would be

$$0.8 \mathscr{D}SD + 0.2 \mathscr{D}S\bar{D} = 0.8 \times 5 + 0.2 \times 100 = 24,$$

whereas since only 55 percent of lenient uncharismatic kings are deposed, the expected utility of lenience in this case would be

$$0.55 \mathscr{D}LD + 0.45 \mathscr{D}L\bar{D} = 0.55 \times 0 + 0.45 \times 80 = 36.$$

[8] We realize that a Bayesian king presented with these data would not ordinarily take on degrees of belief that exactly match the frequencies given in the table; nevertheless, with appropriate prior beliefs and evidence, he would come to have those degrees of belief. Assume that he does.

Thus in either case, lenience is to be preferred, and so by a sure-thing principle, it is rational to prefer lenience in the actual case.

Reoboam's youthful friends argue that on the contrary, sure-thing considerations prescribe severity. Severity is indeed the dominant strategy. There are two possibilities: D, that Reoboam will be deposed, and $\bar{D}$, that he will not be. These two states are stochastically independent of the acts contemplated: both *prob* (D/S) and *prob* (D/L) are 0.48. Therefore, his youthful friends urge, one can without fallacy use the states D and $\bar{D}$ in an argument from dominance. On the assumption that he will be deposed, he prefers to be severe, and likewise on the assumption that he will not be deposed, he prefers to be severe. Thus by dominance, it is rational for him to prefer severity.

Here, then, are two sure-thing arguments which lead to contrary prescriptions. One argument appeals to the finding that charisma is causally independent of the acts contemplated; the other appeals to the finding that being deposed is stochastically independent of the acts. The old advisors and youthful companions are in effect appealing to different versions of a sure-thing principle, one of which requires causal independence and the other of which requires stochastic independence. The two versions lead to incompatible conclusions.

The sure thing principle is this: if a rational agent knows *aut* $(S_1, \ldots, S_n)$ and prefers A to B in each case, then he prefers A to B. If the propositions $S_1, \ldots, S_n$ are required to be states in a matrix formulation of the decision problem, so that each pair of state and act determine a unique outcome, the sure thing principle becomes the principle of dominance to be discussed in Section 8; the principle of dominance is thus a special case of the sure-thing principle. Now the principle of dominance, we have said, requires a proviso that the states in question be act-independent. The sure-thing principle should presumably include the same proviso. The sure-thing principle, then, should be this: If a rational agent knows that precisely one of the propositions $S_1, \ldots, S_n$ holds and prefers act A to act B in each case, and if in addition the propositions $S_1, \ldots, S_n$ are independent of the acts A and B, then he prefers A to B.

The problem in the case of Reoboam is that his two groups of advisors appeal to different kinds of independence to reach opposing conclusions. The older advisors appeal to causal independence; they cite the finding that a king's degree of charisma is unaffected by his adult actions. His youthful companions appeal to stochastic independence; they cite the finding that there is no correlation between severity in kings and revolt. The two appeals yield opposite conclusions.

It seems, then, that the sure-thing principle comes in two different versions, one of which requires that the propositions in question be causally independent of the acts, and the other of which requires the propositions to be stochastically independent of the acts.

The principle to which the youthful companions appeal can be put as follows.

> *Definition.* Act A *is sure against* act B *with stochastic independence of* $S_1, \ldots, S_n$ iff the following hold. The agent knows that independently of the choice between A and B, propositions $S_1, \ldots, S_n$ partition the possibilities; that is to say, $prob(aut(S_1, \ldots, S_n)/A) = 1$ and $prob(aut(S_1, \ldots, S_n)/B) = 1$. The propositions $S_1, \ldots, S_n$ are epistemically independent of the choice between A and B, in the sense that for each, $prob(S_i/A) = prob(S_i/B)$. Finally, for each of these propositions S_i it would be rational to prefer A to B if it were known that S_i held.

Sure-thing with stochastic independence. If act A is sure against act B with stochastic independence, then it is rational to prefer A to B.

The principle to which the older advisors appeal will take longer to formulate. The proviso for this version will be that the propositions $S_1, \ldots, S_n$ be causally independent of the choice between A and B; this can be formulated in terms of counterfactuals. To say that a state S_i is causally independent of the choice between A and B is to say that S_i would hold if A were performed iff S_i would hold if B were performed: $(A \;\square\!\!\rightarrow S_i) \equiv (B \;\square\!\!\rightarrow S_i)$. We now want to suppose that for each state S_i, A would be preferred to B given, in some sense, knowledge of S_i. This knowledge of S_i should not simply be knowledge that S_i holds, but knowledge that S_i holds independently of the choice between A and B: that $(A \;\square\!\!\rightarrow S_i) \;\&\; (B \;\square\!\!\rightarrow S_i)$. We can now state the principle.

> For each S_i let S_i^* be $(A \;\square\!\!\rightarrow S_i) \;\&\; (B \;\square\!\!\rightarrow S_i)$.

> *Definition. A is sure against B with causal independence of* $S_1, \ldots, S_n$ iff the following hold. The agent knows $aut (S_1^*, \ldots, S_n^*)$, and for each S_i it would be rational to prefer A to B if S_i^* were known to hold.[9]

> (Note that since for each S_i, $(A \;\square\!\!\rightarrow S_i) \equiv (B \;\square\!\!\rightarrow S_i)$ follows from $aut (S_i^*, \ldots, S_n^*)$, this guarantees that our agent knows that each S_i is causally independent of the choice between A and B.) We can now state the principle to which the older advisors appeal.

Sure-thing with causal independence. If A is sure against B with causal independence, then it is rational to prefer A to B.

---

[9] Under these conditions, if A and B are the only alternatives, then S_i^* holds if and only if S_i holds. If there are other alternatives, it may be that neither A nor B is performed and S_i holds without either $A \;\square\!\!\rightarrow S_i$ or $B \;\square\!\!\rightarrow S_i$. In that case, what matters is not whether it would be rational to prefer A to B knowing that S_i holds, but whether it would be rational to prefer A to B knowing $(A \;\square\!\!\rightarrow S_i) \;\&\; (B \;\square\!\!\rightarrow S_i)$.

In the case of Reoboam, we have seen, sure-thing with stochastic independence prescribes severity and sure-thing with causal independence prescribes lenience. Now to us it seems clear that the only rational action in this case is that prescribed by sure-thing with causal independence. It is rational for Reoboam to prefer lenience because severity tends to bring about deposition and he wants not to be deposed much more strongly than he wants to be severe. To be guided by sure-thing with stochastic independence in this case is to ignore the finding that severity tends to bring about revolt – to ignore that finding simply because severity is not on balance a *sign* that revolt will occur. To choose to be severe is to act in a way that tends to bring about a dreaded consequence, simply because the act is not a sign of the consequence. That seems to us to be irrational.

The two versions of the sure-thing principle we have discussed correspond to the two kinds of utility discussed earlier. Sure-thing with stochastic independence follows from the principle that an act is rationally preferred to another iff it has greater $\mathcal{V}$-utility, whereas sure-thing with causal independence follows from the principle that an act is rationally preferred to another iff it maximizes $\mathcal{U}$-utility.

Assertion. Suppose that in any possible situation, it is rational to prefer an act A to an act B iff the $\mathcal{U}$-utility of A is greater than that of B. Then sure-thing with causal independence holds.

Proof. Suppose A is sure against B with causal independence of $S1$, ..., S_n, and that in any possible circumstance, it would be rational to prefer A to B iff A's $\mathcal{U}$-utility were greater than B's. The assertion will be proved if we show from these assumptions that $\mathcal{U}(A) > \mathcal{U}(B)$.

Since A is sure against B with causal independence of S_1, ..., S_n, for each S_i it would be rational to prefer A to B if S_i^* were known to hold. Therefore if S_i^* were known to hold, the $\mathcal{U}$-utility of A would be greater than that of B. Now the $\mathcal{U}$-utility that A would have if S_i^* were known is

$$\Sigma_O \ prob(A \ \Box\!\!\rightarrow O/S_i^*)\, \mathcal{D}O.$$

Call this $\mathcal{U}_i^*(A)$, and define $\mathcal{U}_i^*(B)$ in a like manner. We have supposed that for each S_i, $\mathcal{U}_i^*(A) > \mathcal{U}_i^*(B)$. Now by definition of the function $\mathcal{U}$,

$$\mathcal{U}(A) = \Sigma_O \ prob \ (A \ \Box\!\!\rightarrow O)\, \mathcal{D}O;$$

Since A is sure against B with causal independence of S_1, ..., S_n, it is known that $aut(S_1^*, \ldots, S_n^*)$ holds. By the probability calculus, then, for each outcome O

$$prob(A \ \Box\!\!\rightarrow O) = \Sigma_i \ prob(A \ \Box\!\!\rightarrow O/S_i^*)probS_i^*.$$

Therefore

$$\mathcal{U}(A) = \Sigma_O[\Sigma_i \, prob(A \boxminus\!\!\to O/S_i^*) \, probS_i^*] \, \mathcal{D}O$$
$$= \Sigma_i \, prob \, S_i^*[\Sigma_O \, prob(A \boxminus\!\!\to O/S_i^*) \, \mathcal{D}O],$$
$$= \Sigma_i \, \mathcal{U}_i^*(A) prob \, S_i^*.$$

By a like argument,

$$\mathcal{U}(B) = \Sigma_i \, \mathcal{U}_i^*(B) \, probS_i^*.$$

Since for each S_i, $\mathcal{U}_i^*(A) > \mathcal{U}_i^*(B)$, it follows that $\mathcal{U}(A) > \mathcal{U}(B)$, and the Assertion is proved.

Assertion. Suppose that in any possible circumstance, it is rational to prefer an act A to an act B iff the $\mathcal{V}$-utility of A is greater than that of B. Then sure-thing with stochastic independence holds.

Proof. Suppose A is sure against B with causal independence of S_1, ..., S_n, and that in any possible circumstances, it would be rational to prefer A to B iff A's $\mathcal{V}$-utility is greater than B's. The Assertion will be proved if we show from these assumptions that $\mathcal{V}(A) > \mathcal{V}(B)$.

Now since A is sure against B with stochastic independence of S_1, ..., S_n, for each S_i it would be rational to prefer A to B if S_i were known to hold. Therefore, if S_i were known to hold, then the $\mathcal{V}$-utility of A would be greater than that of B. Now the $\mathcal{V}$-utility that A would have if S_i were known to hold is

$$\Sigma_O \, prob(O/AS_i) \, \mathcal{D}O.$$

Call this $\mathcal{V}_i^*(A)$, and define $\mathcal{V}_i^*(B)$ correspondingly. We have that for each S_i, $\mathcal{V}_i^*(A) > \mathcal{V}_i^*(B)$. Now by definition of the function $\mathcal{V}$,

$$\mathcal{V}(A) = \Sigma_O \, prob(O/A) \, \mathcal{D}O.$$

Since A *is sure against* B with stochastic independence of S_1, ..., S_n, we have $prob \, (aut \, (S_1, \, ..., \, S_n)/A) = 1$, and so by the probability calculus, for each O,

$$\mathcal{V} \, prob(O/A) = \Sigma_i \, prob(O/AS_i) \, prob(S_i/A).$$

Hence

$$\mathcal{V}(A) = \Sigma_O[\Sigma_i \, prob(O/AS_i) \, prob(S_i/A)] \, \mathcal{D}O$$
$$= \Sigma_i \, prob(S_i/A)[\Sigma_O \, prob(O/AS_i) \, \mathcal{D}O]$$
$$= \Sigma_i \, prob(S_i/A) \, \mathcal{V}_i^*(A)$$

By a like argument,

$$\mathcal{V}(B) = \Sigma_i \, prob(S_i/B) \, \mathcal{V}_i^*(B).$$

Since for each S_i, $prob(S_i/A) = prob(S_i/B)$ and $\mathcal{V}_i^*(A) > \mathcal{V}_i^*(B)$ it follows that $\mathcal{V}(A) > \mathcal{V}(B)$, and the assertion is proved.

8. Two kinds of dominance

We have said that the principle of dominance is the sure-thing principle restricted to a special case, and that the sure-thing principle has two versions, one of which holds for $\mathcal{U}$-maximization and the other for $\mathcal{V}$-maximization. There should, then, be two versions of the principle of dominance, one for each kind of utility maximization. The principles can be formulated as follows.

> *Definition.* Let $S_1, \ldots, S_n$ be the states of a standard decision matrix, and let A and B be acts. Then A *strongly dominates* B *with respect to* $S_1, \ldots, S_n$ if for each S_i, the outcome of A in S_i is more desirable than the outcome of B in S_i.

Principle of dominance with causal independence. Suppose act A strongly dominates act B with respect to states $S_1, \ldots, S_n$. If for each state S_i, the agent knows that $(A \;\square\!\!\rightarrow S_i) \equiv S_i$ and $(B \;\square\!\!\rightarrow S_i) \equiv S_i$, then it is rational for him to prefer A to B.

Principle of dominance with stochastic independence. Suppose act A strongly dominates act B with respect to states $S_1, \ldots, S_n$. If for each state S_i, $prob\,(S_i/A) = prob\,(S_i) = prob\,(S_i/B)$, then it is rational for him to prefer A to B.

The principle of dominance with causal independence holds if rationality requires maximization of $\mathcal{U}$, and the principle of dominance with stochastic independence holds if rationality requires maximization of $\mathcal{V}$.[10]

Although these two principles are respective consequences of two principles of expected utility maximization which may conflict, they cannot themselves conflict. For suppose A strongly dominates B with respect to some set of states $S_1, \ldots, S_n$. Then the worst outcome of A is more desirable than some outcome of B. For the worst outcome of A is the outcome of A in some state S_i, and since A strongly dominates B with respect to $S_1, \ldots, S_n$, the outcome of A in S_i is more desirable than the outcome of B in S_i. Thus the worst outcome of A is more desirable than the worst outcome of B. It cannot be the case, then, that B strongly dominates A with respect to some other set of states $T_1, \ldots, T_n$. For if that indeed were the case, then, we have seen, the worst outcome of B would be more desirable than the worst outcome of A.

---

[10] Nozick (1969) in effect endorses the principle of dominance with stochastic independence (p. 127), but not $\mathcal{V}$-maximization: in cases of the kind we have been considering, he considers the recommendations of $\mathcal{V}$-maximization "perfectly wild" (p. 126). Nozick also states and endorses the principle of dominance with causal independence (p. 132).

We have seen that if A strongly dominates B with respect to a set of states, then there is no set of states with respect to which B strongly dominates A. For that reason, the two principles of dominance we have stated will never yield conflicting prescriptions for a simple decision problem.

In a weaker form, however, dominance indeed can be exploited to yield conflicting prescriptions.

Definition. Let $S_1, \ldots, S_n$ be the states of a standard decision matrix, and let A and B be acts. *A weakly dominates B with respect to $S_1, \ldots, S_n$* iff for each state S_i, the outcome of A in S_i is at least as desirable as the outcome of B in S_i, and for some state S_i with *prob* $(S_i) > 0$, the outcome of A in S_i is more desirable than the outcome of B in S_i.

We now get two principles of weak dominance by substituting "weakly dominates" for "strongly dominates" in the two principles of dominance stated above.

Case 4. A subject is presented with two boxes, one to the left and one to the right. He must choose between two acts:

A_L Take the box on the left.
A_R Take the box on the right.

The experimenter has already done one of the following.

M_{11} Place a million dollars in each box.
M_{01} Place a million dollars in the box on the right and nothing in the box on the left.
M_{00} Place nothing in either box.

He has definitely not placed money in the left box without placing money in the right box. Now the experimenter has predicted the behavior of the subject, and before making his prediction, he has used a random device to select one of the following three strategies.

1. Reward choice of left box: M_{11} if A_L is predicted; M_{00} if A_R is predicted.
2. Ensure payment: M_{11} if A_L is predicted; M_{01} if A_R is predicted.
3. Ensure non-payment: M_{01} if A_L is predicted; M_{00} if A_R is predicted.

The subject knows all this, and believes in the accuracy of the experimenter's predictions with complete certainty.

The principle of weak dominance with causal independence prescribes taking the box on the right. The three states M_{11}, M_{01}, and M_{00} are causally independent of the act the subject performs. The possible outcomes are shown in the table, where 1 is getting the million dollars and 0 is not getting it.

	M_{11}	M_{01}	M_{00}
A_L	1	0	0
A_R	1	1	0

M_{01} has non-zero probability, since if A_L was predicted it would result from the experimenter's using strategy (3) and if A_R was predicted, it would result from the experimenter's using strategy (2). Thus A_R weakly dominates A_L with respect to M_{11}, M_{01}, M_{00}, and the principle of weak dominance with causal independence prescribes taking the box on the right.

The principle of weak dominance with stochastic independence, in contrast, prescribes taking the box on the left.

The possibilities can be partitioned as follows:

S_1 the experimenter predicts correctly and follows strategy (1).
S_2 S_1 does not hold and the subject wins a million dollars.
S_3 S_1 does not hold and the subject wins nothing.

The payoffs are given in the table.

	S_1	S_2	S_3
A_L	1	1	0
A_R	0	1	0

Now *prob* $(S_1) \neq 0$, and hence A_L weakly dominates A_R with respect to S_1, S_2, S_3. Moreover, the states S_1, S_2, and S_3 are stochastically independent of A_L and A_R. For the subject knows that the experimenter has selected his strategy independently of his prediction, by means of a random device; hence learning that he was about to perform A_L, say, would not affect the probability he ascribes to the experimenter's having had any given strategy. By the subject's probability function, then, which strategy the experimenter has used is stochastically independent of the subject's act. Now the subject believes that the experimenter has predicted correctly and used strategy (1), (2), or (3). Hence he thinks that S_1 holds iff the experimenter has used strategy (1), that S_2 holds iff the experimenter has used strategy (2), and that S_3 holds if the experimenter has used strategy (3). Hence under his probability function, states S_1, S_2, and S_3 are stochastically independent of A_L and A_R. Thus the principle of weak dominance with stochastic independence applies, and it prescribes taking the box on the left.

Some readers may object in Case 4 to the subject's complete certainty that the experimenter has predicted correctly. It is possible to construct a conflict between the two principles of weak dominance without requiring such certainty, but the example becomes more complicated.

Case 5. Same as Case 4, except for the following.

The subject ascribes a probability of 0.8 to the experimenter's having predicted correctly, and this probability is independent of the subject's choice of A_L or A_R. Thus where C is "the experimenter has predicted correctly," $prob(C/A_L) = 0.8$ and $prob(C/A_R) = 0.8$.

The experimenter has chosen among the following three strategies by means of a random device.

1. M_{11} if A_L is predicted; M_{00} if A_R is predicted.
2.* M_{11} if A_L is predicted; M_{01} or M_{00}, with equal probability, if A_R is predicted.
3.* M_{11} or M_{01}, with equal probability, if A_L is predicted; M_{00} if A_R is predicted.

He has followed (1) with a probability 0.5, (2*) with a probability 0.25, and (3*) with a probability 0.25.

In Case 5, as in Case 4, the states M_{11}, M_{01}, and M_{00}, are causally independent of the acts A_R and A_L, and from the principle of weak dominance with causal independence and the facts of the case, it follows that it is rational to prefer A_R to A_L.

Now let states S_1, S_2, and S_3 be as before: S_1 is that the experimenter predicts correctly and follows strategy (1); S_2 is that S_1 does not hold and the subject receives a million dollars; S_3 is that S_1 does not hold and the subject receives nothing. As in Case 4, if S_1, S_2, and S_3 are stochastically independent of A_L and A_R, then from the principle of weak dominance with stochastic independence and the facts of the case, it follows that it is rational to prefer A_L to A_R. It is clear that S_1 is stochastically independent of the acts A_L and A_R; we now show that S_2 and S_3 are as well: that $prob(S_2/A_L) = prob(S_2/A_R)$ and $prob(S_3/A_L) = prob(S_3/A_R)$.

There are two possible acts, two possible experimenter's predictions, and three possible experimenter's strategies, some of which may involve the flip of a coin. Call a combination of act, prediction, experimenter's strategy, and result of coin flip if it matters, a *case*. For each case, the Table 17. 2 shows.

1. The state M_{11}, M_{01}, or M_{00} which would hold in that case.
2. The conditional probability of the case given the act.
3. The outcome in that case: 1 for getting the million dollars, 0 for not.
4. The state S_1, S_2, or S_3 which holds in that case.

The conditional probability $prob(S_2/A_L)$ is then obtained by adding up

Table 17.2

	A_L Performed		A_R Performed	
	A_L Predicted 0.8	A_R Predicted 0.2	A_L Predicted 0.2	A_R Predicted 0.8
Strategy (1) 0.5	M_{11} 0.4 / 1 S_1	M_{00} 0.1 / 0 S_3	M_{11} 0.1 / 1 S_2	M_{00} 0.4 / 0 S_1
Strategy (2*) 0.25		M_{01} 0.025		M_{01} 0.1
	M_{11} 0.2 / 1 S_2	0 S_3	M_{11} 0.05 / 1 S_2	1 S_2
		M_{00} 0.025 / 0 S_3		M_{00} 0.1 / 0 S_3
Strategy (3*) 0.25	M_{11} 0.1 / 1 S_2	M_{00} 0.05 / 0 S_3	M_{11} 0.025 / 1 S_2	M_{00} 0.2 / 0 S_3
	M_{01} 0.1 / 0 S_3		M_{01} 0.025 / 1 S_2	
Totals	$prob(S_2/A_L) = 0.3$ $prob(S_3/A_L) = 0.3$		$prob(S_2/A_R) = 0.3$ $prob(S_3/A_R) = 0.3$	

the conditional probabilities given A_L of cases in which S_2 holds; a like procedure gives $prob(S_3/A_L)$, $prob(S_2/A_R)$, and $prob(S_3/A_R)$.

The conclusion of Table 17.2 is that the states S_1, S_2, and S_3 are indeed epistemically independent of the acts A_L and A_R. Since A_L weakly dominates A_R with respect to states S_1, S_2, and S_3, it follows that A_L weakly dominates A_R with respect to stochastically independent states. We already know that A_R weakly dominates A_L with respect to causally independent states M_{11}, M_{01}, and M_{00}. In Case 5, then, the two principles of weak dominance are in conflict.

9. Act-independence in the Savage formulation

In Section 4, we said that to apply the Savage framework to a decision problem, one must find states of the world that are in some sense act-independent. In the last section, we distinguished two kinds of independence, causal and epistemic. Which kind is needed in the Savage formulation of decision problems?

The answer is that the Savage formulation has both a $\mathscr{U}$-maximizing interpretation and a $\mathscr{V}$-maximizing interpretation. On the $\mathscr{U}$-maximizing interpretation, the states must be causally independent of the acts, whereas on the γ-maximizing interpretation, the states must be epistemically independent of the acts. That is to say, if the states are causally act-independent, then utility as calculated by the Savage method is $\mathscr{U}$-utility, whereas if the states are epistemically act-independent, then utility as calculated by the Savage method is γ-utility. If the states are both causally and epistemically act-independent, then the $\mathscr{U}$-utility of each act equals its γ-utility. Thus the Savage formulation itself is not committed to either kind of utility: the kind of utility it yields depends on the way it is applied to decision problems.

The expected utility of an act A in the Savage theory is

$$\Sigma_S \, prob \, (S) \mathscr{D} O(A, S). \tag{3}$$

If the states S are all known to be causally independent of A, so that for each state S, the agent knows that $(A \,\square\!\!\rightarrow S) \equiv S$, then for each S, we have $prob \, (S) = prob(A \,\square\!\!\rightarrow S)$. (3) thus becomes

$$\Sigma_S \, prob \, (A \,\square\!\!\rightarrow S) \mathscr{D} O(A, S),$$

and this, we said in Section 3, is $\mathscr{U}(A)$. If, on the other hand, the states S are stochastically independent of A, so that for each S, $prob(S) = prob(S/A)$, then (3) becomes

$$\Sigma_S \, prob \, (S/A) \mathscr{D} O(A, S),$$

which is $\gamma(A)$.

10. Newcomb's problem

The Newcomb paradox discussed by Nozick (1969) has the same structure as the case of Solomon discussed in Section 3. Nozick treats it as a conflict between the principle of expected utility maximization and the principle of dominance. On the views we have propounded in this paper, the problem is rather a conflict between two kinds of expected utility maximization. The problem is this. There are two boxes, transparent and opaque; the transparent box contains a thousand dollars. The agent can perform A_1, taking just the contents of the opaque box, or A_2, taking the contents of both boxes. A predictor has already placed a million dollars in the opaque box if he predicted A_1 and nothing if he predicted A_2. The agent knows all this, and he knows the predictor to be highly reliable in that both $prob$(he has predicted A_1/A_1) and $prob$(he has predicted A_2/A_2) are close to one.

To show how the expected utility calculations work, we must add detail to the specification of the situation. Suppose, somewhat unrealistically, that getting no money has a utility of zero, getting $1000 a utility of 10, that getting $1,000,000 has a utility of 100, and that getting $1,001,000 has a utility of 101. Let M be 'there are a million dollars in the opaque box', and suppose $prob$ $(M/A_1) = 0.9$ and $prob$ $(M/A_2) = 0.1$. The calculation of $\mathscr{V}(A_1)$ and $\mathscr{V}(A_2)$ is familiar.

$$\mathscr{V}(A_1) = prob(M/A_1)\,\mathscr{D}\$1{,}000{,}000 + prob(\bar{M}/A_1)\,\mathscr{D}\$0$$
$$= 0.9(100) + 0.1(0) = 90.$$

$$\mathscr{V}(A_2) = prob(M/A_2)\,\mathscr{D}\$1{,}001{,}000 + prob(\bar{M}/A_2)\,\mathscr{D}\$1000$$
$$= 0.1(101) + 0.9(10) = 19.1$$

Maximization of $\mathscr{V}$, as is well known, prescribes taking only the contents of the opaque box.[11]

$\mathscr{U}(A_1)$ and $\mathscr{U}(A_2)$ depend on the probability of M, which in turn depends on the probabilities of A_1 and A_2. For any probability of M, though, we have $\mathscr{U}(A_2) > \mathscr{U}(A_1)$. For let the probability of M be μ; then since M is causally act-independent, $prob(A_1 \;\Box\!\!\rightarrow M) = \mu$ and $prob(A_2 \;\Box\!\!\rightarrow M) = \mu$. Therefore

$$\mathscr{U}(A_1) = prob(A_1 \;\Box\!\!\rightarrow M)\,\mathscr{D}\$1{,}000{,}000 + prob(A_1 \;\Box\!\!\rightarrow \bar{M})\,\mathscr{D}\$0$$
$$= 100\mu + 0(1 - \mu) = 100\mu.$$

$$\mathscr{U}(A_2) = prob(A_2 \;\Box\!\!\rightarrow M)\,\mathscr{D}\$1{,}001{,}000 + prob(A_2 \;\Box\!\!\rightarrow \bar{M})\,\mathscr{D}\$1000$$
$$= 101\mu + 10(1 - \mu) = 91\mu + 10.$$

Thus $\mathscr{U}(A_2) - \mathscr{U}(A_1) = 10 - 9\mu$, and since $\mu \leqslant 1$, this is always positive. Therefore whatever probability M may have, $\mathscr{U}(A_2) > \mathscr{U}(A_1)$, and $\mathscr{U}$-maximization prescribes taking both boxes.

To some people, this prescription seems irrational.[12] One possible argument against it takes roughly the form 'If you're so smart, why ain't you rich?' $\mathscr{V}$-maximizers tend to leave the experiment millionaires whereas $\mathscr{U}$-maximizers do not. Both very much want to be millionaires, and the $\mathscr{V}$-maximizers usually succeed; hence it must be the $\mathscr{V}$-maximizers who are making the rational choice. We take the moral of the paradox to be something else: If someone is very good at predicting behavior and rewards predicted irrationality richly; then irrationality will be richly rewarded.

To see this, consider a variation on Newcomb's story: the subject of the experiment is to take the contents of the opaque box first and learn what it is; he then may choose either to take the thousand dollars in the second box or not to take it. The predictor has an excellent record, and a thoroughly accepted theory to back it up. Most people find

[11] For $\mathscr{V}$-maximizing treatments of Newcomb's problem, see Bar-Hillel and Margalit (1972) and Levi (1975).
[12] Levi (1975) reconstructs Nozick's argument for taking both boxes in a way which uses $prob(M)$ rather than $prob(M/A_1)$ and $prob(M/A_2)$ as the appropriate probabilities for computing expected utility in Newcomb's problem. This agrees with $\mathscr{U}$-maximizing in that the same probabilities are used for computing expected utility for A_1 as for A_2, and results in the same recommendation to take both boxes. Levi is one of the people to whom this recommendation seems irrational.

nothing in the first box and then take the contents of the second box. Of the million subjects tested, 1 percent have found a million dollars in the first box, and strangely enough only 1 percent of these – 100 in 10,000 – have gone on to take the thousand dollars they could each see in the second box. When those who leave the thousand dollars are later asked why they do so, they say things like "If I were the sort of person who would take the thousand dollars in that situation, I wouldn't be a millionaire."

On both grounds of $\mathcal{U}$-maximization and of $\mathcal{V}$-maximization, these new millionaires have acted irrationally in failing to take the extra thousand dollars. They know for certain that they have the million dollars; therefore the $\mathcal{V}$-utility of taking the thousand as well is 101, whereas the $\mathcal{V}$-utility of not taking it is 100. Even on the view of $\mathcal{V}$-maximizers, then, this experiment will almost always make irrational people and only irrational people millionaires. Everyone knows so at the outset.

Return now to the unmodified Newcomb situation, where the subject must take or pass up the thousand dollars before he sees whether the opaque box is full or empty. What happens if the subject knows not merely that the predictor is highly reliable, but that he is infallible? The argument that the $\mathcal{U}$-utility of taking both boxes exceeds that of taking only one box goes through unchanged. To some people, however, it seems especially apparent in this case that it is rational to take only the opaque box and irrational to take both. For in this case the subject is certain that he will be a millionaire if and only if he takes only the opaque box. If in the case where the predictor is known to be infallible it is irrational to take both boxes, then, $\mathcal{U}$-maximization is not always the rational policy.

We maintain that $\mathcal{U}$-maximization is rational even in the case where the predictor is known to be infallible. True, where R is "I become a millionaire", the agent knows in this case that R holds if A_1 holds: he knows the truth-functional proposition $R \equiv A_1$. From this proposition, however, it does not follow that he *would* be a millionaire if he did A_1, or that he *would* be a non-millionaire if he did A_2.

If the subject knows for sure that he will take just the opaque box, then he knows for sure that the million dollars is in the opaque box, and so he knows for sure that he will be a millionaire. But since he knows for sure that the million dollars is already in the opaque box, he knows for sure that even if he were to take both boxes, he would be a millionaire. If, on the other hand, the subject knows for sure that he will take both boxes, then he knows for sure that the opaque box is empty, and so he knows for sure that he will be a non-millionaire. But since in this case he knows for sure that the opaque box is empty, he knows for sure that even if he were to take just the opaque box, he would be a non-millionaire.

If the subject does not know what he will do, then what he knows is this: either he will take just the opaque box and be a millionaire, or he will take both boxes and be a non-millionaire. From this, however, it follows neither that (1) if he took just the opaque box, he would be a millionaire, nor that (2) if he took both boxes he would be a non-millionaire. For (1), the subject knows, is true iff the opaque box is filled with a million dollars, and (2), the subject knows, is true iff the opaque box is empty. Thus, if (1) followed from what the agent knows, he could conclude for certain that the opaque box contains a million dollars, and if (2) followed from what the agent knows, he could conclude that the opaque box is empty. Since the subject, we have supposed, does not know what he will do, he can conclude neither that the opaque box contains a million dollars nor that it is empty. Therefore neither (1) nor (2) follows from what the subject knows.

Rational choice in Newcomb's situation, we maintain, depends on a comparison of what would happen if one took both boxes with what would happen if one took only the opaque box. What the agent knows for sure is this: if he took both boxes, he would get a thousand dollars more than he would if he took only the opaque box. That, on our view, makes it rational for someone who wants as much much as he can get to take both boxes, and irrational to take only one box.

Why, then, does it seem obvious to many people that if the predictor is known to be infallible, it is rational to take only the opaque box and irrational to take both boxes? We have three possible explanations. The first is that a person may have a tendency to want to bring about an indication of a desired state of the world, even if it is known that the act that brings about the indication in no way brings about the desired state itself. Taking just the opaque box would be a sure indication that it contained a million dollars, even though taking just the opaque box in no way brings it about that the box contains a million dollars.

The second possible explanation lies in the force of the argument "If you're so smart, why ain't you rich?" That argument, though, if it holds good, should apply equally well to the modified Newcomb situation, with a predictor who is known to be highly accurate but fallible. There the conclusion of the argument seems absurd: according to the argument, having already received the million dollars, one should pass up the additional thousand dollars one is free to take, on the grounds that those who are disposed to pass it up tend to become millionaires. Since the argument leads to an absurd conclusion in one case, there must be something wrong with it.

The third possible explanation is the fallacious inference we have just discussed, from

Either I shall take one box and be a
millionaire, or I shall take both boxes
and be a non-millionaire

to the conclusion

If I were to take one box, I would be a
millionaire, and if I were to take both
boxes, I would be a non-millionaire.

If, to someone who is free of fallacies, it is still intuitively apparent
that the subject should take only the opaque box, we have no further
arguments to give him. If in addition he thinks the subject should take
only the opaque box even in the case where the predictor is known to
be somewhat fallible, if he also thinks that in the modified Newcomb
situation the subject, on receiving the extra million dollars, should take
the extra thousand, if he also thinks that it is rational for Reoboam to
be severe, and if he also thinks it is rational for Solomon to abstain
from his neighbor's wife, then he may genuinely have the intuitions of
a $\mathcal{V}$-maximizer: $\mathcal{V}$-maximization then provides a systematic account of
his intuitions. If he thinks some of these things but not all of them,
then we leave it to him to provide a systematic account of his views.
Our own views are systematically accounted for by $\mathcal{U}$-maximization.

11. Stability of decision

When a person decides what to do, he has in effect learned what he
will do, and so he has new information. He will adjust his probability
ascriptions accordingly. These adjustments may affect the $\mathcal{U}$-utility of
the various acts open to him.

Indeed, once the person decides to perform an act A, the $\mathcal{U}$-utility of
A will be equal to its $\mathcal{V}$-utility.[13] Or at least this holds if Consequence 1
in Section 2, that $A \supset [(A \;\square\!\!\rightarrow C) \equiv C]$, is a logical truth. For we saw in
the proof of Assertion 1 that if Consequence 1 is a logical truth, then
for any pair of propositions P and Q, $prob(D \;\square\!\!\rightarrow Q/P) = prob(Q/P)$.
Now let $\mathcal{U}_A(A)$ be the $\mathcal{U}$-utility of act A as reckoned by the agent after
he has decided for sure to do A, let $prob$ give the agent's probability
ascriptions before he has decided what to do. Let $prob_A$ give the agent's
probability ascriptions after he has decided for sure to do A. Then for
any proposition P, $prob_A(P) = prob(P/A)$. Thus

$$
\begin{aligned}
\mathcal{U}_A(A) &= \Sigma_O \; prob_A(A \;\square\!\!\rightarrow O)\,\mathcal{D}O \\
&= \Sigma \; prob(A \;\square\!\!\rightarrow O/A)\,\mathcal{D}O \\
&= \Sigma_O \; prob(O/A)\,\mathcal{D}O \\
&= \mathcal{V}(A).
\end{aligned}
$$

The $\mathcal{V}$-utility of an act, then, is what its $\mathcal{U}$-utility would be if the agent
knew he were going to perform it.

[13] We owe this point to Barry O'Neill.

It does not follow that once a person knows what he will do, γ-maximization and $\mathscr{U}$-maximization give the same prescriptions. For although for any act A, $\mathscr{U}_A(A) = \gamma(A)$, it is not in general true that for alternatives B to A, $\mathscr{U}_A(B) = \gamma(B)$. Thus in a case where $\mathscr{U}(A) < \mathscr{U}(B)$ but $\gamma(A) > \gamma(B)$, it is consistent with what we have said to suppose that $\mathscr{U}_A(A) < \mathscr{U}_A(B)$. In such a case, γ-maximization prescribes A regardless of what the agent believes he will do, but even if he believes he will do A, $\mathscr{U}$-maximization prescribes B. The situation is this:

$$\mathscr{U}_A(B) > \mathscr{U}_A(A) = \gamma(A) > \gamma(B).$$

Even though, once an agent knows what he will do, the distinction between the $\mathscr{U}$-utility of that act and its γ-utility disappears, the distinction between $\mathscr{U}$-maximization and γ-maximization remains.

That deciding what to do can affect the $\mathscr{U}$-utilities of the acts open to an agent raises a problem of stability of decision for $\mathscr{U}$-maximizers. Consider the story of the man who met death in Damascus.[14] Death looked surprised, but then recovered his ghastly composure and said, "I am coming for you tomorrow". The terrified man that night bought a camel and rode to Aleppo. The next day, death knocked on the door of the room where he was hiding and said "I have come for you".

"But I thought you would be looking for me in Damascus", said the man.

"Not at all", said death "that is why I was surprised to see you yesterday. I knew that today I was to find you in Aleppo".

Now suppose the man knows the following. Death works from an appointment book which states time and place; a person dies if and only if the book correctly states in what city he will be at the stated time. The book is made up weeks in advance on the basis of highly reliable predictions. An appointment on the next day has been inscribed for him. Suppose, on this basis, the man would take his being in Damascus the next day as strong evidence that his appointment with death is in Damascus, and would take his being in Aleppo the next day as strong evidence that his appointment is in Aleppo.

Two acts are open to him: A, go to Aleppo, and D, stay in Damascus. There are two possibilities: S_A, death will seek him in Aleppo, and S_D, death will seek him in Damascus. He knows that death will find him if and only if death looks for him in the right city, so that, where L is that he lives, he knows $(D \;\square\!\!\rightarrow L) \equiv S_A$ and $(A \;\square\!\!\rightarrow L) \equiv S_D$. He ascribes conditional probabilities $prob(S_A/A) \approx 1$ and $prob(S_D/D) \approx 1$; suppose these are both 0.99 and that $prob(S_D/A) = 0.01$

---

[14] A version of this story quoted from Somerset Maugham's play *Sheppey* (New York, Doubleday 1934) appears on the facing page of John O'Hara's novel *Appointment in Samarra*. (New York, Random House 1934). The story is undoubtedly much older.

and *prob* $(S_A/D) = 0.01$. Suppose $\mathscr{D}(\bar{L}) = -100$ and $\mathscr{D}(L) = 0$. Then where α is *prob*(A), his probability of going to Aleppo, and $1 - \alpha$ is his probability of going to Damascus,

$$prob(A \:\square\!\!\rightarrow L) = prob(S_D) = \alpha \; prob(S_D/A) + (1 - \alpha) \; prob(S_D/D)$$
$$= 0.01\alpha + 0.99(1 - \alpha) = 0.99 - 0.98\alpha$$

$$prob(A \:\square\!\!\rightarrow \bar{L}) = prob(S_A) = 1 - prob(S_D) = 0.01 + 0.98\alpha.$$

Thus

$$\mathscr{U}(A) = prob(A \:\square\!\!\rightarrow L)\,\mathscr{D}(L) + prob(A \:\square\!\!\rightarrow \bar{L})\,\mathscr{D}(\bar{L})$$
$$= (0.01 + 0.98\alpha)\,(-100) = -1 - 98\alpha.$$

By a like calculation, $\mathscr{U}(D) = -99 + 98\alpha$. Thus if $\alpha = 1$, then $\mathscr{U}(D) = -1$ and $\mathscr{U}(A) = -99$, and thus $\mathscr{U}(D) > \mathscr{U}(A)$. If $\alpha = 0$ then $\mathscr{U}(D) = -99$ and $\mathscr{U}(A) = -1$, so that $\mathscr{U}(A) > \mathscr{U}(D)$. Indeed we have $\mathscr{U}(D) > \mathscr{U}(A)$ whenever *prob* $(A) > \frac{1}{2}$, and $\mathscr{U}(A) > \mathscr{U}(D)$ whenever *prob* $(D) > \frac{1}{2}$.

What are we to make of this? If the man ascribes himself equal probabilities of going to Aleppo and staying in Damascus, he has equal grounds for thinking that death intends to seek him in Damascus and that death intends to seek him in Aleppo. If, however, he decides to go to Aleppo, he then has strong grounds for expecting that Aleppo is where death already expects him to be, and hence it is rational for him to prefer staying in Damascus. Similarly, deciding to stay in Damascus would give him strong grounds for thinking that he ought to go to Aleppo: once he knows he will stay in Damascus, he can be almost sure that death already expects him in Damascus, and hence that if he had gone to Aleppo, death would have sought him in vain.

$\mathscr{V}$-maximization does not lead to such instability. What happens to $\mathscr{V}$-utility when an agent knows for sure what he will do is somewhat unclear. Standard probability theory offers no interpretation of $prob_A(O/B)$ where $prob(B/A) = 0$, and so on the standard theory, once an agent knows for sure what he will do, the $\mathscr{V}$-utility of the alternatives ceases to be well-defined. What we can say about $\mathscr{V}$-utility is this: as long as an act's being performed has non-zero probability, its $\mathscr{V}$-utility is independent of its probability and the probabilities of alternatives to it. For the $\mathscr{V}$-utility of an act A depends on conditional probabilities of the form $prob(O/A)$. This is just the probability the agent would ascribe to O on learning A for sure, and that is independent of how likely he now regards A. Whereas, then, the $\mathscr{U}$-utility of an act may vary with its probability of being performed, its $\mathscr{V}$-utility does not. $\mathscr{U}$-maximization, then, may give rise to a kind of instability which $\mathscr{V}$-maximization precludes: in certain cases, an act will be $\mathscr{U}$-maximal if and only if the probability of its performance is low.

Is this a reason for preferring $\mathscr{V}$-maximization? We think not. In the case of death in Damascus, rational decision does seem to be unstable. Any reason the doomed man has for thinking we will go to Aleppo is a reason for thinking he would live longer if he stayed in Damascus, and any reason he has for thinking he will stay in Damascus is reason for thinking he would live longer if he went to Aleppo. Thinking he will do one is reason for doing the other. That there can be cases of unstable $\mathcal{U}$-maximization seems strange, but the strangeness lies in the cases, not in $\mathcal{U}$-maximization: instability of rational decision seems to be a genuine feature of such cases.

12. Applications to game theory

Game theory provides many cases where $\mathcal{U}$-maximizing and $\mathcal{V}$-maximizing diverge; perhaps the most striking of these is the prisoner's dilemma, for which a desirability matrix is shown.

	A_0	A_1
B_0	1 / 1	0 / 10
B_1	10 / 0	9 / 9

Here A_0 and B_0 are respectively A's and B's options of confessing, while A_1 and B_1 are the options of not confessing. The desirabilities reflect these facts: (1) if both confess, they both get long prison terms; (2) if one confesses and the other doesn't, then the confessor gets off while the other gets an even longer prison term; (3) if neither confesses, both get off with very light sentences.

Suppose each prisoner knows that the other thinks in much the same way he does. Then his own choice gives him evidence for what the other will do. Thus, the conditional probability of a long prison term on his confessing is greater than the conditional probability of a long prison term on his not confessing. If the difference between these two conditional probabilities is sufficiently great, then $\mathcal{V}$-maximizing will prescribe not confessing.

The $\mathcal{V}$-utilities of the acts open to B will be as follows:

$$\mathcal{V}(B_0) = prob(A_0/B_0) \times 1 + prob(A_1/B_0) \times 10$$
$$\mathcal{V}(B_1) = prob(A_0/B_1) \times 0 + prob(A_1/B_1) \times 9.$$

If $prob(A_1/B_1) - prob(A_1/B_0)$ is sufficiently great (in this case 1/9 or more), then $\mathcal{V}$-maximizing recommends that B take option B_1 and not confess. If the probabilities for A are similar, then $\mathcal{V}$-maximizing also recommends not confessing for A. The outcome if both $\mathcal{V}$-maximize is A_1B_1, the optimal one of mutual co-operation.[15]

For a $\mathcal{U}$-maximizer, dominance applies because his companion's choice is causally independent of his own. Therefore, $\mathcal{U}$-maximizing yields the classical outcome of the prisoner's dilemma. This suggests that $\mathcal{U}$-maximizing and not $\mathcal{V}$-maximizing corresponds to the kind of utility maximizing commonly assumed in game theory.

[15] Nozick (1969), Brams (1975), Grofman (1975) and Rapoport (1975), have all suggested a link between Newcomb's problem and the Prisoner's Dilemma. Brams, Grofman and Rapport all endorse co-operative solutions, Rapoport (1975, p. 619) appears to endorse $\mathcal{V}$-maximizing.

18. Causal decision theory

David Lewis

1. Introduction

Decision theory in its best-known form (as presented, for instance, in Jeffrey, 1965) manages to steer clear of the thought that what's best to do is what the agent believes will most tend to cause good results. Causal relations and the like go unmentioned. The theory is simple, elegant, powerful, and conceptually economical. Unfortunately it is not quite right. In a class of somewhat peculiar cases, called Newcomb problems, this noncausal decision theory gives the wrong answer. It commends an irrational policy of managing the news so as to get good news about matters which you have no control over.

I am one of those who have concluded that we need an improved decision theory, more sensitive to causal distinctions. Noncausal decision theory will do when the causal relations are right for it, as they very often are, but even then the full story is causal. Several versions of causal decision theory are on the market in the works of Gibbard and Harper (1978, Chapter 17, this volume), Skyrms (1980), and Sobel (1978), and I shall put forward a version of my own. But also I shall suggest that we causal decision theorists share one common idea, and differ mainly on matters of emphasis and formulation. The situation is not the chaos of disparate approaches that it may seem.

Of course there are many philosophers who understand the issues

This paper is based on a talk given at a conference on Conditional Expected Utility at the University of Pittsburgh in November 1978. It has benefited from discussions and correspondence with Nancy Cartwright, Allan Gibbard, William Harper, Daniel Hunter, Frank Jackson, Richard Jeffrey, Gregory Kavka, Reed Richter, Brian Skyrms, J. Howard Sobel, and Robert Stalnaker.

Reprinted from *Australasian Journal of Philosophy*, 59 (1981), pp. 5–30, by kind permission of the author and the publisher.

very well, and yet disagree with me about which choice in a Newcomb problem is rational. This paper is about a topic that does not arise for them. Noncausal decision theory meets their needs and they want no replacement. I will not enter into debate with them, since that debate is hopelessly deadlocked and I have nothing new to add to it. Rather, I address myself to those who join me in presupposing that Newcomb problems show the need for some sort of causal decision theory, and in asking what form that theory should take.

2. Preliminaries: credence, value, options

Let us assume that a (more or less) rational agent has, at any moment, a *credence* function and a *value* function. These are defined in the first instance over single possible worlds. Each world W has a credence $C(W)$, which measures the agent's degree of belief that W is the actual world. These credences fall on a scale from zero to one, and they sum to one. Also each world W has a value $V(W)$, which measures how satisfactory it seems to the agent for W to be the actual world. These values fall on a linear scale with arbitrary zero and unit.

We may go on to define credence also for sets of worlds. We call such sets *propositions*, and we say that a proposition *holds* at just those worlds which are its members. I shall not distinguish in notation between a world W and a proposition whose sole member is W, so all that is said of propositions shall apply also to single worlds. We sum credences: for any proposition X,

$$C(X) =^{df} \Sigma_{W \epsilon X} C(W).$$

We define conditional credences as quotients of credences, defined if the denominator is positive:

$$C(X/Y) =^{df} C(XY)/C(Y),$$

where XY is the conjunction (intersection) of the propositions X and Y. If $C(Y)$ is positive, then $C(-/Y)$, the function that assigns to any world W or proposition X the value $C(W/Y)$ or $C(X/Y)$, is itself a credence function. We say that it *comes from* C *by conditionalising on* Y. Conditionalising on one's total evidence is a rational way to learn from experience. I shall proceed on the assumption that it is the only way for a fully rational agent to learn from experience; however, nothing very important will depend on that disputed premise.

We also define (expected) value for propositions. We take credence-weighted averages of values of worlds: for any proposition X,

$$V(X) =^{df} \Sigma_W C(W/X)V(W) = \Sigma_{W \epsilon X} C(W)V(W)/C(X).$$

A *partition* (or a *partition of* X) is a set of propositions of which exactly one holds at any world (or at any X-world). Let the variable Z range over any partition (in which case the XZ's, for fixed X and varying Z, are a partition of X). Our definitions yield the following *Rules of Additivity* for credence, and for the product of credence and expected value:

$$C(X) = \Sigma_Z C(XZ),$$
$$C(X)V(X) = \Sigma_Z C(XZ)V(XZ). \tag{1}$$

This *Rule of Averaging* for expected values follows:

$$V(X) = \Sigma_Z C(Z/X)V(XZ). \tag{2}$$

Thence we can get an alternative definition of expected value. For any number v, let $[V = v]$ be the proposition that holds at just those worlds W for which $V(W)$ equals v. Call $[V = v]$ a *value-level proposition*. Since the value-level propositions are a partition,

$$V(X) = \Sigma_v C([V = v]/X)v. \tag{3}$$

I have idealized and oversimplified in three ways, but I think the dodged complications make no difference to whether, and how, decision theory ought to be causal. First, it seems most unlikely that any real person could store and process anything so rich in information as the C and V functions envisaged. We must perforce make do with summaries. But it is plausible that someone who really did have these functions to guide him would not be so very different from us in his conduct, apart from his supernatural prowess at logic and mathematics and *a priori* knowledge generally. Second, my formulation makes straightforward sense only under the fiction that the number of possible worlds is finite. There are two remedies. We could reformulate everything in the language of standard measure theory, or we could transfer our simpler formulations to the infinite case by invoking nonstandard summations of infinitesimal credences. Either way the technicalities would distract us, and I see little risk that the fiction of finitude will mislead us. Third, a credence function over possible worlds allows for partial beliefs about the way the world is, but not for partial beliefs about who and where and when in the world one is. Beliefs of the second sort are distinct from those of the first sort; it is important that we have them; however they are seldom very partial. To make them partial we need either an agent strangely lacking in self-knowledge, or else one who gives credence to strange worlds in which he has close duplicates. I here ignore the decision problems of such strange agents.[1]

Let us next consider the agent's options. Suppose we have a

---

[1] I consider them in (1979a, especially p. 534). There, however, I ignore the causal aspects of decision theory. I trust there are no further problems that would arise from merging the two topics.

partition of propositions that distinguish worlds where the agent acts differently (he or his counterpart, as the case may be). Further, he can act at will so as to make any one of these propositions hold; but he cannot act at will so as to make any proposition hold that implies but is not implied by (is properly included in) a proposition in the partition. The partition gives the most detailed specifications of his present action over which he has control. Then this is the partition of the agents' alternative *options*.[2] (Henceforth I reserve the variable A to range over these options.) Say that the agent *realises* an option iff he acts in such a way as to make it hold. Then the business of decision theory is to say which of the agent's alternative options it would be rational for him to realise.

All this is neutral ground. Credence, value, and options figure both in noncausal and in causal decision theory, though of course they are put to somewhat different uses.

3. Noncausal decision theory

Noncausal decision theory needs no further apparatus. It prescribes the rule of V-maximising, according to which a rational choice is one that has the greatest expected value. An option A is V-*maximal* iff $V(A)$ is not exceeded by any $V(A')$, where A' is another option. The theory says that to act rationally is to realise some V-maximal option.

Here is the guiding intuition. How would you like to find out that A holds? Your estimate of the value of the actual world would then be $V(A)$, if you learn by conditionalising on the news that A. So you would like best to find out that the V-maximal one of the A's holds (or one of the V-maximal ones, in case of a tie). But it's in your power to find out that whichever one you like holds, by realising it. So go ahead – find out whichever you'd like best to find out! You make the news, so make the news you like best.

This seeking of good news may not seem so sensible, however, if it turns out to get in the way of seeking good results. And it does.

4. Newcomb problems

Suppose you are offered some small good, take it or leave it. Also you may suffer some great evil, but you are convinced that whether you suffer it or not is entirely outside your control. In no way does it depend causally on what you do now. No other significant payoffs are at stake. Is it rational to take the small good? Of course, say I.

[2] They are his narrowest options. Any proposition implied by one of them might be called an option for him in a broader sense, since he could act at will so as to make it hold. But when I speak of options, I shall always mean the narrowest options.

I think enough has been said already to settle that question, but there is some more to say. Suppose further that you think that some prior state, which may or may not obtain and which also is entirely outside your control, would be conducive both to your deciding to take the good and to your suffering the evil. So if you take the good, that will be evidence that the prior state does obtain and hence that you stand more chance than you might have hoped of suffering the evil. Bad news! But is that any reason not to take the good? I say not, since if the prior state obtains, there's nothing you can do about it now. In particular, you cannot make it go away by declining the good, thus acting as you would have been more likely to act if the prior state had been absent. All you accomplish is to shield yourself from the bad news. That is useless. (*Ex hypothesi*, dismay caused by the bad news is not a significant extra payoff in its own right. Neither is the exhilaration or merit of boldly facing the worst.) To decline the good lest taking it bring bad news is to play the ostrich.

The trouble with noncausal decision theory is that it commends the ostrich as rational. Let G and $-G$ respectively be the propositions that you take the small good and that you decline it; suppose for simplicity that just these are your options. Let E and $-E$ respectively be the propositions that you suffer the evil and that you do not. Let the good contribute g to the value of a world and let the evil contribute $-e$, suppose the two to be additive, and set an arbitrary zero where both are absent. Then by averaging,

$$V(-G) = C(E/-G)V(E-G) + C(-E/-G)V(-E-G)$$
$$= -eC(E/-G) \tag{4}$$

$$V(G) = C(E/G)V(EG) + C(-E/G)V(-EG) = -eC(E/G) + g$$

That means that $-G$, declining the good, is the V-maximal option iff the difference $(C(E/G) - C(E/-G))$, which may serve as a measure of the extent to which taking the good brings bad news, exceeds the fraction g/e. And that may well be so under the circumstances considered. If it is, noncausal decision theory endorses the ostrich's useless policy of managing the news. It tells you to decline the good, though doing so does not at all tend to prevent the evil. If a theory tells you that, it stands refuted.

In Newcomb's original problem, (see Nozick, 1969) verisimilitude was sacrificed for extremity. $C(E/G)$ was close to one and $C(E/-G)$ was close to zero, so that declining the good turned out to be V-maximal by an overwhelming margin. To make it so, we have to imagine someone with the mind-boggling power to detect the entire vast combination of causal factors at some earlier time that would cause you to decline the good, in order to inflict the evil if any such combination is present.

Some philosophers have refused to learn anything from such a tall story.

If our aim is to show the need for causal decision theory, however, a more moderate version of Newcomb's problem will serve as well. Even if the difference of C(E/G) and C(E/−G) is quite small, provided that it exceeds g/e, we have a counterexample. More moderate versions can also be more down-to-earth, as witness the medical Newcomb problems.[3] Suppose you like eating eggs, or smoking, or loafing when you might go out and run. You are convinced, contrary to popular belief, that these pleasures will do you no harm at all. (Whether you are right about this is irrelevant.) But also you think you might have some dread medical condition: a lesion of an artery, or nascent cancer, or a weak heart. If you have it, there's nothing you can do about it now and it will probably do you a lot of harm eventually. In its earlier stages, this condition is hard to detect. But you are convinced that it has some tendency, perhaps slight, to cause you to eat eggs, smoke, or loaf. So if you find yourself indulging, that is at least some evidence that you have the condition and are in for big trouble. But is that any reason not to indulge in harmless pleasures? The V-maximising rule says yes, if the numbers are right. I say no.

So far, I have considered pure Newcomb problems. There are also mixed problems. You may think that taking the good has some tendency to produce (or prevent) the evil, but also is a manifestation of some prior state which tends to produce the evil. Or you may be uncertain whether your situation is a Newcomb problem or not, dividing your credence between alternative hypotheses about the causal relations that prevail. These mixed cases are still more realistic, yet even they can refute noncausal decision theory.

However, no Newcomb problem, pure or mixed, can refute anything if it is not possible. The Tickle Defence of noncausal decision theory[4] questions whether Newcomb problems really can arise. It runs as follows: "Supposedly the prior state that tends to cause the evil also tends to cause you to take the good. The dangerous lesion causes you to choose to eat eggs, or whatever. How can it do that? If you are fully rational your choices are governed entirely by your beliefs and desires so nothing can influence your choices except by influencing your beliefs and desires. But if you are fully rational, you know your own mind. If the lesion produces beliefs and desires favourable to eating

---

[3] Discussed in Skyrms (1980a), Nozick (1969), and Jeffrey (1980a, b) I discuss another sort of moderate and down-to-earth Newcomb problem in (1979b).

[4] Discussed in Skyrms (1980a), and most fully presented in Ellery Eells (1981). Eells argues that Newcomb problems are stopped by assumptions of rationality and self-knowledge somewhat weaker than those of the simple Tickle Defence considered here, but even those weaker assumptions seem to me unduly restrictive.

eggs, you will be aware of those beliefs and desires at the outset of deliberation. So you won"t have to wait until you find yourself eating eggs to get the bad news. You will have it already when you feel that tickle in the tastebuds – or whatever introspectible state it might be – that manifests your desire for eggs. Your consequent choice tells you nothing more. By the time you decide whether to eat eggs, your credence function already has been modified by the evidence of the tickle. Then $C(E/G)$ does not exceed $C(E/-G)$, their difference is zero and so does not exceed g/e, $-G$ is not V-maximal, and noncausal decision theory does not make the mistake of telling you not to eat the eggs."

I reply that the Tickle Defence does establish that a Newcomb problem cannot arise for a fully rational agent, but that decision theory should not be limited to apply only to the fully rational agent.[5] Not so, at least, if rationality is taken to include self-knowledge. May we not ask what choice would be rational for the partly rational agent, and whether or not his partly rational methods of decision will steer him correctly? A partly rational agent may very well be in a moderate Newcomb problem, either because his choices are influenced by something besides his beliefs and desires or because he cannot quite tell the strengths of his beliefs and desires before he acts. ("How can I tell what I think till I see what I say?" – E. M. Forster.) For the dithery and the self-deceptive, no amount of *Gedankenexperimente* in decision can provide as much self-knowledge as the real thing. So even if the Tickle Defence shows that noncausal decision theory gives the right answer under powerful assumptions of rationality (whether or not for the right reasons), Newcomb problems still show that a general decision theory must be causal.

5. Utility and dependency hypotheses

Suppose someone knows all there is to know about how the things he cares about do and do not depend causally on his present actions. If something is beyond his control, so that it will obtain – or have a certain chance of obtaining – no matter what he does, then he knows that for certain. And if something is within his control, he knows that for certain; further, he knows the extent of his influence over it and he

[5] In fact, it may not apply to the fully rational agent. It is hard to see how such an agent can be uncertain what he is going to choose, hence hard to see how he can be in a position to deliberate. See Richard C. Jeffrey (1977a). Further, the "fully rational agent" required by the Tickle Defence is, in one way, not so very rational after all. Self-knowledge is an aspect of rationality, but so is willingness to learn from experience. If the agent's introspective data make him absolutely certain of his own credences and values, as they must if the Defence is to work, then no amount of evidence that those data are untrustworthy will ever persuade him not to trust them.

knows what he must do to influence it one way or another. Then there can be no Newcomb problems for him. Whatever news his actions may bring, they cannot change his mind about the likely outcomes of his alternative actions. He knew it all before.

Let us call the sort of proposition that this agent knows – a maximally specific proposition about how the things he cares about do and do not depend causally on his present actions – a *dependency hypothesis* (for that agent at that time). Since there must be some truth or other on the subject, and since the dependency hypotheses are maximally specific and cannot differ without conflicting, they comprise a partition. Exactly one of them holds at any world, and it specifies the relevant relations of causal dependence that prevail there.

It would make no difference if our know-it-all didn't really know. If he concentrates all his credence on a single dependency hypothesis, whether rightly or wrongly, then there can be no Newcomb problems for him. His actions cannot bring him news about which dependency hypothesis holds if he already is quite certain which one it is.

Within a single dependency hypothesis, so to speak, V-maximising is right. It is rational to seek good news by doing that which, according to the dependency hypothesis you believe, most tends to produce good results. That is the same as seeking good results. Failures of V-maximising appear only if, first, you are sensible enough to spread your credence over several dependency hypotheses, and second, your actions might be evidence for some dependency hypotheses and against others. That is what may enable the agent to seek good news not in the proper way, by seeking good results, but rather by doing what would be evidence for a good dependency hypothesis. That is the recipe for Newcomb problems.

What should you do if you spread your credence over several dependency hypotheses? You should consider the expected value of your options under the several hypotheses; you should weight these by the credences you attach to the hypotheses; and you should maximise the weighted average. Henceforth I reserve the variable K to range over dependency hypotheses (or over members of partitions that play a parallel role in other versions of causal decision theory). Let us define the (*expected*) *utility* of an option A by:

$$U(A) =^{df} \Sigma_K C(K)V(AK).$$

My version of causal decision theory prescribes the rule of U-*maximising* according to which a rational choice is one that has the greatest expected utility. Option A is U-maximal iff $U(A)$ is not exceeded by any $U(A')$, and to act rationally is to realise some U-maximal option.

In putting this forward as the rule of rational decision, of course I speak for myself; but I hope I have found a neutral formulation which fits not only my version of causal decision theory but also the versions proposed by Gibbard and Harper, Skyrms, and Sobel. There are certainly differences about the nature of dependency hypotheses; but if I am right, these are small matters compared to our common advocacy of utility maximising as just defined.

In distinguishing as I have between V and U – value and utility – I have followed the notation of Gibbard and Harper. But also I think I have followed the lead of ordinary language, in which "utility" means much the same as "usefulness". Certainly the latter term is causal. Which would you call the useful action: the one that tends to produce good results? Or the one that does no good at all (or even a little harm) and yet is equally welcome because it is a sign of something else that does produce good results? (Assume again that the news is not valued for its own sake.) Surely the first – and that is the one with greater utility in my terminology, though both may have equal value.

It is essential to define utility as we did using the unconditional credences $C(K)$ of dependency hypotheses, not their conditional credences $C(K/A)$. If the two differ, any difference expresses exactly that news-bearing aspect of the options that we meant to suppress. Had we used the conditional credences, we would have arrived at nothing different from V. For the Rule of Averaging applies to any partition; and hence to the partition of dependency hypotheses, giving

$$V(A) = \Sigma_K C(K/A)V(AK). \tag{5}$$

Let us give noncausal decision theory its due before we take leave of it. It works whenever the dependency hypotheses are probabilistically independent of the options, so that all the $C(K/A)$'s equal the corresponding $C(K)$'s. Then by (5) and the definition of U, the corresponding $V(A)$'s and $U(A)$'s also are equal. V-maximising gives the same right answers as U-maximising. The Tickle Defence seems to show that the K's must be independent of the A's for any fully rational agent. Even for partly rational agents, it seems plausible that they are at least close to independent in most realistic cases. Then indeed V-maximising works. But it works because the agent's beliefs about causal dependence are such as to make it work. It does not work for reasons which leave causal relations out of the story.

I am suggesting that we ought to undo a seeming advance in the development of decision theory. Everyone agrees that it would be ridiculous to maximise the "expected utility" defined by

$$\Sigma_Z C(Z)V(AZ)$$

where Z ranges over just any old partition. It would lead to different answers for different partitions. For the partition of value-level propositions, for instance, it would tell us fatalistically that all options are equally good! What to do? Savage suggested, in effect, that we make the calculation with unconditional credences, but make sure to use only the right sort of partition (Savage, 1954, p. 15; the suggestion is discussed by Jeffrey, 1977*b*). But what sort is that? Jeffrey responded that we would do better to make the calculation with conditional credences, as in the right hand side of (2). Then we need not be selective about partitions, since we get the same answer, namely V(*A*), for all of them. In a way, Jeffrey himself was making decision theory causal. But he did it by using probabilistic dependence as a mark of causal dependence, and unfortunately the two need not always go together. So I have thought it better to return to unconditional credences and say what sort of partition is right.

As I have formulated it, causal decision theory is causal in two different ways. The dependency hypotheses are causal in their content: they class worlds together on the basis of likenesses of causal dependence. But also the dependency hypotheses themselves are causally independent of the agent's actions. They specify his influence over other things, but over them he has no influence. (Suppose he did. Consider the dependency hypothesis which we get by taking account of the ways the agent can manipulate dependency hypotheses to enhance his control over other things. This hypothesis seems to be right no matter what he does. Then he has no influence over whether this hypothesis or another is right, contrary to our supposition that the dependency hypotheses are within his influence.) Dependency hypotheses are "act-independent states" in a causal sense, though not necessarily in the probabilistic sense. If we say that the right sort of partition for calculating expected utility is a causally act-independent one, then the partition of dependency hypotheses qualifies. But I think it is better to say just that the right partition is the partition of dependency hypotheses, in which case the emphasis is on their causal content rather than their act-independence.

If any of the credences C(*AK*) is zero, the rule of U-maximising falls silent. For in that case V(*AK*) becomes an undefined sum of quotients with denominator zero, so U(*A*) in turn is undefined and *A* cannot be compared in utility with the other options. Should that silence worry us? I think not, for the case ought never to arise. It may seem that it arises in the most extreme sort of Newcomb problem: suppose that taking the good is thought to make it absolutely certain that the prior state obtains and the evil will follow. Then if *A* is the option of taking the good and *K* says that the agent stands a chance of escaping the evil, C(*AK*) is indeed zero and U(*A*) is indeed undefined. What should you do in such an extreme Newcomb problem? V-maximise after all?

No; what you should do is not be in that problem in the first place. Nothing should ever be held as certain as all that, with the possible exception of the testimony of the senses. Absolute certainty is tantamount to a firm resolve never to change your mind no matter what, and that is objectionable. However much reason you may get to think that option A will not be realised if K holds, you will not if you are rational lower $C(AK)$ quite to zero. Let it by all means get very, very small; but very, very small denominators do not make utilities go undefined.

What of the partly rational agent, whom I have no wish to ignore? Might he not rashly lower some credence $C(AK)$ all the way to zero? I am inclined to think not. What makes it so that someone has a certain credence is that its ascription to him is part of a systematic pattern of ascriptions, both to him and to others like him, both as they are and as they would have been had events gone a bit differently, that does the best job overall of rationalising behaviour.[6] I find it hard to see how the ascription of rash zeros could be part of such a best pattern. It seems that a pattern that ascribes very small positive values instead always could do just a bit better, rationalising the same behaviour without gratuitously ascribing the objectionable zeros. If I am right about this, rash zeros are one sort of irrationality that is downright impossible.[7]

6. Reformulations

The causal decision theory proposed above can be reformulated in various equivalent ways. These will give us some further understanding of the theory, and will help us in comparing it with other proposed versions of causal decision theory.

Expansions: We can apply the Rule of Averaging to expand the $V(AK)$'s that appear in our definition of expected utility. Let Z range over any partition. Then we have

$$U(A) = \Sigma_K \Sigma_Z C(K)C(Z/AK)V(AKZ). \tag{6}$$

(If any $C(AKZ)$ is zero we may take the term for K and Z as zero,

---

[6] See my (1974). I now think that discussion is too individualistic, however, in that it neglects the possibility that one might have a belief or desire entirely because the ascription of it to him is part of a systematic pattern that best rationalises the behaviour of *other* people. On this point, see my discussion of the madman in (1980a).

[7] Those who think that credences can easily fall to zero often seem to have in mind credences conditional on some background theory of the world which is accepted, albeit tentatively, in an all-or-nothing fashion. While I don't object to this notion, it is not what I mean by credence. As I understand the term, what is open to reconsideration does not have a credence of zero or one; these extremes are not to be embraced lightly.

despite the fact that V(*AKZ*) is undefined.) This seems only to make a simple thing complicated; but if the partition is well chosen, (6) may serve to express the utility of an option in terms of quantities that we find it comparatively easy to judge.

Let us call a partition *rich* iff, for every member *S* of that partition and for every option *A* and dependency hypothesis *K*, V(*AKS*) equals V(*AS*). That means that the *AS*'s describe outcomes of options so fully that the addition of a dependency hypothesis tells us no more about the features of the outcome that matter to the agent. Henceforth I reserve the variable *S* to range over rich partitions. Given richness of the partition, we can factor the value terms in (6) part way out, to obtain

$$U(A) = \Sigma_S(\Sigma_K C(K)C(S/AK))V(AS). \tag{7}$$

Equation (7) for expected utility resembles equation (2) for expected value, except that the inner sum in (7) replaces the conditional credence C(*S/A*) in the corresponding instance of (2). As we shall see, the analogy can be pushed further. Two examples of rich partitions to which (7) applies are the partition of possible worlds and the partition of value-level propositions [V = v].

Imaging: Suppose we have a function that selects, for any pair of a world *W* and a suitable proposition *X*, a probability distribution W_X. Suppose further that W_X assigns probability only to *X*-worlds, so that $W_X(X)$ equals one. (Hence at least the empty proposition must not be "suitable".) Call the function an *imaging function*, and call W_X the *image of W on X*. The image might be sharp, if W_X puts all its probability on a single world; or it might be blurred, with the probability spread over more than one world.

Given an imaging function, we can apply it to form images also of probability distributions. We sum the superimposed images of all the worlds, weighting the images by the original probabilities of their source worlds. For any pair of a probability distribution C and a suitable proposition *X*, we define C_X, the *image of C on X*, as follows. First, for any world *W'*.

$$C_X(W') =^{\mathrm{df}} \Sigma_W C(W)W_X(W');$$

think of C(*W*) $W_X(W')$ as the amount of probability that is moved from *W* to *W'* in making the image. We sum as usual: for any proposition *Y*,

$$C_X(Y) =^{\mathrm{df}} \Sigma_{W \in Y} C_X(W).$$

It is easy to check that C_X also is a probability distribution; and that it assigns probability only to *X*-worlds, so that $C_X(X)$ equals one. Imaging is one way – conditionalising is another – to revise a given probability distribution so that all the probability is concentrated on a given proposition.[8]

For our present purposes, what we want are images of the agent's credence function on his various options. The needed imaging function can be defined in terms of the partition of dependency hypotheses: let

$$W_A(W') =^{df} C(W'/AK_W)$$

for any option A and worlds W and W', where K_W is the dependency hypothesis that holds at W. In words: move the credence of world W over to the A-worlds in the same dependency hypothesis, and distribute it among those worlds in proportion to their original credence. (Here again we would be in trouble if any of the $C(AK)$'s were zero, but I think we needn't worry.) It follows from the several definitions just given that for any option A and proposition Y,

$$C_A(Y) = \Sigma_K C(K)C(Y/AK). \tag{8}$$

The inner sum in (7) therefore turns out to be the credence, imaged on A, of S. So by (7) and (8) together,

$$U(A) = \Sigma_S C_A(S)V(AS). \tag{9}$$

Now we have something like the Rule of Averaging for expected value, except that the partition must be rich and we must image rather than conditionalising. For the rich partition of possible worlds we have

$$U(A) = \Sigma_W C_A(W)V(W) \tag{10}$$

which resembles the definition of expected value. For the rich partition of value-level propositions we have something resembling (3):

$$U(A) = \Sigma_v C_A([V = v])v. \tag{11}$$

7. Primitive imaging: Sobel

To reformulate causal decision theory in terms of imaging, I proceeded in two steps. I began with the dependency hypotheses and used them to define an imaging function; then I redefined the expected utility of an option in terms of imaging. We could omit the first step and leave

---

[8] Sharp imaging by means of a Stalnaker selection function is discussed in my (1976, especially pp. 309–311). This generalisation to cover blurred imaging as well is due to Peter Gärdenfors (1982). A similar treatment appears in Donald Nute (1980, chap. 6). What is technically the same idea, otherwise motivated and under other names, appears in my (1973c, section 8), Pollock (1976, pp. 219–236), and Sobel (1978). The possibility of deriving an imaging function from a partition was suggested by Brian Skyrms in discussion of a paper by Robert Stalnaker at the 1979 annual meeting of the American Philosophical Association, Eastern Division.

the dependency hypotheses out of it. We could take the imaging function as primitive, and go on as I did to define expected utility by means of it. That is the decision theory of J. Howard Sobel (1978).

Sobel starts with the images of worlds, which he calls *world-tendencies*. (He considers images on all propositions possible relative to the given world, but for purposes of decision theory we can confine our attention to images on the agent's options.) Just as we defined C_A in terms of the W_A's, so Sobel goes on to define images of the agent's credence function. He uses these in turn to define expected utility in the manner of (10), and he advocates maximising the utility so defined rather than expected value.

Sobel unites his decision theory with a treatment of counterfactual conditionals in terms of closest antecedent-worlds.[9] If $W_A(W')$ is positive, then we think of W' as one of the A-worlds that is in some sense closest to the world W. What might be the case if it were the case that A, from the standpoint of W, is what holds at some such closest A-world; what would be the case if A, from the standpoint of W, is what holds at all of them. Sobel's apparatus gives us quantitative counterfactuals intermediate between the mights and the woulds. We can say that if it were that A, it would be with probability p that X; meaning that $W_A(X)$ equals p, or in Sobel's terminology that X holds on a subset of the closest A-worlds whose tendencies, at W and on the supposition A, sum to p.

Though Sobel leaves the dependency hypotheses out of his decision theory, we can perhaps bring them back in. Let us say that worlds *image alike* (on the agent's options) iff, for each option, their images on that option are exactly the same. Imaging alike is an equivalence relation, so we have the partition of its equivalence classes. If we start with the dependency hypotheses and define the imaging function as I did, it is immediate that worlds image alike iff they are worlds where the same dependency hypothesis holds; so the equivalence classes turn out to be just the dependency hypotheses.

The question is whether dependency hypotheses could be brought into Sobel's theory by defining them as equivalence classes under the relation of imaging alike. Each equivalence class could be described, in Sobel's terminology, as a maximally specific proposition about the tendencies of the world on all alternative suppositions about which option the agent realises. That sounds like a dependency hypothesis to me. Sobel tells me (personal communication, 1980) that he is inclined to agree, and does regard his decision theory as causal; though it is

---

[9] As in my (1973a), without the complications raised by possible infinite sequences of closer and closer antecedent-worlds.

hard to tell that from his written presentation, in which causal language very seldom appears.

If the proposal is to succeed technically, we need the following thesis: if K_W is the equivalence class of W under the relation of imaging alike (of having the same tendencies on each option) then, for any option A and world W', $W_A(W')$ equals $C(W'/AK_W)$. If so, it follows that if we start as Sobel does with the imaging function, defining the dependency hypotheses as equivalence classes, and thence defining an imaging function as I did, we will get back the same imaging function that we started with. It further follows, by our results in Section 6, that expected utility calculated in my way from the defined dependency hypotheses is the same as expected utility calculated in Sobel's way from the imaging function. They must be the same, if the defined dependency hypotheses introduced into Sobel's theory are to play their proper role.

Unfortunately, the required thesis is not a part of Sobel's theory; it would be an extra constraint on the imaging function. It does seem a very plausible constraint, at least in ordinary cases. Sobel suspends judgement about imposing a weaker version of the thesis (Connection Thesis 1, discussed in his Section 6.7). But his reservations, which would carry over to our version, entirely concern the extraordinary case of an agent who thinks he may somehow have foreknowledge of the outcomes of chance processes. Sobel gives no reason, and I know of none, to doubt either version of the thesis except in extraordinary cases of that sort. Then if we assume the thesis, it seems that we are only setting aside some very special cases – cases about which I, at least, have no firm views. (I think them much more problematic for decision theory than the Newcomb problems.) So far as the remaining cases are concerned, it is satisfactory to introduce defined dependency hypotheses into Sobel's theory and thereby render it equivalent to mine.

8. Factors outside our influence: Skyrms

Moving on to the version of causal decision theory proposed by Brian Skyrms (1980a), we find a theory that is formally just like mine. Skyrms' definition of *K-expectation* – his name for the sort of expected utility that should be maximised – is our equation (6). From that, with a trivial partition of Z's, we can immediately recover my first definition of expected utility. Skyrms introduces a partition of hypotheses – the K's which give K-expectation its name – that play just the same role in his calculation of expected utility that the dependency hypotheses play in mine. (Thus I have followed Skyrms in notation.) So the only difference, if it is a difference, is in how the K's are characterised.

Skyrms describes them at the outset as maximally specific specifications of the factors outside the agent's influence (at the time of decision) which are causally relevant to the outcome of the agent's action. He gives another characterisation later, but let us take the first one first.

I ask what Skyrms means to count as a "factor". Under a sufficiently broad construal, I have no objection to Skyrms' theory and I think it no different from mine. On a narrower and more literal construal, I do not think Skyrms' theory is adequate as a general theory of rational decision, though I think that in practice it will often serve. Insofar as Skyrms is serving up a general theory rather than practical rules of thumb, I think it is indeed the broad construal that he intends.

(I also ask what Skyrms means by "relevant to the outcome". I can't see how any factor, broadly or narrowly construed, could fail to be relevant to some aspect of the outcome. If the outcome is that I win a million dollars tomorrow, one aspect of this outcome may be that it takes place just one thousand years after some peasant felled an oak with ninety strokes of his axe. So I suppose Skyrms' intent was to include only factors relevant to those features of the outcome that the agent cares about, as opposed to those that are matters of indifference to him. That would parallel a like exclusion of matters of indifference in my definition of dependency hypotheses. In neither case is the exclusion important. Richer hypotheses, cluttered with matters of indifference, ought to give the same answers.)

On the broad construal, a "factor" need not be the sort of localised particular occurrence that we commonly think of as causing or being caused. It might be any matter of contingent fact whatever. It might indeed be some particular occurrence. It might be a vast dispersed pattern of occurrences throughout the universe. It might be a law of nature. It might be a dependency hypothesis. On the broad construal, Skyrms is saying only that the K's are maximally specific propositions about matters outside the agent's influence and relevant to features of the outcome that the agent cares about.

A dependency hypothesis is outside the agent's influence. It is relevant to features of the outcome that he cares about. (*Causally* relevant? – Not clear, but if we're construing "factor" broadly, we can let that by as well.) Any specification of something outside the agent's influence is included in a dependency hypothesis – recall that they cover what doesn't depend on the agent's actions as well as what does – unless it concerns something the agent doesn't care about. I conclude that on the broad construal, Skyrms' K's are nothing else than the dependency hypotheses. In that case his theory is the same as mine.

On the narrow construal, a "factor" must be the sort of localised occurrence – event, state, omission, etc. – that we normally think of as a cause. In the medical Newcomb problems, for instance, the lesion or the nascent cancer or the weak heart is a causal factor narrowly and literally. In motivating his theory, it is factors like these that Skyrms considers.

Our topic is rational decision according to the agent's beliefs, be they right or wrong. So it seems that we should take not the factors which really are outside his influence, but rather those he thinks are outside his influence. But what if he divides his credence between several hypotheses as to which factors are outside his influence, as well he might? Skyrms responds to this challenge by redescribing his partition of hypotheses. On his new description, each hypothesis consists of two parts: (i) a preliminary hypothesis specifying which of the relevant causal factors are outside the agent's influence, and (ii) a full specification of those factors that are outside his influence according to part (i).

That is a welcome amendment, but I think it does not go far enough. Influence is a matter of degree, so shouldn't the hypotheses say not just that the agent has some influence over a factor or none, but also how much? And if the hypothesis says that the agent has influence over a factor, shouldn't it also say which way the influence goes? Given that I can influence the temperature, do I make it cooler by turning the knob clockwise or counterclockwise? Make Skyrms' amendment and the other needed amendments, and you will have the dependency hypotheses back again.

To illustrate my point, consider an agent with eccentric beliefs. He thinks the influence of his actions ramifies but also fades, so that everything in the far future is within his influence but only a little bit. Perhaps he thinks that his actions raise and lower the chances of future occurrences, but only very slightly. Also he thinks that time is circular, so that the far future includes the present and the immediate past and indeed all of history. Then he gives all his credence to a single one of Skyrms' two-part hypotheses: the one saying that no occurrence whatever – no factor, on the narrow construal – is entirely outside his influence. That means that on Skyrms' calculation his $U(A)$'s reduce to the corresponding $V(A)$'s, so V-maximising is right for him. That's wrong. Since he thinks he has very little influence over whether he has the dread lesion, his decision problem about eating eggs is very little different from that of someone who thinks the lesion is entirely outside his influence. V-maximising should come out wrong for very much the same reason in both cases.

No such difficulty threatens Skyrms' proposal broadly construed. The agent may well wonder which of the causal factors narrowly construed are within his influence, but he cannot rationally doubt that the dependency hypotheses are entirely outside it. On the broad construal, Skyrms' second description of the partition of hypotheses is a gloss on the first, not an amendment. The hypotheses already specify which of the (narrow) factors are outside the agent's influence, for that is itself a (broad) factor outside his influence. Skyrms notes this, and that is why I think it must be the broad construal that he intends. Likewise the degrees and directions of influence over (narrow) factors are themselves (broad) factors outside the agent's influence, hence already specified according to the broad construal of Skyrms' first description.

Often, to be sure, the difference between the broad and narrow construals will not matter. There may well be a correlation, holding throughout the worlds which enjoy significant credence, between dependency hypotheses and combinations of (narrow) factors outside the agent's influence. The difference between good and bad dependency hypotheses may in practice amount to the difference between absence and presence of a lesion. However, I find it rash to assume that there must always be some handy correlation to erase the difference between the broad and narrow construals. Dependency hypotheses do indeed hold in virtue of lesions and the like, but they hold also in virtue of the laws of nature. It would seem that uncertainty about dependency hypotheses might come at least partly from undertainty about the laws.

Skyrms is sympathetic, as am I,[10] to the neo-Humean thesis that every contingent truth about a world – law, dependency hypothesis, or what you will – holds somehow in virtue of that world's total history of manifest matters of particular fact. Same history, same everything. But that falls short of implying that dependency hypotheses hold just in virtue of causal factors, narrowly construed; they might hold partly in virtue of dispersed patterns of particular fact throughout history, including the future and the distant present. Further, even if we are inclined to accept the neo-Humean thesis, it still seems safer not to make it a presupposition of our decision theory. Whatever we think of the neo-Humean thesis, I conclude that Skyrms' decision theory is best taken under the broad construal of "factor" under which his K's are the dependency hypotheses and his calculation of utility is the same as mine.[11]

---

[10] Although sympathetic, I have some doubts; see my (1980b, pp. 290–292).

[11] The decision theory of Nancy Cartwright (1979) is, as she remarks, "structurally identical" to Skyrms' theory for the case where value is a matter of reaching some all-or-nothing goal. However, hers is not a theory of subjectively rational decision in the single case, like Skyrms' theory and the others considered in this paper, but instead is a theory of objectively effective generic strategies. Since the subject matters are different, the structural identity is misleading. Cartwright's theory might somehow imply a single-case theory having more than structure in common with Skyrms' theory, but that would take principles she does not provide; inter alia,

9. Counterfactual dependence: Gibbard and Harper

If we want to express a dependency hypothesis in ordinary language, it is hard to avoid the use of counterfactual conditionals saying what would happen if the agent were to realise his various alternative options. Suppose that on a certain occasion I'm interested in getting Bruce to purr. I could try brushing, stroking, or leaving alone; pretend that these are my narrowest options. Bruce might purr loudly, softly, or not at all; pretend that these alternatives are a rich partition. (Those simplifying pretences are of course very far from the truth.) Much of my credence goes to the dependency hypothesis given by these three counterfactuals:

I brush Bruce $\Box\!\!\rightarrow$ he purrs loudly;
I stroke Bruce $\Box\!\!\rightarrow$ he purrs softly;
I leave Bruce alone $\Box\!\!\rightarrow$ he doesn't purr.

($\Box\!\!\rightarrow$ is used here as a sentential connective, read "if it were that ... it would be that ...". I use it also as an operator which applies to two propositions to make a proposition; context will distinguish the uses.) This hypothesis says that loud and soft purring are within my influence – they depend on what I do. It specifies the extent of my influence, namely full control. And it specifies the direction of influence, what I must do to get what. That is one dependency hypothesis. I give some of my credence to others, for instance this (rather less satisfactory) one:

I brush Bruce $\Box\!\!\rightarrow$ he doesn't purr;
I stroke Bruce $\Box\!\!\rightarrow$ he doesn't purr;
I leave Bruce alone $\Box\!\!\rightarrow$ he doesn't purr.

That dependency hypothesis says that the lack of purring is outside my influence, it is causally independent of what I do. Altogether there are twenty-seven dependency hypotheses expressible in this way, though some of them get very little credence.

Note that it is the pattern of counterfactuals, not any single one of them, that expresses causal dependence or independence. As we have seen, the same counterfactual

I leave Bruce alone $\Box\!\!\rightarrow$ he doesn't purr

figures in the first hypothesis as part of a pattern of dependence and in the second as part of a pattern of independence.

principles relating generic causal conduciveness to influence in the single case. So it is not clear that Cartwright's decision theory, causal though it is, falls under my claim that "We causal decision theorists share one common idea".

It is clear that not just any counterfactual could be part of a pattern expressing causal dependence or independence. The antecedent and consequent must specify occurrences capable of causing and being caused, and the occurrences must be entirely distinct. Further, we must exclude "back-tracking counterfactuals" based on reasoning from different supposed effects back to different causes and forward again to differences in other effects. Suppose I am convinced that stroking has no influence over purring, but that I wouldn't stroke Bruce unless I were in a mood that gets him to purr softly by emotional telepathy. Then I give credence to

I stroke Bruce $\Box\!\rightarrow$ he purrs softly

taken in a back-tracking sense, but not taken in the sense that it must have if it is to be part of a pattern of causal dependence or independence.

Let us define *causal counterfactuals* as those that can belong to patterns of causal dependence or independence. Some will doubt that causal counterfactuals can be distinguished from others except in causal terms; I disagree, and think it possible to delimit the causal counterfactuals in other terms and thus provide noncircular counterfactual analyses of causal dependence and causation itself. But that is a question for other papers (see Lewis, 1973d, 1979c). For present purposes, it is enough that dependency hypotheses can be expressed (sometimes, at least) by patterns of causal counterfactuals. I hope that much is adequately confirmed by examples like the one just considered. And that much can be true regardless of whether the pattern of counterfactuals provides a noncircular analysis.

Turning from language to propositions, what we want are causal counterfactuals $A \Box\!\rightarrow S$, where A is one of the agent's options and S belongs to some rich partition. The rich partition must be one whose members specify combinations of occurrences wholly distinct from the actions specified by the agent's options. It seems a safe assumption that some such rich partition exists. Suppose some definite one to be chosen (it should make no difference which one). Define a *full pattern* as a set consisting of exactly one such counterfactual proposition for each option. I claim that the conjunction of the counterfactuals in any full pattern is a dependency hypothesis.

Conjunctions of different full patterns are contraries, as any two dependency hypotheses should be. For if S and S' are contraries, and A is possible (which any option is), then also $A \Box\!\rightarrow S$ and $A \Box\!\rightarrow S'$ are contraries;[12] and any two full patterns must differ by at least one such contrary pair.

---

[12] Here and henceforth, I make free use of some fairly uncontroversial logical principles for counterfactuals: namely, those given by the system CK + ID + MP of Brian F. Chellas (1975).

What is not so clear is that some full pattern or other holds at any world, leaving no room for any other dependency hypotheses besides the conjunctions of full patterns. We shall consider this question soon. But for now, let us answer it by fiat. Assume that there is a full pattern for every world, so that the dependency hypotheses are all and only the conjunctions of full patterns.

That assumption yields the causal decision theory proposed by Allan Gibbard and William Harper (1978), following a suggestion of Robert Stalnaker. My statement of it amounts to their Savage-style formulation with conjunctions of full patterns of counterfactuals as act-independent states; and their discussion of consequences in their Section 6 shows that they join me in regarding these conjunctions as expressing causal dependence or independence. Although they do not explicitly distinguish causal counterfactuals from others, their Section 2 sketches a theory of counterfactuals which plainly is built to exclude back-trackers in any ordinary situation. This is essential to their purpose. A theory which used counterfactuals in formally the same way, but which freely admitted back-trackers, would not be a causal decision theory. Its conjunctions of full patterns including back-trackers would not be causal dependency hypotheses, and it would give just those wrong answers about Newcomb problems that we causal decision theorists are trying to avoid. (Such a theory is defended in Horgan, 1981.)

Consider some particular A and S. If a dependency hypothesis K is the conjunction of a full pattern that includes $A \,\square\!\!\rightarrow\, S$, then AK implies S and $C(S/AK)$ equals one. If K is the conjunction of a full pattern that includes not $A \,\square\!\!\rightarrow\, S$ but some contrary $A \,\square\!\!\rightarrow\, S'$, then AK contradicts S and $C(S/AK)$ equals zero. *Ex hypothesi*, every dependency hypothesis K is of one kind or the other. Then the K's for which $C(S/AK)$ equals one comprise a partition of $A \,\square\!\!\rightarrow\, S$, while $C(S/AK)$ equals zero for all other K's. It follows by the Rule of Additivity for credence that
Substituting (12) into (7) we have

$$C(A \,\square\!\!\rightarrow\, S) = \Sigma_K C(K)C(S/AK).\tag{12}$$

(Comparing (12) with (8), we find that our present assumptions equate $C(A \,\square\!\!\rightarrow\, S)$ with $C_A(S)$, the credence of S imaged on the option A.)

$$U(A) = \Sigma_S C(A \,\square\!\!\rightarrow\, S)V(AS),\tag{13}$$

which amounts to Gibbard and Harper's defining formula for the "genuine expected utility" they deem it rational to maximise.[13]

We have come the long way around to (13), which is not only simple but also intuitive in its own right. But (13) by itself does not display the

---

[13] To get exactly their formula, take their "outcomes" as conjunctions. AS, with "desirability" given by $V(AS)$; and bear in mind (i) that $A \,\square\!\!\rightarrow\, AS$ is the same as $A \,\square\!\!\rightarrow\, S$, and (ii) that if A and A' are contraries, $A \,\square\!\!\rightarrow\, A'S$ is the empty proposition with credence zero.

causal character of Gibbard and Harper's theory, and that is what makes it worthwhile to come at it by way of dependency hypotheses. No single $C(A \,\square\!\!\rightarrow S)$ reveals the agent's causal views, since it sums the credences of hypotheses which set $A \,\square\!\!\rightarrow S$ in a pattern of dependence and others which set $A \,\square\!\!\rightarrow S$ in a pattern of independence. Consequently the roundabout approach helps us to appreciate what the theory of Gibbard and Harper has in common with that of someone like Skyrms who is reluctant to use counterfactuals in expressing dependency hypotheses.

10. Counterfactual dependence with chancy outcomes

The assumption that there is a full pattern for each world is a consequence of Stalnaker's principle of Conditional Excluded Middle,[14] which says that either $X \,\square\!\!\rightarrow Y$ or $X \,\square\!\!\rightarrow\, - Y$ holds at any world (where $- Y$ is the negation of Y). It follows that if $Y, Y', \ldots$ are a partition and X is possible, then $X \,\square\!\!\rightarrow Y$, $X \,\square\!\!\rightarrow Y'$, $\ldots$ also are a partition. The conjunctions of full patterns are then a partition because, for any option A, the counterfactuals $A \,\square\!\!\rightarrow S$, $A \,\square\!\!\rightarrow S'$, $\ldots$ are a partition.

Conditional Excluded Middle is open to objection on two counts, one more serious than the other. Hence so is the decision theory of Gibbard and Harper, insofar as it relies on Conditional Excluded Middle to support the assumption that there is a full pattern for each world. Gibbard and Harper themselves are not to be faulted, for they tell us that their "reason for casting the rough theory in a form which gives these principles is that circumstances where these can fail involve complications which it would be best to ignore in preliminary work." (Gibbard & Harper, 1978, p. 128; pp. 344–345 in this volume). Fair enough; still, we have unfinished business on the agenda.

The first objection to Conditional Excluded Middle is that is makes arbitrary choices. It says that the way things would be on a false but possible supposition X is no less specific than the way things actually are. Some single, fully specific possible world is the one that would be actualised if it were that X. Since the worlds $W, W', \ldots$ are a partition, so are the counterfactuals $X \,\square\!\!\rightarrow W$, $X \,\square\!\!\rightarrow W'$, $\ldots$ saying exactly how things would be if X. But surely some questions about how things would be if X have no nonarbitrary answers: if you had a sister, would she like blintzes?

The less specific the supposition, the less it settles; the more far-fetched it is, the less can be settled by what carries over from actuality; and the less is settled otherwise, the more must be settled arbitrarily

---

[14] Robert C. Stalnaker (1968) gives a semantical analysis in which Conditional Excluded Middle follows from ordinary Excluded Middle applied to the selected antecedent-world.

or not at all. But the supposition that an agent realises one of his narrowest options is neither unspecific nor far-fetched. So the Arbitrariness Objection may be formidable against the general principle of Conditional Excluded Middle, yet not formidable against the special case of it that gives us a full pattern for each world.

Further, Bas van Fraassen has taught us a general method for tolerating arbitrariness.[15] When forced to concede that certain choices would be arbitrary, we leave those choices unmade and we ask what happens on all the alternative ways of making them. What is constant over all the ways of making them is determinate, what varies is indeterminate. If the provision of full patterns for certain worlds is partly arbitrary, so be it. Then indeed some arbitrary variation may infect the C(K)'s, C(S/AK)'s, C(A $\square\!\!\rightarrow$ S)'s, and even the U(A)'s. It might even infect the set of U-maximal options. Then indeed it would be (wholly or partly) indeterminate which options the Gibbard-Harper theory commends as rational. All of that might happen, but it needn't. The arbitrary variation might vanish part way through the calculation, leaving the rest determinate. The less arbitrary variation there is at the start, of course, the less risk that there will be any at the end.

I conclude that the Arbitrariness Objection by itself is no great threat to Gibbard and Harper's version of causal decision theory. We can well afford to admit that the theory might fail occasionally to give a determinate answer. Indeed, I admit that already, for any version, on other grounds: I think there is sometimes an arbitrary element in the assignment of C and V functions to partly rational agents. No worries, so long as we can reasonably hope that the answers are mostly determinate.

Unfortunately there is a second, and worse, objection against Conditional Excluded Middle and the Gibbard-Harper theory. In part it is an independent objection; in part an argument that van Fraassen's method of tolerating arbitrariness would be severely overloaded if we insisted on providing full patterns all around (and *a fortiori* if we insisted on saving Conditional Excluded Middle generally), and we could not reasonably hope that the answers are mostly determinate. Suppose the agent thinks – as he should if he is well-educated – that the actual world may very well be an indeterministic one, where many things he cares about are settled by chance processes. Then he may give little of his credence to worlds where full patterns hold. In fact he may well give little credence to any of the A $\square\!\!\rightarrow$ S counterfactuals that make up these patterns.

---

[15] See Bas van Fraassen (1966). Use of van Fraassen's method to concede and tolerate arbitrariness in counterfactuals was suggested to me by Stalnaker in 1971 (personal communication) and is discussed in my (1973a, pp. 81–83).

Consider again my problem of getting Bruce to purr. I think that Bruce works by firing of neurons, I think neurons work by chemical reactions, and I think the making or breaking of a chemical bond is a chance event in the same way that the radioactive decay of a nucleus is Mayble I still give some small credence to the twenty-seven full patterns considered in Section 9 – after all, I might be wrong to think that Bruce is chancy. But mostly I give my credence to the denials of all the counterfactuals that appear in those patterns, and to such counterfactuals as

I brush Bruce □→ a chance process goes on in him which has certain probabilities of eventuating in his purring loudly, softly, or not at all;

and likewise for the options of stroking and leaving alone. A diehard supporter of the Gibbard-Harper theory (not Gibbard or Harper, I should think) might claim that I give my credence mostly to worlds where it is arbitrary which one of the twenty-seven full patterns holds, but determinate that some one of them holds. If he is right, even this easy little decision problem comes out totally indeterminate, for the arbitrary variation he posits is surely enough to swing the answer any way at all. Nor would it help if I believe that whichever I did, all the probabilities of Bruce's purring loudly, softly, or not at all would be close to zero or one. Nor would a more realistic decision problem fare any better: unless the agent is a fairly convinced determinist, the answers we want vanish into indeterminacy. The diehard destroys the theory in order to save it.

Anyway, the diehard is just wrong. If the world is the chancy way I mostly think it is, there's nothing at all arbitrary or indeterminate about the counterfactuals in the full patterns. They are flatly, determinately false. So is their disjunction; the diehard agrees that it is determinate in truth value, but the trouble is that he thinks it is determinately true.

Unlike the Arbitrariness Objection, the Chance Objection seems to me decisive both against Conditional Excluded Middle generally and against the assumption that there is a full pattern for each world. Our conception of dependency hypotheses as conjunctions of full patterns is too narrow. Fortunately, the needed correction is not far to seek.

I shall have to assume that anyone who gives credence to indeterministic worlds without full patterns is someone who – implicitly and in practice, if not according to his official philosophy – distributes his credence over contingent propositions about single-case, objective chances. Chance is a kind of probability that is neither frequency nor credence, though related to both. I have no analysis to offer, but I am convinced that we do have this concept and we don't have any substitute for it.[16]

---

[16] For a fuller discussion of chance and its relations to frequency and credence, see Lewis (1980b).

Suppose some rich partition to be chosen which meets the requirement of distinct occurrences laid down in Section 9. Let the variable p range over candidate probability distributions for this rich partition: functions assigning to each S in the partition a number $p(S)$ in the interval from zero to one, such that the $p(S)$'s sum to one. Let $[P = p]$ be the proposition that holds at just those worlds where the chances of the S's, as of the time when the agent realises his chosen option, are correctly given by the function p. Call $[P = p]$ a *chance proposition*, and note that the chance propositions are a partition. Now consider the causal counterfactuals $A \;\Box\!\!\rightarrow\; [P = p]$ from the agent's options to the chance propositions. Define a *probabilistic full pattern* as a set containing exactly one such counterfactual for each option. I claim that the conjunction of the counterfactuals in any probabilistic full pattern is a causal dependency hypothesis. It specifies plain causal dependence or independence of the chances of the S's on the A's, and thereby it specifies a probabilistic kind of causal dependence of the S's themselves on the A's.

Here, for example, are verbal expressions of three chance propositions.

$[P = p_1]$ The chance that Bruce purrs loudly is 50 percent; the chance that he purrs softly is 40 percent; and the chance that he purrs not at all is 10 percent.

$[P = p_2]$ (similar, but with 30 percent, 50 percent, 20 percent).

$[P = p_3]$ (similar, but with 10 percent, 10 percent, 80 percent).

(The chance is to be at the time of my realising an option; the purring or not is to be at a certain time shortly after.) And here is a dependency hypothesis that might get as much of my credence as any:

I brush Bruce $\Box\!\!\rightarrow$ $[P = p_1]$ holds;
I stroke Bruce $\Box\!\!\rightarrow$ $[P = p_2]$ holds;
I leave Bruce alone $\Box\!\!\rightarrow$ $[P = p_3]$ holds.

Observe that this hypothesis addresses itself not only to the question of whether loud and soft purring are within my influence, but also to the question of the extent and the direction of my influence.

If a chance proposition says that one of the S's has a chance of one, it must say that the others all have chances of zero. Call such a chance proposition *extreme*. I shall not distinguish between an extreme chance proposition and the S that it favours. If they differ, it is only on worlds where something with zero chance nevertheless happens. I am inclined to think that they do not differ at all, since there are no worlds where anything with zero chance happens; the contrary opinion comes of mistaking infinitesimals for zero. But even if there is a difference between extreme chance propositions and their favoured S's, it will not matter to calculations of utility so let us neglect it. Then our previous dependency hypotheses, the conjunctions of full patterns, are subsumed under the conjunctions of probabilistic full patterns. So are the conjunctions of mixed full patterns that consist partly of $A \;\Box\!\!\rightarrow S$'s and partly of $A \;\Box\!\!\rightarrow [P = p]$'s.

Dare we assume that there is a probabilistic full pattern for every world, so that on this second try we have succeeded in capturing all the dependency hypotheses by means of counterfactuals? I shall assume it, not without misgivings. That means accepting a special case of Conditional Excluded Middle, but (i) the Chance Objection will not arise again[17], (ii) there should not be too much need for arbitrary choice on other grounds, since the options are quite specific suppositions and not far-fetched, and (iii) limited arbitrary choice results in nothing worse than a limited risk of the answers going indeterminate.

So my own causal decision theory consists of two theses. My main thesis is that we should maximise expected utility calculated by means of dependency hypotheses. It is this main thesis that I claim is implicitly accepted also by Gibbard and Harper, Skyrms, and Sobel. My subsidiary thesis, which I put forward much more tentatively and which I won't try to foist on my allies, is that the dependency hypotheses are exactly the conjunctions of probabilistic full patterns.

(The change I have made in the Gibbard-Harper version has been simply to replace the rich partition of S's by the partition of chance propositions $[P = p]$ pertaining to these S's. One might think that perhaps that was no change at all: perhaps the S's already were the chance propositions for some other rich partition. However, I think it at least doubtful that the chance propositions can be said to "specify combinations of occurrences" as the S's were required to do. This question would lead us back to the neo-Humean thesis discussed in Section 8.)

Consider some particular A and S. If a dependency hypothesis K is the conjunction of a probabilistic full pattern, then for some p, K implies $A \;\Box\!\!\rightarrow [P = p]$. Then AK implies $[P = p]$; and $C(S/AK)$ equals $p(S)$, at least in any ordinary case.[18] For any p, the K's that are conjunctions of probabilistic full pattern including $A \;\Box\!\!\rightarrow [P = p]$ are a partition of $A \;\Box\!\!\rightarrow [P = p]$. So we have

---

[17] Chances aren't chancy; if $[P = p]$ pertains to a certain time, its own chance at that time of holding must be zero or one, by the argument of Lewis (1980b, pp. 276–277).

[18] That follows by what I call the Principal Principle connecting chance and credence, on the assumption that (i) AK holds or fails to hold at any world entirely in virtue of the history of that world up to action time together with the complete theory of chance for that world, and (ii) the agent gives no credence to worlds where the usual asymmetries of time break down. Part (ii) fails in the case which we have already noted in Section 7 as troublesome, in which the agent thinks he may have foreknowledge of the outcomes of chance processes. See Lewis (1980b, pp. 266–276).

$$\Sigma_p C(A \mathrel{\Box}\!\!\!\rightarrow [P = p])p(S) = \Sigma_K C(K)C(S/AK). \tag{14}$$

Substituting (14) into (7) gives us a formula defining expected utility in terms of counterfactuals with chance propositions as consequents:

$$U(A) = \Sigma_S \Sigma_p C(A \mathrel{\Box}\!\!\!\rightarrow [P = p])p(S)V(AS). \tag{15}$$

For any S and any number q from zero to one, let $[P(S) = q]$ be the proposition that holds at just those worlds where the chance of S, at the time when the agent realises his option, is q. It is the disjunction of those $[P = p]$'s for which $p(S)$ equals q. We can lump together counterfactuals in (14) and (15) to obtain reformulations in which the consequents concern chances of single S's:

$$\Sigma_q C(A \mathrel{\Box}\!\!\!\rightarrow [P(S) = q])q = \Sigma_K C(K)C(S/AK), \tag{16}$$

$$U(A) = \Sigma_S \Sigma_q C(A \mathrel{\Box}\!\!\!\rightarrow [P(S) = q])qV(AS). \tag{17}$$

There are various ways to mix probabilities and counterfactuals. I have argued that when things are chancy, it isn't good enough to take credences of plain $A \mathrel{\Box}\!\!\!\rightarrow S$ counterfactuals. The counterfactuals themselves must be made probabilistic. I have made them so by giving them chance propositions as consequents. Sobel makes them so in a different way: as we noted in Section 7, he puts the probability in the connective. Under our present assumptions (and setting aside extraordinary worlds where the common asymmetries of time break down), the two approaches are equivalent. Sobel's quantitative counterfactual with a plain consequent

If it were that A, it would be with probability q that S

holds at W iff $W_A(S)$ equals q. Given my derivation of the imaging function from the dependency hypotheses, that is so iff $C(S/AK_W)$ equals q. That is so (setting aside the extraordinary worlds) iff K_W implies $A \mathrel{\Box}\!\!\!\rightarrow [P(S) = q]$. Given that there is a probabilistic full pattern for each world, that is so iff $A \mathrel{\Box}\!\!\!\rightarrow [P(S) = q]$ holds at W. Hence the Sobel quantitative counterfactual with a plain consequent is the same proposition as the corresponding plain counterfactual with a chance consequent. If ever we must retract the assumption that there is a probabilistic full pattern for each world (or if we want to take the extraordinary worlds into account), the two approaches will separate and we may need to choose; but let us cross that bridge if we come to it.

11. The Hunter-Richter problem

That concludes an exposition and survey of causal decision theory. In this final section, I wish to defend it against an objection raised by

Daniel Hunter and Reed Richter (1978, especially pp. 257–259). Their target is the Gibbard-Harper version; but it depends on nothing that is special to that version, so I shall restate it as an objection against causal decision theory generally.

Suppose you are one player in a two-person game. Each player can play red, play white, play blue, or not play. If both play the same colour, each gets a thousand dollars; if they play different colours, each loses a thousand dollars; if one or both don't play, the game is off and no money changes hands. Value goes by money; the game is played only once; there is no communication or prearrangement between the players; and there is nothing to give a hint in favour of one colour or another – no "Whites rule OK!" sign placed where both can see that both can see it, or the like. So far, this game seems not worthwhile. But you have been persuaded that you and the other player are very much alike psychologically and hence very likely to choose alike, so that you are much more likely to play and win than to play and lose. Is it rational for you to play?

Yes. So say I, so say Hunter and Richter, and so (for what it is worth) says noncausal decision theory. But causal decision theory seems to say that it is not rational to play. If it says that, it is wrong and stands refuted. It seems that you have four dependency hypotheses to consider, corresponding to the four ways your partner might play:

K_1 Whatever you do, he would play red;
K_2 Whatever you do, he would play white;
K_3 Whatever you do, he would play blue;
K_4 Whatever you do, he would not play.

By the symmetry of the situation, K_1 and K_2 and K_3 should get equal credence. Then the expected utility of not playing is zero, whereas the expected utilities of playing the three colours are equal and negative. So we seem to reach the unwelcome conclusion that not playing is your U-maximal option.

I reply that Hunter and Richter have gone wrong by misrepresenting your partition of options. Imagine that you have a servant. You can play red, white, or blue; you can not play; or you can tell your servant to play for you. The fifth option, delegating the choice, might be the one that beats not playing and makes it rational to play. Given the servant, each of our previous dependency hypotheses splits in three. For instance K_1 splits into:

$K_{1,1}$ Whatever you do, your partner would play red, and your servant would play red if you delegated the choice;

$K_{1,2}$ Whatever you do, your partner would play red, and your servant would play white if you delegated the choice;

$K_{1,3}$ Whatever you do, your partner would play red, and your servant would play blue if you delegated the choice.

(If you and your partner are much alike, he too has a servant, so we can split further by dividing the case in which he plays red, for instance, into the case in which he plays red for himself and the case in which he delegates his choice and his servant plays red for him. However, that difference doesn't matter to you and is outside your influence, so let us disregard it.) The information that you and your partner (and your respective servants) are much alike might persuade you to give little credence to the dependency hypotheses $K_{1,2}$ and $K_{1,3}$ but to give more to $K_{1,1}$, and likewise for the subdivisions of K_2 and K_3. Then you give your credence mostly to dependency hypotheses according to which you would either win or break even by delegating your choice. Then causal decision theory does not tell you, wrongly, that it is rational not to play. Playing by delegating your choice is your U-maximal option.

But you don't have a servant. What of it? You must have a tie-breaking procedure. There must be something or other that you do after deliberation that ends in a tie. Delegating your choice to your tie-breaking procedure is a fifth option for you, just as delegating it to your servant would be if you had one. If you are persuaded that you will probably win if you play because you and your partner are alike psychologically, it must be because you are persuaded that your tie-breaking procedures are alike. You could scarcely think that the two of you are likely to coordinate *without* resorting to your tie-breaking procedures, since *ex hypothesi* the situation plainly *is* a tie! So you have a fifth option, and as the story is told, it has greater expected utility than not playing. This is not the option of playing red, or white, or blue, straightway at the end of deliberation, although if you choose it you will indeed end up playing red or white or blue. What makes it a different option is that it interposes something extra – something other than deliberation – after you are done deliberating and before you play.

19. Ratifiability and stability

Wlodzimierz Rabinowicz

1. Background

In this paper I am going to discuss Richard Jeffrey's maxim of ratifiability – a maxim that enjoins the agent to make decisions that he can ratify once he has made them. (Cf. the second edition of *The Logic of Decision*, Jeffrey, 1983. See also Jeffrey, 1981.) I shall also consider a number of decision-theoretical proposals according to which ratifiability or "stability", which is a closely related property, constitutes at least a necessary, if not a sufficient, condition on rational options. I want to oppose this recent trend. I am going to suggest that stability considerations are, in a sense, irrelevant. They do not provide the agent with reasons for or against different options. The present paper is a successor to Rabinowicz (1985). There I suggested that stability considerations, while not being decisive, should at least be given some weight in decision-making – that they at least provide the agent with *some* reasons to act. I don't believe this any longer.[1]

First, a short background. In the first edition of his *Logic of Decision* (1965), Jeffrey advised us to choose so as to maximise the expected desirability of our actions. The expected desirability of an action was defined as the weighted sum of that action's desirability-values under different states (or, to use Jeffrey's own terminology, under different possible "conditions"). As his weights, Jeffrey used the conditional probabilities for the different states given the action in question. Thus, Jeffrey's weights were of the form: $P(S/A)$, where S stands for a possible states, A represents the action in question, and $P(S/A)$ – the conditional probability for S given A – is defined in the standard way,

---

[1] I am greatly indebted to Howard Sobel for his stimulating suggestions and generous criticism.

as the ratio between the probabilities for the conjunction of S and A and for A: $P(SA)/P(A)$. If $P(A)$ equals zero, $P(S/A)$ is left undefined.

Strictly speaking, A and S stand for the propositions to the effect that a given action (state) is performed (obtains): Jeffrey assumes that both P and the desirability function operate on propositions. For simplicity, however, I shall often talk about "actions" and "states" even when I mean to refer to the corresponding propositions.

The relevant probabilities are here given a "subjective" interpretation: the function P is supposed to measure the strength of the agent's beliefs just before the choice.

Jeffrey's approach from 1965 may be compared with the classical proposal developed by Leonard Savage in *The Foundations of Statistics* (1954). Savage chooses as his weights the *un*conditional probabilities for states: $P(S)$. In what follows, I shall talk about an action's *Savage-value* in order to refer to its expected value calculated in accordance with Savage's criterion in terms of the unconditional probabilities.

Jeffrey defended his own approach against Savage by pointing out that, in some cases, the states might be causally dependent on actions. And the agent might be aware of this dependence relation. In such cases, it would be obviously irrational of him to make his choice in terms of the unconditional probabilities for states.

However, there is every reason to suppose that in 1965 Jeffrey would have considered Savage's criterion as perfectly unassailable if it were restricted to those cases in which the agent, prior to the choice, is certain that the different states are causally *in*dependent of the available actions. Jeffrey would have thought, I believe, that such a *restricted Savage criterion* was just a special case of his own principle of choice. He would have assumed that firm beliefs about causal act-independence imply the corresponding independence in the probability assignments:

The Independence Assumption. If, just before the choice, the agent is certain that a state S is causally independent of an action A available to him, then

$$P(S/A) = P(S).^2$$

However, in the years that followed, the Independence Assumption has been rejected by several philosophers, including Jeffrey himself. [3]

---

[2] This assumption may be seen as a special case of a more general presupposition that seems to underlie Jeffrey's original approach. The differences between conditional probabilities for states given different actions are assumed to reflect *nothing* but the differences in those actions' expected causal influence on states. For a possible formulation of this general presupposition see note 6.

[3] See, for example, Nozick (1969), Gibbard and Harper (1978) [Chapter 17, this volume], Sobel (1978), Skyrms (1980a), Lewis (1981) [Chapter 18, this volume], and Jeffrey (1983).

The assumption fails in all situations in which, before the choice, the agent is certain that the states are outside his influence, but, at the same time, he considers his coming action as an *indication* of the obtaining state. Thus, the agent ascribes to his actions purely evidentiary bearings on states – perhaps because he believes that different states might cause him to perform different actions or because he thinks that his coming action and the obtaining state can both be traced back to the same underlying common cause. I shall call such cases *Newcomblike*, since the most famous case of this kind has come to be known as "Newcomb's Problem".[4] In Newcomblike cases, there is at least one possible state S such that, for some action A, $P(S/A)$ differs from $P(S)$, even though the agent is convinced that the state in question is causally act-independent.

It is obvious that, in Newcomblike cases, Jeffrey's original principle of choice may easily come into conflict with the restricted Savage criterion. The agent who follows Jeffrey's advice may choose an action simply because of its high "evidentiary value" – simply because its performance would indicate that the world is in an advantageous state. This may happen even when the action's expected *causal* effects would be worse than the effects the agent may expect to achieve by performing some alternative action instead. And, in fact, this is what does happen in Newcomb's Problem itself and in all the other Newcomblike cases mentioned in note 3. In all of them, it is the restricted Savage criterion and not Jeffrey's principle that leads to the right prescriptions. This is quite understandable: whenever the states are firmly believed to be causally independent of actions, the Savage-value of the action – the weighted sum of its values under different

---

On the other hand, the Independence Assumption is defended by Eells (1982) and, in a somewhat different way, by Eells (1984b). According to Eells, the Independence Assumption is going to hold if the agent is rational and self-knowing. For such an agent, all information about the relevant states that can be gained from learning how he is going to act is already available before he makes his choice. If the agent is rational, his choice and the ensuing action are determined by his beliefs and desires. If the agent is self-knowing, he will know the determining factors and the determining mechanism itself before the choice. In Eells (1984b), this line of reasoning is supported by a description of a dynamic process of ideal deliberation with continual recalculation of expected desirabilities and corresponding readjustment of probability assignments – a process which eventually leads the agent to a system of beliefs satisfying the Independence Assumption. Obviously, Eells' defense makes very severe demands on the agent's rationality and self-knowledge. Furthermore, as has been argued by Sobel (1986b), Eells' recipe for ideal deliberation sometimes seems to lead to intuitively unreasonable choices. If I am not mistaken, Sobel's argument equally well applies to yet another defense of the Independence Assumption to be found in Price (1986).

[4] This problem was first presented in Nozick (1969). Here belong also certain versions of the Prisoner's Dilemma, in which each prisoner "sees his own choice as a strong clue to the other's" (Jeffrey, 1983, p. 16; see also Lewis, 1979b, and Sobel, 1986b), Smoker's Choice (Skyrms, 1980a, Solomon Case (Gibbard and Harper, 1978), Popcorn Problem (Sobel, 1986b, c), and many other more or less fanciful cases.

states, with weights being the prior probabilities for the states in question – seems to constitute a proper measure of that action's expected causal efficacy. Its "Jeffrey-value", on the other hand, measures only its "value as news". It seems clear that the choice should go by the former and not by the latter.

When confronted with Newcomblike cases, Jeffrey (1983) decided to revise his original proposal and suggested instead a new principle of choice: the maxim of ratifiability. Therefore, it seems appropriate to compare this new maxim with the restricted Savage criterion. Are the two compatible with each other or do they sometimes lead to conflicting prescriptions? This is the question that I am going to consider in sections 3 and 4, after a short presentation of the ratifiability idea in section 2. In section 5, I am going to discuss what happens when we replace ratifiability with stability.

It should be noted that a principle essentially identical with the restricted Savage criterion has been endorsed by Gibbard and Harper (1978). They advise the agent to act so as to maximize the "U-value", where the U-value of an action A is the weighted sum of its desirability values under different states, with weights being the agent's probabilities for corresponding subjunctive conditionals. Thus, instead of using conditional probabilities, $P(S/A)$, we are to use probabilities for conditionals, $P(A \mathbin{\square\!\!\rightarrow} S)$. This conception of U-value presupposes the validity of "Conditional Excluded Middle":

$$(A \mathbin{\square\!\!\rightarrow} S) \vee (A \mathbin{\square\!\!\rightarrow} -S).^{[5]}$$

At the same time, Gibbard and Harper provide an explication of certain causal act-independence in terms of subjunctive conditionals. Assuming Conditional Excluded Middle, the agent may be said to be certain that S is causally independent of A whenever he is certain that $A \mathbin{\square\!\!\rightarrow} S$ holds iff S obtains. Thus, whenever

$$P((A \mathbin{\square\!\!\rightarrow} S) \longleftrightarrow S) = 1.^{[6]}$$

It is easily seen that $P(A \mathbin{\square\!\!\rightarrow} S)$ coincides with $P(S)$ when the agent is certain that S is causally independent of A. Thus, given certain causal act-independence, the U-value of an action turns out to be just its Savage-value.

---

[5] For the theories that avoid this assumption and replace probabilities of conditionals with a more complicated notion of "probable chances", see Sobel (1978, 1986c). Cf. also Lewis (1981) [Chapter 18, this volume]. The theory of Lewis is compared with Sobel's in Rabinowicz (1983).

[6] Gibbard and Harper [Chapter 17, this volume] also provide a formulation of the general presupposition mentioned in note 2. The tacit assumption behind Jeffrey's approach from 1965 seems to be that, for every state S and every available action A, the agent's probability for $A \mathbin{\square\!\!\rightarrow} S$ coincides with his conditional probability for $A \mathbin{\square\!\!\rightarrow} S$ given A. From this, together with Gibbard and Harper's subjunctive explication of certain causal independence, the Independence Assumption follows as a special case.

It should, however, be noted that Gibbard and Harper endorse maximization of U-value only when the partition of states is chosen with care: only when, for each state S and each available action A, the conjunction of S and A determines the outcome in all value-relevant respects – in all respects that the agent cares about. This qualification, while important, is not relevant for our discussion. In all the examples that we are going to consider, it may be safely assumed that nothing of substance depends on our partitioning of states.

2. Ratificationism

In the second edition of *The Logic of Decision*, Jeffrey points out that *decisions* to act usually are just as reliable indicators of states as the actions themselves. In other words, the evidentiary value of the action is preempted by the corresponding decision. This is something that we can profitably use, according to Jeffrey. When we confront a choice, we should first make a purely hypothetical assumption about our coming decision to act and only then – on the basis of this assumption – should we compare the different actions that are available to us with respect to their expected desirability. The idea is that conditionalizing expected desirability on a hypothetical assumption about our decision to act *screens off*, so to speak, the differences between the actions that depend only on their different evidentiary value. And we want to screen them off, since purely evidentiary differences between actions should not be allowed to determine the choice.

Let me now describe Jeffrey's proposal in some more detail. Let "dA" stand for the proposition that A is the action that the agent will finally decide to perform. Observe that, according to Jeffrey, a decision is final if it is not going to be changed or retracted. On the other hand, implementation failures are always possible. It is always possible that the agent will fail to carry out his final decision and that he will perform some other action instead. Therefore, Jeffrey assumes that, for any two actions A and B, the conditional probability for A given dB is always higher than zero, even though this probability may be extremely small. [7]

If A and B are alternative options, then *the expected desirability of A given dB* is defined as the expected desirability value assigned to A on the hypothetical assumption that the agent's final decision will be to

---

[7] This assumption is necessary if Jeffrey's ratificationism is to work (see below, note 8), but its plausibility may well be disputed. It is one thing to admit that I can always fail to carry out my final decision, and quite another to allow that such to failure may in principle result in my performing a totally different action instead. Can I really fail to carry out my final decision to stay at home tonight in such a way that I instead go to a movie? It would be different if I "changed my mind", but then my decision to stay at home would not, after all, be final!

perform the action B. This value is calculated in the same way as the expected desirability of A on Jeffrey's original proposal, but the weights are now different. All the weights of the form: $P(S/A)$, are now conditionalized on the hypothetical assumption dB. Thus, the new weights look as follows:

$P(S/A\&dB)$.

Now we can explain what Jeffrey means by ratifiability. The intuitive idea is simple: a decision to perform an action A is ratifiable iff, on the assumption that A is the action that the agent will finally decide to perform, A exhibits maximal expected desirability. Or, to be more precise,

a decision to perform A is *ratifiable* iff the expected desirability of A given dA is at least as high as the expected desirability of any alternative action B given dA.

Note that, in this definition, we conditionalize each time on the same hypothetical decision.

In what follows, I am going to talk sometimes about ratifiable *actions*. An action is said to be ratifiable if the decision to perform it is ratifiable.

Jeffrey's maxim of ratifiability prescribes that the agent should make ratifiable decisions. Thus, ratifiability is taken to be a necessary and sufficient condition on choiceworthy actions.[8]

How does this maxim take care of Newcomblike cases? I have already sketched Jeffrey's answer to this question. By conditionalizing on a decision to act we can usually screen off the purely evidentiary differences between actions: the decision to act constitutes an equally reliable indicator of the obtaining state as the action itself.

In Newcomblike cases, in which the agent is certain that the states are outside his influence, this "screening assumption" amounts to the demand that conditionalizing on a decision to act should make the probability of each state independent of which action the agent is going to perform:

The Screening Assumption. If, just before the choice, the agent is certain that a state S is causally independent of an action A available to him, then, for any alternative action B,

---

[8] As was mentioned above, Jeffrey assumes that $P(A/dB) > 0$, for all alternative actions A and B. This amounts to the condition that $P(A\&dB)$ is always higher than zero. It can now easily be seen why Jeffrey needs this assumption. If $P(A\&dB) = 0$, then all the weights of the form $P(S/A\&dB)$ will be undefined. Therefore, it will not be possible to define the expected desirability of A given dB and, consequently, it will not be possible to apply the ratifiability maxim to the action B. We will not be able to compare B with A in accordance with Jeffrey's recipe, and the decision to perform B will be neither ratifiable nor unratifiable.

$$P(S/A\&dB) = P(S/dB).^9$$

Note that, given the Screening Assumption and given certain causal act-independence, the formula for expected desirability of A given dB can be simplified. Jeffrey's weights of the form: $P(S/A\&dB)$, can now be replaced by

$$P(S/dB).$$

Thus, the reference to A can be avoided. This is due to the fact that the agent is certain that the performance of A cannot have any causal effects on S and, at the same time, having already conditionalized on dB, he does not ascribe to A any further evidentiary bearings on S (Screening Assumption!).

With purely evidentiary differences screened off by the hypothesis about the agent's decision for A, it becomes possible to compare A with its alternatives with respect to their expected causal value. And this is the only kind of value that really should count when it comes to choice. Thus, *if* the Screening Assumption is satisfied, the ratifiability approach seems to work in Newcomblike cases. Or, at least, this is what Jeffrey thought in 1983.

In what follows, I shall not discuss the obvious objection against the Screening Assumption that has been made by Bas van Fraassen [cf. Jeffrey (1983); see also Sobel (1986b)]: Though a decision to act might be a good indicator of the obtaining state, the action itself might sometimes turn out to be an even better indicator. In such cases, the ratifiability approach will not be successful in screening off the mere "news values" of the actions. Jeffrey himself admits that this objection is well-taken and he therefore explicitly restricts his ratifiability maxim to cases in which the Screening Assumption is satisfied. For the argument's sake, I shall follow him in that respect.

Jeffrey needs the Screening Assumption because he conditionalizes the expected desirability of an action A on a *decision* to perform another action B. He cannot close the gap between decisions and acts and talk about the expected desirability of A given B *itself*. This is impossible because the agent's probability for A given B equals zero. (The agent is assumed to know that alternative actions are mutually incompatible.) In consequence, the expected desirability of A given B cannot be defined.[10]

---

[9] In the general case, when we allow that the states might be causally act-dependent, the Screening Assumption could be expressed as the following demand: for any S, and for any available actions A and B, $P(S/A\&dB) = P(A \; \Box\!\!\rightarrow \; S/dB)$.

[10] To solve this problem, Jeffrey might try to use *non-standard* conditional probabilities – probabilities that are defined even when the conditions are assigned probability zero. Thus, Sobel (1986b) writes:

The situation becomes radically different when we move to 'causal' decision theories, which explicitly make use of causal or subjunctive notions in the definition of expected value. Thus, to give an example, Gibbard and Harper can talk about *the U-value of A given B*, which is then simply the weighted sum of A's desirability-values under different possible states, with weights being of the form:

$$P(A \ \square\!\rightarrow S/B).$$

Clearly, such weights may have definite values even when the agent's probability for A given B equals zero. Note also that, when the agent is certain that the states are outside his influence, these weights can be simplified: Given Gibbard and Harper's explication of certain causal act-independence, $P(A \ \square\!\rightarrow S/B)$ will then equal $P(S/B)$. Thus, under these circumstances, the U-value of A given B will coincide with A's Savage-value given B.

On this approach, there is no need to rely on the Screening Assumption. Conditionalizing on the decisions to act gives way to

"Given such non-standard conditional probabilities, ratifiability could be defined in terms of whether or not an act excels in Desirability, on the supposition that it *itself* will take place. Decisions as things distinct from acts would not come into it."

Let us try to develop this suggestion a little in order to see where it leads. If P is the agent's probability function, and X is any proposition describing a possible state of affairs, then P_X shall stand for the hypothetical probability function that codifies the agent's beliefs upon becoming certain that X. Clearly, $P_X(X)$ equals 1. P_X is taken to be defined even when $P(X)$ equals zero, but we assume that, whenever $P(X)$ is higher than zero, P_X coincides with the standard conditional probability function $P(/X)$. Note that conditionalizing may be iterated. Thus, we can talk about such probability functions as $(P_X)_Y$ and so on. Observe that $(P_X)_Y$ does not have to coincide with $P_{X\&Y}$. In fact the latter function may be undefined. This will happen when X&Y describes an impossible state of affairs. (For further discussion, see Gärdenfors (1987), chapter 5.)

Now, instead of talking about the expected desirability of A given the *decision* for B, we may try to close the gap between acts and decisions by working with the expected desirability of A given the action B *itself*, where the latter notion would be defined in terms of non-standard conditional weights of the form: $(P_B)_A(S)$. P stands, as usual, for the agent's probability assignment just before the choice.

However, this maneuver is in fact much less promising than one might expect. There is very little reason to think that such non-standard conditional weights would satisfy the screening assumption. It does not seem reasonable to suppose that, whenever the agent is certain that the states are outside his causal influence, $(P_B)_A(S)$ will equal $P_B(S)$, for all states S and all alternative actions A and B.

In fact, the opposite seems to be the case. Note that, insofar as the agent, before the choice, assigns to B a non-zero probability, $P_B(S)$ equals $P(S/B)$. At the same time, since the agent knows that A and B are incompatible actions, it is very natural to expect that $(P_B)_A(S)$ will normally simply coincide with $P(S/A)$ and that it will therefore differ from $P(S/B)$ in the Newcomblike cases. To put it differently, if the agent were to become certain of A after first becoming certain of B, then A's evidentiary bearings would *not* be screened off by the agent's prior belief in B. Upon becoming certain of A, the agent's prior belief in B would not be able to function as a 'screen' since it would then simply disappear!

conditionalizing on the actions themselves – ratifiability gives way to "stability":

> An action A is *stable* iff, given A, A's U-value is at least as high as the U-value of any other alternative action given the same assumption.

I shall return to the discussion of stability in the last section.

3. Jeffrey meets Savage – first round

What happens when none of the available options is ratifiable? Jeffrey's maxim will then not give us any guidance. Jeffrey suggests that a case like that can only arise for an irrational agent: When none of the options is ratifiable, there is reason to suspect that the agent's beliefs or preferences must be pathological in some way. Why does Jeffrey think so? I must admit I have no idea. In Rabinowicz (1985), I describe a clearly non-pathological Newcomblike case in which the agent lacks a ratifiable option. And, as a matter of fact, a similar case had already been described by Gibbard and Harper way back in the seventies. (Cf. the "story of the man who met death in Damascus", Gibbard and Harper, 1978, section 11, pp. 372–375, this volume.) The existence of such cases points to an important difference between the ratifiability maxim and the restricted Savage criterion: The latter provides guidance even in the absence of ratifiable options. Here, however, I prefer to consider a Newcomblike case in which the maxim of ratifiability does yield a definite prescription. It will turn out, however, that this prescription conflicts with the one that would follow from the Savage criterion. (Similar examples have been presented in Rabinowicz, 1985; Skyrms, 1984; see also Harper, 1986.)

Crossroads

We imagine two agents, X and Y, who want to meet each other. They have started from different directions and now each of them has come to a crossroad. Each of them has to choose between going left (L), right (R) or straight ahead (S). The situation looks as follows:

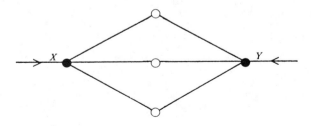

The points show X and Y's present positions, while the circles represent possible meeting-places. For simplicity, we assume that neither of the agents can just take it easy and wait for the other. They must both keep on the move.

As easily seen, they will meet only if they both go straight ahead or if one goes right while the other goes left. The road straight ahead, being quite winding, will take just as long as the other two. Also, whoever takes it will have to pay a rather high road-tax. (If we want, we may imagine that X and Y are two medieval merchants and that the road straight ahead passes through the domains of a powerful baron.) Furthermore, we assume that, for each of them, the road to the left is more convenient than the other two and that they both know about it. All these factors together determine their desirability-values for the different outcomes. Here, I present only the value matrix for X. Y's matrix is assumed to be exactly analogous.

		Y		
		L	S	R
	L	1	1	11
X	S	−5	5	−5
	R	10	0	0

The best event (11) is to meet the other agent after having taken the convenient road to one's left. But even if the meeting takes place after one has chosen the less convenient road to the right, there will still be reason to rejoice (10). Paying the road-tax in order to meet is not too attractive (5), but it is better than not meeting at all – independent of whether the road was otherwise convenient (1) or not (0). And the worst event (−5) is, of course, to pay the road-tax and still miss the meeting!

No communication between the agents is possible. The distance is too great and there are no telephones available. I also assume that neither of the agents can make his choice by tossing a coin or a die: Mixed strategies are not allowed. (Perhaps they will be severely punished if they try to randomize, or perhaps their religion forbids leaving important choices to chance.)

X and Y are assumed to be very similar to each other psychologically. They are more or less psychological twins and they know that. Furthermore, they have quite often confronted similar coordination problems before and they have always or nearly always acted in the same way: left-left, right-right, or straight ahead-straight ahead.

However, fearing the road-tax and seeking comfortable travel, they have usually chosen to go to their left. Thus, they have tended to play it safe. The worst possible outcome if one goes to the left (1) is better than the worst outcomes for the other two options (-5 and 0, respectively).

At this point, the reader might want to point out that X and Y should have long ago coordinated their behaviour through a suitable agreement. For example: "In the future, I (X) shall go to the left, while you (Y) will take the less convenient road to your right, for which you will receive a suitable compensation." Yes, they should have done that, but now it is too late. They stand at the crossroads and they must do the best they can.

Each agent knows all that we know and this knowledge determines his probability assignments. In fact, their probability judgments are quite analogous. If we therefore concentrate on X and let P be his probability function just before the choice, then the following holds:

1. P(Y goes to the left) is high, say, higher than .5;
2. P(Y goes in the direction α/d(X goes in the direction α)) is very high, say, higher than .8.

Assumption (1) depends on X's knowledge about Y's tendency to play it safe; (2) holds for every direction α may stand for (to the left, to the right, or straight ahead) because X considers his coming final decision to act, whatever it may be, as a reliable indication of how Y, his psychological twin, is going to behave. (If this is to be true, then X's belief in Y's playing it safe should be less "robust" than his confidence in Y's being his psychological replica. Cf. Levi, 1975, for a similar observation in connection with Newcomb's Problem.) Let us also, for the argument's sake, presuppose that the Screening Assumption is satisfied. This completes my description of the Crossroads Case.

Clearly, the Crossroads Case is Newcomblike: X is convinced that the relevant states (Y's actions) are outside his influence (X and Y act independently), even though he considers his own actions and decisions to act as reliable state-indicators.

The restricted Savage criterion prescribes that X should go *to the right*. As we know, the unconditional probability for Y's going to the left is high [cf. (1) above], and if Y goes to the left, the agents will meet only if X goes to the right.

The maxim of ratifiability prescribes, on the other hand, that X should go *straight ahead*, this being the only ratifiable option. Let me explain. Assume, first, that X decides to go straight ahead. On this hypothetical assumption, it is very probable (from X's point of view) that Y will go straight ahead. According to (2), this conditional probability exceeds .8. If Y goes straight ahead, they will meet only if X does likewise. In fact, it is easy to calculate that, given the assumption that X decides to go straight ahead, his expected desirability for this option exceeds 3 [$.8 \times 5 + .2 \times (-5)$]. Given the same assumption, however, the expected desirability of going to the left is lower than 3 (at the very best, it is still lower than $.8 \times 1 + .2 \times 11$). The corresponding value for going to the right is lower than 2 ($.8 \times 0 + .2 \times 10$). Thus, going straight ahead is clearly ratifiable. On the other hand, suppose that X decides to go to the right. On *this* hypothetical assumption, it becomes highly probable that Y will go to the right [cf. (2) above]. Then X, by going to the right, will miss the meeting. The meeting will take place only if X, contrary to his decision, goes to the left instead. It is therefore obvious that going to the right is not a ratifiable option. An analogous argument shows that going to the left is not ratifiable either.[11]

That there exists this conflict beween the two principles is easy to understand. When the agent is convinced that the states are outside his control, the ratifiability maxim demands that the agent should make his choice in terms of his *posterior* probabilities for states – the probabilities of the form: $P(S/dA)$. Assuming, as Jeffrey does, the conditionalization model for probability kinematics, we might put this demand as follows: The choice is to be made in terms of the probabilities that the agent would assign to the states after the choice. The restricted Savage criterion recommends, on the other hand, using one's *prior* probabilities – the ones that the agent assigns to the states before the choice. It is obvious that, in the Newcomblike cases, the two kinds of probabilities will differ from each other. In such cases, the agent's choice is assumed to have evidentiary bearings on states. It is

---

[11] Were the options "to the left" and "to the right" fully symmetric in value-relevant respects (which would be the case if the two alternatives were equally convenient), then, as Peter Gärdenfors has pointed out to me, the case for their unratifiability would be much weaker. Thus, let r be the road that lies to X's right and to Y's left. Now, if X decides to take r, then, knowing that Y is his psychological twin, he *might* consider it highly probable that Y will take r as well, in which case the meeting between them *is* going to take place! Thus, it would seen that a given option might appear as ratifiable or unratifiable depending on how the agent describes it to himself. How is he to decide which description is the appropriate one?

By introducing the asymmetry in the relevant outcomes, we make this description task much easier. It is plausible to suppose that psychological twins make *relevantly* similar decisions – decisions similar with respect to what they *care* about. Thus, on the hypothetical assumption that X decides to choose the less convenient road (to his right), it is to be expected that Y will do likewise. And for Y, road r is the one that is *more* convenient!

In that respect, the Crossroads Case is clearly more satisfactory than the symmetrical Vexing Coordination Problem in Rabinowicz (1985).

therefore not astonishing that the two principles of choice will sometimes lead to conflicting prescriptions.[12]

I shall now consider two objections that might be leveled against
Savage by someone who finds Jeffrey's advice attractive in the
Crossroads Case.

Objection 1. According to Savage, X should go to the right. Since the
two agents are symmetrically placed, the same advice must apply to Y.
But if both agents follow Savage's prescription, they will never meet.
They would do better if they followed the ratifiability maxim
and went straight ahead instead. Thus, the restricted Savage criterion
recommends to each agent a way of acting such that its combined
result is worse for each agent than the combined result of an
alternative way of acting. Doesn't this show that such a principle of
choice cannot be upheld?

Savage can respond to this objection by pointing out that the same
difficulty may equally well arise for Jeffrey. Given the Screening
Assumption, it can be shown that, under the conditions of certain
causal act-independence, the ratifiability maxim, just as the restricted
Savage criterion, always recommends the agent to perform the
dominant option, if such an option exists. (An option is dominant if
under every possible state it gives the agent more than the alternative
options.) This is why the ratifiability approach, unlike Jeffrey's original
principle of choice, gives the right advice in Newcomblike cases
containing the dominant option. (All the famous Newcomblike cases,
including Newcomb's Problem itself, are of this kind. On the other
hand, in the Crossroads Case, no action is dominant.) But this also
means that, in all Prisoner's Dilemmas, in which the combined result
of the agents' dominant actions is bad for each agent, Jeffrey's
principle of ratifiability will lead to the same unsettling effect as does
the restricted principle of Savage in the Crossroads Case. Each agent
would be better off if all of them, instead of following the principle,
chose an alternative way of acting.

That his maxim works that way in Prisoner's Dilemmas is, of course,
no news for Jeffrey and, in fact, nothing that he considers to be a

---

[12] As Harper (1986) points out in connection with Skyrms' example and the earlier
example of my own, the situation would contain more than one ratifiable alternative if
we allowed the agents to use mixed strategies. Thus, if we removed the penalties on
randomizing and if we assumed that X's decision to use a given mixed strategy would
make it highly probable that Y, his psychological twin, is going to randomize in the
same way, then X's tossing a coin in order to determine whether he is to go to the left
or to the right would not only be ratifiable but it would also exhibit a higher
Savage-value than going straight ahead. (To be more precise, if the coin tossing is to
be ratifiable, the coin will have to be slightly biased toward going to the left: going to
the left will have to get nine chances out of sixteen.) Harper is right, of course. But
even in such circumstances, the restricted Savage criterion would still conflict with the
ratifiability approach. Going to the right would still be the only alternative that
maximizes Savage-value. And going to the right is unratifiable.

weakness. Quite the contrary (cf. Jeffrey, 1983, chap. 1). He probably would never have criticized Savage in the way I have suggested. On the other hand, the objection that follows is much more congenial to Jeffrey's approach.

Objection 2. If X decided to go to the right, as the Savage criterion prescribes, then, having made this decision, he would come to believe that Y, his psychological twin, was also going to the right. This belief on X's part would be perfectly rational under such circumstances. But if Y goes to the right, X, by going to the right, will miss the meeting. Doesn't this show that the decision to go to the right is irrational? This decision itself would provide the agent with a reason against carrying it out. It would lead X to a belief in light of which the chosen action must appear as clearly disadvantageous. Shouldn't X try to avoid this (predictable) disappointment and instead make his decision in terms of his hypothetical *after-choice* probabilities?

Perhaps Savage would respond as follows: It's true that the decision to go to the right would lead X to believe that Y is also going to the right and it's true that this belief would be perfectly rational under these (hypothetical) circumstances. *But,* before the choice, X believes that Y will go to the *left.* What's more, X is sure that his own choice, whatever it might be, cannot influence Y's behaviour. This means that X, before the choice, considers his hypothetical future belief as most probably false. Rational (under the hypothetical circumstances), but still false. What reason is there for X to take such (most probably mistaken) hypothetical beliefs into consideration when he makes his decision? None at all.

Let me develop this argument in some more detail. As we know,

1. P(Y goes to the left) is high.

X is certain that Y's behaviour is causally independent of what X himself does or decides to do. Applying Gibbard and Harper's subjunctive interpretation of certain causal independence (see above, section 2), we get:

2. $P((d(\text{X goes to the right}) \,\square\!\!\rightarrow\, \text{Y goes to the left}) \leftrightarrow \text{Y goes to the left}) = 1$.

Given (2), (1) implies:

3. $P(d(\text{X goes to the right}) \,\square\!\!\rightarrow\, \text{Y goes to the left})$ is high.

To put this conclusion less formally:

Prior to the choice, X deems it highly probable that Y would go to the left (even) if X himself were to decide to go to the right.

This means that deciding to go to the right should lead X to adopt beliefs about Y's behaviour that, in terms of X's pre-choice probabilities, most probably would be mistaken. Certainly, X has no reason to take such hypothetical beliefs into consideration when he makes his choice. He has no reason to abstain from choosing the road to his right, even though this option is unratifiable. It is true that his decision to perform this option would *give* him a reason against carrying it out. But he doesn't *have* this reason when he makes his decision!

4. Savage meets Jeffrey – second round

The trouble with Jeffrey's original principle of choice from 1965 was that it advised that agent to engage in "wishful acting" – to perform an action not because of its expected causal effects but because of its hypothetical evidentiary bearings – because performing it would indicate that the world is in a state advantageous to the agent.

The ratifiability approach is a more subtle version of such wishful acting. The maxim of ratifiability advises the agent to choose an action not because of its actually expected causal effects but because choosing it would indicate that the world is in a state advantageous to the agent who makes that choice and carries it out. If this is not "wishful acting" then it is at least "wishful choosing".

Jeffrey rejected his original principle of choice when he confronted Newcomblike cases "with dominance". In such cases, it may well happen that Jeffrey's old principle picks out a dominated action – an action that gives less to the agent under each possible state – even though the states are assumed to be certainly causally act-independent. Clearly, the restricted Savage criterion would never lead to such a perverse prescription and it is a virtue of Jeffrey's new principle – the ratifiability maxim – that it coincides with the Savage criterion in this respect.

However, the maxim of ratifiability, while respecting dominance, does not respect *weak* dominance. Sometimes, when all the states prior to the choice are positively probable and known to be causally act-independent, Jeffrey's maxim picks out an option A and forbids another option B even though B weakly dominates A – even though B never gives less to the agent and sometimes gives more.

Admittedly, such things may happen only under extreme and therefore quite unrealistic assumptions about the agent's conditional probability assignments, only when certain *ex ante* positively probable states are assigned probability zero, given certain decisions to act.[13]

On the other hand, no such extreme assumptions are needed in order to show that the maxim of ratifiability sometimes violates a closely related and intuitively very plausible condition of *Indifference*.

---

[13] Consider a case with three available options A_1–A_3, and three certainly causally act-independent states, S_1–S_3, where all the states are positively probable prior to the choice and where, for each $A_i (1 \leq i \leq 3)$, the decision to perform A_i makes the corresponding state S_i more probable than not: $P(S_i/dA_i) > \frac{1}{2}$. In addition, assume that $P(S_3/dA_1)$ equals zero. (This is the extreme probability assignment that I have

Indifference: If the available actions *A* and *B* exhibit the same desirability-values under each possible state, and if the states are certainly causally act-independent, then *A* is choiceworthy iff *B* is choiceworthy.

It is easily seen that the restricted Savage criterion implies Indifference as a special case. That Indifference is violated by the ratifiability maxim is shown by the example that follows.

Open boxes

Suppose that our old friends, X and Y, are placed in separate rooms, without being able to communicate with each other. In each room there are two boxes, one red and one white, both of which are open. Each box is seen to contain one hundred dollars. Each agent has to choose among three options: taking the red box, taking the white box, or abstaining. The agents have been informed that, if either abstains, he will receive nothing *unless* the other agent takes his red box. In the latter case, but only then, the agent who abstains is going to receive the contents of both boxes: two hundred dollars. Assuming that their desirabilities go by money and that each of the agents cares only about his own profit, the desirability matrix looks like this:

		Y		
		Red	White	Abstain
	Red	1,1	1,1	1,2
X	White	1,1	1,1	1,0
	Abstain	2,1	0,1	0,0

been referring to. Note that this assignment is compatible with Jeffrey's demand that each action is to be positively probable given any final decision to act: on the hypothetical assumption dA_1, A_3 may have positive probability even though S_3 would have probability zero. Suppose that the desirability-matrix looks as follows:

	S_1	S_2	S_3
A_1	1	0	0
A_2	1	0	1
A_3	0	1	0

It is easily seen that A_1 is the only ratifiable option even though A_2 weakly dominates A_1. Thus, in this case, the maxim of ratifiability violates weak dominance.

(In each cell, the first number represents X's desirability value, while the second gives the corresponding value for Y.)

Since each agent knows that the other agent is his psychological replica, he considers his coming decision and action, whatever they will be, as reliable indications of the obtaining state (the other agent's behaviour). Further, we suppose that each agent deems it at least as probable that the other agent is going to opt for White or abstain as that he is going to opt for Red. Concentrating on X, we get the following probability assignments (Y's probabilities are, of course, analogous):

1. $P(Y \text{ takes White or abstains}) \geq P(Y \text{ takes Red})$;
2. $P(Y \text{ performs } \alpha/d(X \text{ performs } \alpha)) > .5$, for any action α (Red, White, Abstain).

Since each agent is convinced that the states are causally act-independent, the situation is clearly Newcomblike. We also assume that the Screening Assumption is satisfied.

Given (1), the probability of getting two hundred dollars if one abstains does not exceed one-half; thus, the Savage-value of abstaining does not exceed 1. In consequence, Red and White, which give 1 for certain, both maximize Savage-value. But only White is ratifiable. Choosing Red would make it more probable than not that abstaining is a better choice, though choosing to abstain would make both Red and White more attractive. On the other hand, choosing White would indicate that abstaining is not to be preferred. Thus, White is the only ratifiable option even though Red would lead to exactly the same results under each possible state: the two boxes contain exactly the same amounts of money!

The ratifiability approach, while satisfying Dominance, still violates Indifference. This is what happens when you engage in wishful choosing.

5. Stability

Advice to perform a *stable* action – the "maxim of stability" – could be seen as a causal theorist's analogue of Jeffrey's maxim of ratifiability. Such a maxim would avoid the problems arising in connection with the Screening Assumption, but it would still invite the agent to engage in wishful activities. An action is stable if its performance indicates that the world is in a state advantageous to the agent who performs that action. The maxim of stability would advise the agent to perform an action not because of its actually expected effects but because of its own hypothetical evidentiary bearings on its own effects.

Insofar as a decision to act and the ensuing action itself are equally good state-indicators, the maxim of stability and the maxim of ratifiability are going to lead to the same prescriptions. Thus, in the Crossroads Case, going straight ahead is the only stable action even though its Savage-value is less than maximal. (To get this result, we only have to replace the assumption about the evidentiary bearings of X's decisions to act on Y's actions by the corresponding assumption about the evidentiary bearings of X's actions themselves. As we remember, when the agent is certain that the states are causally act-independent, his conditional probabilities for states given an action constitute appropriate weights when it comes to determining whether the action in question is stable or not. Cf. above, section 2.)

Analogously (and once again assuming that the evidentiary bearings of the agent's decisions to act transfer to the actions themselves), White turns to be the only stable option in Open Boxes, even though White and Red would lead to exactly the same results under every possible state. Thus, the maxim of stability would violate Indifference. It would also violate Weak Dominance given extreme values for conditional probabilities of states given actions.

The maxim of stability was suggested by Weirich (1985), but he is no longer prepared to defend it [cf. Weirich (1986)]. Instead, the existing proposals combine, in various ways, stability with maximization of Savage-value – or, in the general case, when the states are no longer assumed to be certainly causally act-independent, with something like maximization of U-value. Stability is seen as a necessary but not as a sufficient condition of rationality and choiceworthiness. (For an excellent review of these various proposals, see Sobel, 1986a.)

To give some examples, Harper (1986) argues for a lexicographical approach, in which stability comes prior to maximization. An action is rational if it is stable and if, in comparison with *other stable options*, its Savage-value (or, in the general case, its U-value) is maximal. In other words, first remove the unstable options and then maximize within the remaining set!

Eells (1985b) suggests a more complicated procedure which starts with maximization and then continues until a stable option is reached. This "ideally persistent maximization theory" can perhaps be described as follows. (Both the label and the description are due to Sobel, 1986a.)

Let us say that there is a *path* from option A to option B iff there is a sequence of options starting with A and ending with B such that, for any option C in the sequence, distinct from B, the Savage-value (U-value, in the general case) of C *given* C is not greater than the Savage-value of the next option in the sequence given the *same* assumption. [For this notion of conditionalized Savage-value (U-value) see above, the end of section 2.] To put it metaphorically, there is a path from A to B if we can reach B from A without loss of expected value, provided that, in the process, we continually adjust our expectations in view of changing assumptions about the action that we are going to perform.

According to the ideally persistent maximization theory, an action B is rational iff B is stable and there is a path to B from some action A that maximizes Savage-value (U-value). In other words, first maximize and then, starting from that point, continue to readjust your expectations until you reach a stable solution!

Finally, we have Sobel (1983, 1986a), whose proposal is, in a way, the simplest of the three: stability and maximization are taken to be necessary and jointly sufficient conditions of rationality. Thus, in order to be rational, an action should maximize in terms of *both* the agent's actual pre-choice probability assignments *and* his hypothetical after-choice probabilities.

Since all these "mixed" proposals treat stability as a necessary condition on rational actions, thereby making a virtue of wishful acting, and since, as I have been arguing, the agent has no reason to abstain from an unstable action just because its performance (or a final decision to perform it) would *give* him such a reason, I must reject all of them. In my view, they all have unacceptable consequences. All of them violate Indifference in Open Boxes and, in the Crossroads Case, they all deny that going to the right is the rational thing to do. (Sobel, however, unlike the others, would say that going straight ahead is not rational either.)[14]

Sobel (1986a) admits that the instability of an option does not constitute any reason against its performance, and that, in particular, the agent need not take into consideration his hypothetical after-choice beliefs which he, before the choice, considers as most probably mistaken. Sobel thinks, however, that this criticism of mine does not touch his own theory. He writes (1986a, section on "Arguments and Perspectives"):

this theory is offered not as a guide to making rational decisions or in a prescriptive spirit as a system for generating practical directives and recommendations. It is intended in a descriptive-explanatory spirit as a statement concerning how rational actions are related to beliefs and preferences.

Turning to specific points in or suggested by Rabinowicz's argument, I do *not* think that the instability of a decision (to act) should be *taken into consideration* by the agent, or that it gives him a *reason* against deciding (for an action). I claim only that if an agent is rational he will not make a decision he could not maintain on reflection – *that rational processes for translating beliefs and*

---

[14] Just as it was with ratifiability (see above, the beginning of section 3), it is easy to show that certain Newcomblike situations will lack a stable solution. Thus, like the ratifiability approach and unlike the straightforward Savage-value (or U-value) maximization, mixed theories are "incomplete". They entail that, in some situations no option is rational. However, in my view, this is only a relatively minor difficulty; for a theory of rational action, completeness is a plus but not a must.

desires into actions cannot issue in ideally unstable final decisions.... Maximization-cum-stability is proposed as a theory of how rational actions are "externally" related to beliefs and preferences, and not as part of a theory of reasons, of what things 'figure in' reasons for action.

If this is what his theory of rational action is about, then Sobel is right: stability *is* important. Thus, in Open Boxes, a decision to perform the unstable Red would not be "maintained to reflection" – such a tentative decision would lead a reflective agent to change his beliefs and, in consequence, to change his choice. Thus, for a reflective agent, decisions for unstable actions would not be final (at least, providing that, as in Open Boxes, he has at his disposal some other action that is both stable and maximizing). On the other hand, as Sobel would probably agree, the agent in Open Boxes has just as good reasons for Red as for the stable White. Thus, were he to decide finally for Red and act accordingly, he would have had a perfectly good reason for what he has done (even though he would then prove himself not to be sufficiently reflective).

Unlike Sobel, I am primarily interested in a theory of practical rationality that studies the agent's reasons for his decisions and actions. From *this* perspective, stability is *not* a virtue. Being a virtue is a context-dependent feature.

References

Aczel, J. *Lectures on Functional Equations and Their Applications*, Academic Press, New York, 1966.

Allais, M. "Le comportement de l'homme rationnel devant le risque: Critique des postulats et axioms de l'ecole americaine," *Econometrica, 21* (1953), 503–546.

Allais, M. "The foundations of a positive theory of choice involving risk and criticism of the postulates and axioms of American school," (translation of "Fondements d'une théorie positive des choix comportant un risque et critique des postulats et axiomes de l'ecole americaine," Paris, CNRS), in *Expected Utility Hypotheses and the Allais Paradox*, ed. by M. Allais and O. Hagen, Reidel, Dordrecht, 1979a, 27–144.

Allais, M. "The so-called Allais paradox and rational decision under uncertainty," in *Expected Utility Hypotheses and the Allais Paradox*, ed. by M. Allais and O. Hagen, Reidel, Dordrecht, 1979b, 437–663.

Allais, M., and Hagen, O. (eds.). *Expected Utility Hypotheses and the Allais Paradox*, Reidel, Dordrecht, 1979.

Anderson, N. H., and Shanteau, J. C. "Information integration in risky decision making," *Journal of Experimental Psychology, 84* (1970), 441–451.

Anscombe, G. E. M. *Intention*, Blackwell, Oxford, 1957; 2nd ed., 1963.

Arrow, K. J. *Social Choice and Individual Values*, Wiley, New York, 1951a; 2nd revised edition, 1963.

Arrow, K. J. "Alternative approaches to the theory of choice in risk-taking situations," *Econometrica, 19* (1951b), 404–437.

Arrow, K. J. "Hurwicz's optimality criterion for decision-making under ignorance," *Technical Report 2*, Department of Economics and Statistics, Stanford University, 1953.

Arrow, K. J. *Essays in the Theory of Risk-Bearing*, Markham, Chicago, 1971.

Arrow, K. J. "Exposition of the theory of choice under conditions of uncertainty," in *Decision and Organization*, ed. by C. B. McGuire and R. Radner, North-Holland, 1972, 19–55.

Arrow, K. J. "Risk perception in psychology and economics," *Economic Inquiry, 20* (1982), 1–9.

Austin, J. L. *Sense and Sensibilia*, Oxford University Press, Oxford, 1962.

Bar-Hillel, M. "The paradox of ideal evidence and the concept of relevance," a translation of a paper appearing in *IYYUN*, a Hebrew Philosophical Quarterly, *27* (1979).

Bar-Hillel, M., and Margalit, A. "Newcomb's paradox revisited," *British Journal for the Philosophy of Science, 23* (1972), 295–304.

Barnes, J. D., and Reinmuth J. E. "Comparing imputed and actual utility functions in a competitive bidding setting," *Decision Sciences, 7* (1976), 801–812.

Becker, S. W., and Brownson, F. O. "What price ambiguity? Or the role of ambiguity in decision making," *Journal of Political Economy, 72* (1964), 62–73.

Becker, G. M., DeGroot, M. H., and Marschak, J. "An experimental study of some stochastic models for wagers," *Behavioral Science, 8* (1963), 199–202.

Bernoulli, D. "Specimen theoriae novae de mensura sortis," *Commentarii academiae scientiarum imperialis Petropolitanae* (for 1730 and 1731), *5* (1738), 175–192.

Bernoulli, J. *Ars Conjectandi*, Basel, 1713.

Bernstein, A. R., and Wattenberg, F. "Non-standard measure theory," in *Applications of Model Theory of Algebra, Analysis and Probability*, ed. by W. A. J. Luxemburg, Holt, Rinehart & Winston, New York, 1969, 171–185.

Bolker, E. *Functions Resembling Quotients of Measures*, Ph.D. dissertation, Harvard University, 1965.

Bolker, E. "Functions resembling quotients of measures," *Transactions of the American Mathematical Society, 124* (1966), 293–312.

Bolker, E. "A simultaneous aximatization of utility and subjective probability," *Philosophy of Science, 34* (1967), 333–340.

Brams, S. J. "Newcomb's problem and prisoners' dilemma," *Journal of Conflict Resolution, 19* (1975), 597–612.

Brehmer, B. "In one word: Not from experience," *Acta Psychologica, 45* (1980), 223–241.

Brewer, K. R. W. "Decisions under uncertainty: Comment," *Quarterly Journal of Economics, 77* (1963), 159–161.

Brewer, K. R. W., and Fellners, W. "The slanting of subjective probabilities – agreement on some essentials," *Quarterly Journal of Economics, 79* (1965), 657–663.

Campbell, N. *Physics: The Elements*, Cambridge University Press, Cambridge, 1920.

Campbell, R., and Sowden, L. *Paradoxes of Rationality and Cooperation: Prisoner's Dilemma and Newcomb's Problem*, University of British Columbia Press, Vancouver, 1985.

Carnap, R. *Meaning and Necessity*, University of Chicago Press, 1947.

Carnap, R. *Logical Foundations of Probability*, University of Chicago Press, 1950; 2d ed., 1962.

Carnap, R. "Inductive logic and rational decisions," in *Studies in Inductive Logic and Probability*, ed. by R. Carnap and R. C. Jeffrey, UCLA Press, Berkeley, 1971a.

Carnap, R. "A basic system of inductive logic," in *Studies in Inductive Logic and Probability*, ed. by R. Carnap and R. C. Jeffrey, UCLA Press, Berkeley, 1971b.

Carroll, Lewis. "What the tortoise said to Achilles," *Mind*, 4 (1895), 278–280.

Cartwright, N. "Causal laws and effective strategies," *Noûs*, 13 (1979), 419–437.

Chadwick, J. A. "Logical constants," *Mind*, 36 (1927), 1–11.

Chellas, B. F. "Basic conditional logic," *Journal of Philosophical Logic*, 4 (1975), 133–153.

Chernoff, H. "Remarks on a rational selection of a decision function," *Cowles Commission Discussion Paper: Statistics, No. 326A*, (1949).

Chernoff, H. "Rational selection of decision functions," *Econometrica*, 22 (1954), 422–443.

Chew S. H., and MacCrimmon, K. "Alpha-nu choice theory: A generalization of expected utility theory," University of British Columbia Faculty of Commerce and Business Adminstration, Working Paper no. 669, 1979.

Chipman, J.S. "Stochastic choice and subjective probability," in *Decisions, Values and Groups*, ed. by D. Willner, Pergamon Press, New York, 1960.

C.N.R.S. *Fondements et applications de la théorie du risque en économetrie*, Centre National de la Recherche Scientifique, Paris, 1954.

Coombs, C. H. "Portfolio theory and measurement of risk," in *Human Judgment and Decision Processes*, ed. by M. F. Kaplan and S. Schwartz, Academic Press, New York, 1975, 63–85.

Courant, R. *What is Mathematics?*, Oxford University Press, New York, 1941.

Cox, R. T. *The Algebra of Probable Inference*, Johns Hopkins Press, Baltimore, 1961.

David, F. N. *Gods, Games and Scholars*, London, 1962.

Davidson, D., and Suppes, P. "A finite axiomatization of subjective probability and utility," *Econometrica*, 24 (1956), 264–275.

Davidson, D., Suppes, P., and Siegel, S. *Decision-making: An Experimental Approach*, Stanford University Press, Stanford, 1957.

Debreu, G. *Theory of Value: An Axiomatic Analysis of General Equilibrium*, Yale University Press, New Haven, 1959.

Debreu, G. "Topological methods in cardinal utility theory," in *Mathematical Methods in the Social Sciences*, ed. by K. J. Arrow, S. Karlin, and P. Suppes, Stanford Unversity Press, Stanford, 1960, 16–26.

de Finetti, B. "La prévision: see lois logiques, ses sources subjectives," *Annales de l'Institut Henri Poincaré*, vol. 7, 1937.

de Finetti, B. "Recent suggestions for reconciliation of theories of probability," in *Proceedings of the Second Berkeley Symposium on Mathematical Statistics and Probability*, Berkeley, 1951, 217–226.

de Finetti, B. "Foresight: Its logical laws, its subjective sources," in *Studies in Subjective Probability* (translation of "La prévision: ses lois logiques, ses sources subjectives," 1937), ed. by H. E. Kyburg, Jr., and H. Smokler, John Wiley, New York, 1964. and R. E. Krieger, New York, 1980, 53–118.

de Finetti, B. *Probability, Induction and Statistics*, John Wiley & Sons, New York, 1972.

de Finetti, B. *Theory of Probability*, vols. 1, 2, John Wiley & Sons, New York, 1979.

Dempster, A. P. "A subjectivist look at robustness," *Bulletin of the International Statistical Institute*, 46 (1962), 349–374.

Dempster, A. P. "Upper and lower probabilities induced by multivalued mapping," *Annals of Mathematical Statistics, 38* (1967) 325–339.

Diaconis, P., and Zabell, S.L. "Updating subjective probability," *Journal of the American Statistical Association, 77* (1982), 822–829.

Domotor, Z. "Axiomatization of Jeffrey utilities," *Synthese, 38* (1978), 165–210.

Dreyfus, H. L., and Dreyfus, S. E. "Inadequacies in the decision analysis model of rationality," in *Foundations and Applications of Decision Theory,* vol. I, ed. by C. A. Hooker, J. J. Leach, and E. F. McClennen, Reidel, Dordrecht, 1978, 115–124.

Dreze, J. "Axiomatic theories of choice, cardinal utility and subjective probability: A Review," in *Allocation Under Uncertainty: Equilibrium and Optimality,* ed. by J. Dreze, Wiley, New York, 1974.

Econometrie (XL), Centre National de la Recherche Scientifique, Paris, 1953.

Edman, M. "Adding independent pieces of evidence," in *Modality, Morality and Other Problems of Sense and Nonsense: Essays dedicated to Sören Halldén,* C. W. K. Gleerup, Lund, 1973, 180–191.

Edwards, W. "The predictions of decisions among bets," *Journal of Experimental Psychology, 50* (1955), 201–214.

Edwards, W. "Subjective probabilities inferred from decisions," *Psychological Review, 69* (1962), 109–135.

Edwards, W., and Tversky, A. (eds.) *Decision Making,* Penguin, Harmondsworth, 1967.

Eells, E. "Causality, utility, and decision," *Synthese, 48* (1981), 295–329.

Eells, E. *Rational Decisions and Causality,* Cambridge University Press, Cambridge, 1982.

Eells, E. "Newcomb's many solutions," *Theory and Decision, 16* (1984a), 59–105.

Eells, E. "Metatickles and the dynamics of deliberation," *Theory and Decision, 17* (1984b), 71–95.

Eells, E. "Causal decision theory," in *PSA 1984,* vol. 2, ed. by P. D. Asquith and P. Kitcher, East Lansing, Michigan, 1985a, 171–200.

Eells, E. "Weirich on decision instability," *Australasian Journal of Philosophy, 63* (1985b), 473–478.

Einhorn, H. J., and Hogarth, R. M. "Ambiguity and uncertainty in probabilistic inference," *Psychological Review, 92* (1985), 433–461.

Ekelöf, P. O. *Rättegång,* P. A. Norstedt & Söner, Stockholm, 1977.

Ellis, B. "The logic of subjective probability," *British Journal for the Philosophy of Science, 24* (1973), 125–152.

Ellis, B. *Rational Belief Systems,* Blackwell, Oxford, 1979.

Ellsberg, D. "Classic and current notions of 'measurable utility'," *The Economic Journal, 64* (1954), 528–556.

Ellsberg, D. "Risk, ambiguity, and the Savage axioms," *Quarterly Journal of Economics, 75* (1961), 643–669. (Reprinted in this volume, as Chap. 13.)

Farquhar, P. "Utility assessments methods," University of California, Davis, Graduate School of Administration, Working Paper no. 81–5, 1982.

Fellner, W. "Distortion of subjective probabilities as a reaction to uncertainty," *Quarterly Journal of Economics, 75* (1961), 670–689.

Fellner, W. "Slanted subjective probabilities and randomization: Reply to Howard Raiffa and K. R. W. Brewer," *Quarterly Journal of Economics, 77* (1963), 676–690.

Fellner, W. *Probability and Profit – A Study of Economic Behavior Along the Bayesian Lines*, Homewood, R. D. Irwin, Illinois, 1965.

Field, H. "A note on Jeffrey conditionalization," *Philosophy of Science, 45* (1978), 361–367.

Fine, T. L. *Theories of Probability*, Academic Press, New York, 1973.

Fischhoff, B., Lichtenstein, S., Slovic, P., Derby, S. L., and Keeney, R. L. *Acceptable Risk*, Cambridge University Press, New York, 1981.

Fishburn, P. C. *Decision and Value*, Wiley, New York, 1964.

Fishburn, P. C. "Mean-risk analysis with risk associated with below-target returns," *American Economic Review, 67* (1977), 116–126.

Fishburn, P. C. *The Foundations of Expected Utility Theory*, Reidel, Dordrecht, 1982*a*.

Fishburn, P. C. "Nontransitive measurable utility," *Journal of Mathematical Psychology, 26* (1982*b*), 31–67.

Fishburn, P. C. "Transitive measurable utility," *Journal of Economic Theory, 31* (1983), 293–317.

Fishburn, P. C., and Kochenberger, G. A. "Two-piece von Neumann-Morgenstern utility functions," *Decision Sciences, 10* (1979), 503–518.

Freeling, A. "Fuzzy sets and decision analysis," *IEEE Transactions on Systems, Man and Cybernetics*, SMC–10 (1980), 341–354.

Friedman, M., and Savage, L. J. "The utility analysis of choices involving risks," *Journal of Political Economy, 56* (1948), 279–304. Reprinted in *Reading in Price Theory*, ed. by G. Stigler and K. Boulding, Richard D. Irwin, Chicago, 1952.

Friedman, M., and Savage, L. J. "The expected-utility hypothesis and the measurability of utility," *Journal of Political Economy, 60* (1952), 463–474.

Fuchs, V. R. "From Bismark to Woodcock: The 'irrational' pursuit of national health insurance," *Journal of Law and Economics, 19* (1976), 347–359.

Galanter, E., and Pliner, P. "Cross-modality matching of money against other continua," in *Sensation and Measurement*, ed. by H. R. Moskowitz, et al., Reidel, Dordrecht, 1974, 65–76.

Gärdenfors, P. "Forecasts, decisions and uncertain probabilities," *Erkenntnis, 14* (1979), 159–181.

Gärdenfors, P. "Imaging and conditionalization," *Journal of Philosophy, 79* (1982), 747–760.

Gärdenfors, P. *Knowledge, in Flux: Modeling the Dynamics of Epistemic States*, forthcoming as a Bradford Book, MIT Press, Cambridge, Mass., 1988.

Gärdenfors, P., Hansson, B., and Sahlin, N.-E. (eds.). *Evidentiary Value: Philosophical, Judicial and Psychological Aspects of a Theory: Essays dedicated to Sören Halldén on his sixtieth birthday*, C W K Gleerup, Lund, 1983.

Gärdenfors, P., and Sahlin, N.-E. "Unreliable probabilities, risk taking, and decision making," *Synthese, 53* (1982), 361–386. (Reprinted in this volume as Chap. 16.)

Gärdenfors, P., and Sahlin, N.-E. "Decision making with unreliable probabilities," *The British Journal of Mathematical and Statistical Psychology, 36* (1983), 240–251.

Georgescu-Roegen, N. "Choice, expectation and measurability," *Quarterly Journal of Economics, 68* (1954).

Georgescu-Roegen, N. "The nature of expectation and uncertainty," in *Expectation, Uncertainty and Business Behavior,* ed. by M. Bowman, Social Science Research Council, New York, 1958.

Gibbard, A., and Harper, W. L. "Counterfactuals and two kinds of expected value," in *Foundations and Applications of Decision Theory,* ed. by C. A. Hooker, J. J. Leach and E. F. McClennen, Reidel, Dordrecht, 1978, 125–162. (Reprinted in this volume as Chap. 17.)

Gillies, D. A. "The subjective theory of probability," *British Journal for the Philosophy of Science, 23* (1972), 138–156.

Ginet, C. "Can the will be caused?", *Philosophical Review 71* (1962), 49–55.

Goldman, A. I. *A Theory of Human Action,* Prentice-Hall, Englewood Cliffs, 1970.

Goldsmith, R. W., and Sahlin, N.-E. "The role of second-order probabilities in decision making," in *Analysing and Aiding Decision Processes,* ed. by P. C. Humphreys, O. Svenson, and A. Vari, North-Holland, Amsterdam, 1982, 455–467.

Good, I. J. *Probability and the Weighing of Evidence,* Charles Griffin and Company, London, 1950.

Good, I. J. "Rational decisions," *Journal of the Royal Statistical Society,* Ser. B, *14* (1952), 107–114.

Good, I. J. "Subjective probability as a measure of a non-measurable set," in *Logic, Methodology and Philosophy of Science, Proceedings of the 1960 International Congress,* ed. by E. Nagel, P. Suppes, and A. Tarski, Stanford University Press, Stanford, 1962, 319–329.

Good, I. J. "On the principle of total evidence," *British Journal for the Philosophy of Science, 18* (1967), 319–321.

Good, I. J. *Good Thinking: The Foundations of Probability and Its Applications,* University of Minnesota Press, Minneapolis, 1983.

Grayson, C. J. *Decisions under Uncertainty: Drilling Decisions by Oil and Gas Operators,* Graduate School of Business, Harvard University, Cambridge, Massachusetts, 1960.

Green, P. E. "Risk attitudes and chemical investment decisions," *Chemical Engineering Progress, 59* (1963), 35–40.

Grether, D. M. "Recent psychological studies of behavior under uncertainty," *American Economic Review Papers and Proceedings, 68* (1978), 70–74.

Grether, D. M., and Plott, C. R. "Economic theory of choice and the preference reversal phenomenon," *American Economic Review, 69* (1979), 623–638.

Grofman, B. "A comment on 'Newcomb's problem and prisoners' dilemma'," manuscript, September, 1975.

Hacking, I. *Logic of Statistical Inference,* Cambridge, New York, 1965.

Hacking, I. "Possibility," *Philosophical Review, 76* (1967), 143–168.

Hacking, I. "Slightly more realistic personal probabilities," *Philosophy of Science, 34* (1967), 311–325. (Reprinted in this volume as Chap. 7)

432   REFERENCES

Hacking, I. *The Emergence of Probability*, Cambridge University Press, Cambridge, 1975.

Hadar, J., and Russell, W. "Rules for ordering uncertainty prospects," *American Economic Review, 59* (1969) 25–34.

Hagen, O. "Towards a positive theory of preferences under risk," in *Expected Utility Hypotheses and the Allais Paradox*, ed. by M. Allais and O. Hagen, Reidel, Dordrecht, 1979, 271–302.

Halldén, S. *On the Logic of "Better"*, C W K Gleerup, Lund, 1957.

Halldén, S. "Indiciemekanismer," *Tidskrift for Rettsvitenskap, 86* (1973), 55–64.

Halldén, S. *The Foundations of Decision Logic*, C W K Gleerup, Lund, 1980.

Halldén, S. *The Strategy of Ignorance*, Thales, Stockholm, 1986.

Halter, A. N., and Dean, G. W. *Decisions under Uncertainty*, South Western Publishing Co., Cincinnati, 1971.

Hampshire, S. *Thought and Action*, Chatto and Windus, London, 1959.

Handa, J. "Risks, probability, and a new theory of cardinal utility," *Journal of Political Economy, 85* (1977), 97–122.

Hansson, B. "The appropriateness of the expected utility model," *Erkenntnis, 9* (1975), 175–193.

Hansson, B. "The decision game – The conceptualisation of risk and utility," in *Ethics – Foundations, Problems, and Applications*, ed. by E. Morscher and R. Stranzinger, Hölder-Pichler-Tempsky, Wien, 1981, 187–193.

Harper, W. L. "Mixed strategies and ratifiability in causal decision theory," *Erkenntnis, 24* (1986), 25–36.

Harper, W. L., and Hooker, C. A. (eds.). *Foundations of Probability Theory, Statistical Inference, and Statistical Theories of Science*, vols. I–III, Reidel, Dordrecht, 1976.

Harsanyi, J. C. "On the rationale of the Bayesian approach: Comments on Professor Watkins's paper," in *Foundational Problems in the Special Sciences*, ed. by R. E. Butts and J. Hintikka, Reidel, Dordrecht, 1977, 381–392.

Helson, H. *Adaptation-Level Theory*, Harper, New York, 1964.

Herstein, I., and Milnor, J. "An axiomatic approach to measurable utility," *Econometrica, 47* (1953), 291–297.

Hintikka, J. *Knowledge and Belief*, Cornell, Ithaca, 1962.

Hodges, J. L., and Lehman, E. L. "The uses of previous experience in reaching statistical decisions," *Annals of Mathematical Statistics, 23* (1952), 396–407.

Hooker, C. A., Leach, J. J., and McClennen, E. F. (eds.). *Foundations and Applications of Decision Theory*, vols. I–II, Reidel, Dordrecht, 1978.

Horgan, T. "Counterfactuals and Newcomb's problem," *Journal of Philosophy, 78* (1981), 331–356.

Hunter, D., and Richter R. "Counterfactuals and Newcomb's paradox," *Synthese, 39* (1978), 249–261.

Hurwicz, L. "Optimality Criteria for Decision Making Under Ignorance," *Cowles Commission Discussion Paper, Statistics*, no. 370, 1951a (mimeographed).

Hurwicz, L. "Some specification problems and applications to econometric models," (abstract) *Econometrica, 19* (1951b), 343–344.

Jeffrey, R. C. *The Logic of Decision*, McGraw-Hill, New York, 1965; 2nd revised edition, University of Chicago Press, Chicago, 1983.

Jeffrey, R. C. "Probable knowledge," in *The Problem of Inductive Logic*, ed. by I. Lakatos, North-Holland, Amsterdam, 1968, 166–180. (Reprinted in this volume as Chap. 5)

Jeffrey, R. C. "Dracula meets Wolfman: Acceptance versus partial belief," in *Induction, Acceptance and Partial Belief*, ed. by M. Swain, Reidel, Dordrecht, 1970, 157–185.

Jeffrey, R. C. "Preference among preferences," *Journal of Philosophy*, 71, (1974) 377–391.

Jeffrey, R. C. "A note on the kinematics of preference," *Erkenntnis 11* (1977a), 135–141.

Jeffrey, R. C. "Savage's omelet," in *PSA 1976*, vol. 2, ed. by F. Suppe and P. D. Asquith, East Lansing, Michigan, 1977b, 361–371.

Jeffrey, R. C. "Choice, chance and credence," in *Philosophy of Logic*, ed. by G. H. von Wright and G. Fløistad, M. Nijhoff, Dordrecht, 1980a.

Jeffrey, R. C. "How is it reasonable to base preference on estimates of chance?" in *Science, Belief and Behaviour: Essays in Honour of R. B. Braithwaite*, ed. by D. H. Mellor, Cambridge University Press, Cambridge, 1980b.

Jeffrey, R. C. "The Logic of Decision defended," *Synthese, 48* (1981), 473–492.

Jeffreys, H. *The Theory of Probability*, Oxford University Press, Oxford, 1939.

Kahneman, D., and Tversky, A. "Prospect theory: An analysis of decision under risk," *Econometrica, 47* (1979), 263–291. (Reprinted in this volume as Chap. 11.)

Kahneman, D., Slovic, P., and Tversky, A. (eds.). *Judgment Under Uncertainty: Heuristics and Biases*, Cambridge University Press, New York, 1982.

Karmarkar, U. "The effect of probabilities on the subjective evaluation of lotteries," Massachusetts Institute of Technology Sloan School of Management, Working Paper no. 698–74, 1974.

Karmarkar, U. "Subjectively weighted utility: A descriptive extension of the expected utility model," *Organizational Behavior and Human Performance, 21* (1978), 61–72.

Keeney, R. L., and Raiffa, H. *Decisions with Multiple Objectives: Preferences and Value Tradeoffs*, Wiley, New York, 1976.

Kemeny, J. G. "Fair bets and inductive probabilities," *The Journal of Symbolic Logic, 20* (1955), 263–273.

Keynes, J. M. *A Treatise on Probability*, Macmillan, London, 1921.

Knight, F. H. *Risks, Uncertainty and Profit*, Houghton Mifflin, Boston, 1921.

Koopman, B. O. "The bases of probability," *Bulletin of the American Mathematical Society, 46* (1940), 763–774.

Krantz, D. H., Luce, D. R., Suppes, P., and Tversky, A. *Foundations of Measurement*, Academic Press, New York, 1971.

Kunreuther, H., Ginsberg, R., Miller, L., Sagi, P., Slovic, P., Borkan, B., and Katz, N. *Disaster Insurance Protection: Public Policy Lessons*, Wiley, New York, 1978.

Kyburg, H. E. *Probability and the Logic of Rational Belief*, Wesleyan University Press, Middletown, Conn., 1961.

Kyburg, H. E. "Bets and beliefs," *American Philosophical Quarterly*, no.5 (1968), 54–63. (Reprinted in this volume as Chap. 6)

Kyburg, H. E. "Rational belief," *The Behavioral and Brain Sciences, 6* (1983*a*), 231–273.

Kyburg, H.E. *Epistemology and Inference*, University of Minnesota Press, Minneapolis, 1983*b*.

Larson, J. R. "Exploring the external validity of a subjective weighted utility model of decision making," *Organizational Behavior and Human Performance, 26* (1980), 293–304.

Lehman, R. S. "On confirmation and rational betting," *The Journal of Symbolic Logic, 20* (1955), 251–262.

Levi, I. "On potential surprise," *Ratio, 8* (1966), 107–29.

Levi, I. "Probability and evidence," in *Induction, Acceptance and Rational Belief*, ed. by M. Swain, Reidel, Dordrecht, 1970, 134–156.

Levi, I. "Potential surprise in the context of inquiry," in *Uncertainty and Expectations in Economics*, ed. by C. F. Carter and J.F. Ford, Blackwell, Oxford, 1972, 213–236.

Levi, I. "On indeterminate probabilites," *Journal of Philosophy 71* (1974), 391–418. (Reprinted in this volume as Chap. 15.)

Levi, I. "Newcomb's many problems," *Theory and Decision, 6* (1975), 161–175.

Levi, I. "Irrelevance," in *Foundations and Applications of Decision Theory*, vol. 1, ed. by C. A. Hooker, J. J. Leach, and E. F. McClennen, Reidel, Dordrecht, 1978. 263–273.

Levi, I. *The Enterprise of Knowledge*, MIT Press, Cambridge, Mass., 1980.

Levi, I. *Decisions and Revisions: Philosophical Essays on Knowledge and Value*, Cambridge University Press, Cambridge, 1984.

Levi, I. "Imprecision and indeterminacy in probability judgment," *Philosophy of Science, 52* (1985), 390–409.

Levi, I. *Hard Choices*, Cambridge University Press, Cambridge, 1986.

Lewis, C. I. *An Analysis of Knowledge and Valuation*, La Salle, Illinois, 1946.

Lewis, C. I. and Langford, C. H. *Symbolic Logic*, New York, 1932.

Lewis, D. K. *Counterfactuals*, Blackwell, Oxford, 1973*a*.

Lewis, D. K. "The counterfactual analysis of causation," *Journal of Philosophy, 70* (1973*b*), 556–567.

Lewis, D. K. "Counterfactuals and comparative possibility," *Journal of Philosophical Logic, 2* (1973*c*), 418–446.

Lewis, D. K. "Causation," *Journal of Philosophy, 76* (1973*d*) 556–567.

Lewis, D. K. "Radical interpretation," *Synthese, 23* (1974), 331–334.

Lewis, D. K. "Probabilities of conditionals and conditional probabilities," *Philosophical Review, 85* (1976), 297–315.

Lewis, D. K. "Attitudes De Dicto and De Se," *The Philosophical Review, 88* (1979*a*), 513–543.

Lewis, D. K. "Prisoners' dilemma is a Newcomb problem," *Philosophy and Public Affairs, 8* (1979*b*), 235–240.

Lewis, D. K. "Counterfactual dependence and time's arrow," *Noûs, 13* (1979*c*), 455–476.

Lewis, D. K. "Mad pain and Martian pain," in *Readings in Philosophy of Psychology*, vol. 1, ed. by N. Block, Harvard University Press, Cambridge, Mass., 1980*a*.

Lewis, D. K. "A subjectivist's guide to objective chance," in *Studies in Inductive Logic and Probability*, vol.2, ed. by R. C. Jeffrey, University of California Press, Berkeley and Los Angeles, 1980*b*.

Lewis D. K. "Causal decision theory," *Australasian Journal of Philosophy*, 59 (1981), 5–30. (Reprinted in this volume as Chap. 18.)

Lichtenstein, S., and Slovic, P. "Reversal of preference between bids and choices in gambling decisions," *Journal of Experimental Psychology*, 89 (1971), 46–55.

Lindley, D. V. *Introduction to Probability and Statistics from a Bayesian Viewpoint*, vols. 1–2, Cambridge University Press, Cambridge, 1965.

Lindley, D. V. *Making Decisions*, Wiley, New York, 1971.

Luce, R. D. *Individual Choice Behavior*, Wiley, New York, 1959.

Luce, R. D., and Krantz, D. H. "Conditional expected utility," *Econometrica, 39* (1971), 253–271.

Luce, R. D., and Raiffa, H. *Games and Decisions*, Wiley, New York, 1957. (Partly reprinted in this volume as Chap. 3.)

Luce, R. D. and Suppes, P. "Preference, utility, and subjective probability," in *Handbook of Mathematical Psychology*, vol. III, ed. by R. D. Luce, R. R. Bush, and E. Galanter, Wiley, New York, 1965, 249–410.

MacCrimmon, K. R. "Descriptive and normative implications of the decision theory postulates," in *Risk Under Uncertainty*, ed. by K. Borch and J. Mossin, Macmillan & Co., London, 1968.

MacCrimmon, K. R., and Larsson, S. "Utility theory: Axioms versus paradoxes," in *Expected Utility Hypothesis and the Allais Paradox*, ed. by M. Allais and O. Hagen, Reidel, Dordrecht, 1979, 333–409.

Machina, M. "'Rational' decision making versus 'rational' decision modeling?," *Journal of Mathematical Psychology*, 24 (1981). 163–175.

Machina, M. "'Expected utility' analysis without the independence axiom," *Econometrica, 50* (1982*a*), 277–323.

Machina, M. "A stronger characterization of declining risk aversion," *Econometrica, 50* (1932*b*), 1069–1079.

Machina, M. "Generalized expected utility analysis and the nature of observed violations of the independence axiom," in *Foundations of Utility and Risk Theory with Applications*, B. P. Stigum and F. Wenstøp, Reidel, Dordrecht, 1983, 117–136. (Reprinted in this volume as Chap. 12)

Machina, M. "Temporal risk and nature of induced preferences," *Journal of Economic Theory, 33* (1984), 199–231.

Markowitz, H. "The utility of wealth," *Journal of Political Economy, 60* (1952), 151–158.

Markowitz, H. *Portfolio Selection*, Wiley, New York, 1959.

Marschak, J. "Why 'should' statisticians and businessmen maximize moral expectation?", in Proceedings of the *Second Berkeley Symposium on Mathematical Statistics and Probability*, ed. by J. Neyman, University of California Press, Berkeley and Los Angeles, 1951, 493–506.

Marschak, J. "Decision making: Economic aspects," *International Encyclopedia of the Social Sciences*, vol. 4, Macmillan Co. & The Free Press, New York, 1968, 42–55.

Marschak, J. "Personal probabilities of probabilities," *Theory and Decision, 6* (1975), 121–153.

McClennen, E. "Sure-thing doubts," in *Foundations of Utility and Risk Theory with Applications*, ed. by B. P. Stigum and F. Wenstøp, Reidel, Dordrecht, 1983, 117–136. (Reprinted in this volume as Chap. 10.)

McCord, M., and de Neufville, R. "Fundamental deficiencies of expected utility decision analysis," Massachusetts Institute of Technology, manuscript, 1982.

McGlothlin, W. H. "Stability of choice among uncertain alternatives," *American Journal of Psychology, 69* (1954), 604–615.

Mellor, D. H. *The Matter of Chance*, Cambridge University Press, Cambridge, 1971.

Milnor, J.W. "Games against nature," *Research Memorandum*, RM–679, The RAND Corporation, Santa Monica, 1951.

Milnor, J.W. "Games against nature," in *Decision Processes*, ed. by R. M. Thrall, C. H. Coombs, and R. L. Davis, Wiley, New York, 1954, 49–60.

Morrison, D. "On the consistency of preferences in Allais' paradox," *Behavioral Sciences, 12* (1967), 373–383.

Moskowitz, H. "Effects of problem representation and feedback on rational behavior in Allais and Morlat-type problems," *Decision Sciences, 5* (1974), 225–242.

Mosteller, F. *Fifty Challenging Problems in Probability*, Reading, Mass., Palo Alto, and London, 1965.

Mosteller, F., and Nogee, P. "An experimental measurement of utility," *American Journal of Political Economy, 59* (1951), 371–404.

Nagel, E. "Principles of the theory of probability," *International Encyclopedia of Unified Sciences*, vol. I, no. 6, University of Chicago Press, Chicago, 1939.

Newell, A., and Simon, H. A. *Human Problem Solving*, Prentice-Hall, Englewood Cliffs, 1972.

Nozick, R. "Newcomb's problem and two principles of choice," in *Essays in Honor of Carl G. Hempel*, ed. by N. Rescher, Reidel, Dordrecht, Holland, 1969, 107–133.

Nute, D. *Topics in Conditional Logic*, Reidel, Dordrecht, 1980.

Pears, D. "Predicting and deciding," in *Studies in the Philosophy of Thought and Action*, ed. by P. F. Strawson, Oxford, London, 1968.

Peirce, C. S. *Chance, Love and Logic*, ed. by M. Cohen, Harcourt Brace and Co., New York, 1923.

Peirce, C. S. *Collected Papers*, ed. by C. Hartshorne and P. Weiss, Belknap Press, Cambridge, Mass., 1932.

Pollock, J. L. *Subjunctive Reasoning*, Reidel, Dordrecht, 1976.

Popper, K. R. "Indeterminism in quantum physics and in classical physics," *British Journal for the Philosophy of Science, 1* (1950), 117–133 and 173–195.

Popper, K. R. *The Logic of Scientific Discovery*, Hutchinson, London, 1959/1974.

Pratt, J. W. "Risk aversion in the small and in the large," *Econometrica, 32* (1964), 122–136.

Price, H. "Against causal decision theory," *Synthese, 67* (1986), 195–212.

Quine, W. V. O. "Two dogmas of empiricism," *Philosophical Review, 60* (1951), 20–43.

Quine, W. V. O. *From a Logical Point of View*, Harvard University Press, Cambridge, 1953.

Rabinowicz, W. "Two causal decision theories: Lewis vs. Sobel," in *Philosophical Essays Dedicated to Lennart Åqvist*, ed. by T. Pauli, Department of Philosophy, Uppsala, 1983, 299–321.

Rabinowicz, W. "Ratificationism without ratification: Jeffrey meets Savage," *Theory and Decision, 19* (1985), 171–200.

Raiffa, H. "Risk, ambiguity, and the Savage axioms: Comment," *Quarterly Journal of Economics, 75* (1961), 690–694.

Raiffa, H. *Decision Analysis: Introductory Lectures on Choices Under Uncertainty*, Addison-Wesley, Reading, Mass., 1968.

Raiffa, H., and Schlaifer, R. *Applied Statistical Decision Theory*, Harvard University Press, Boston, 1961.

Ramsey, F. P. "Truth and Probability," in *The Foundations of Mathematics and Other Logical Essays*, ed. by R. B. Braithwaite, Routledge and Kegan Paul, London, 1931, 156–198, and in *Foundations: Essays in Philosophy, Mathematics and Economics*, ed. by D. H. Mellor, Routledge and Kegan Paul, London 1978, 58–100. (Reprinted in this volume as Chap. 2)

Rapoport, A. "Comment on Bram's discussion of Newcomb's paradox," *Journal of Conflict Resolution, 19* (1975), 4.

Reichenbach. *The Theory of Probability*, University of California Press, Berkeley and Los Angeles, 1949. (First German edition, 1934.)

Ritchie, A. D. "Induction and probability," *Mind*, (1926).

Rosenkrantz, R. D. "Probability magic unmasked," *Philosophy of Science, 40* (1973), 227–233.

Rosenkrantz, R. D. *Inference, Method, & Decision*, Reidel, Dordrecht, 1977.

Roskies, R. "A measurement axiomatization of essentially multiplicative representation of two factors," *Journal of Mathematical Psychology, 2* (1965), 266–276.

Rothschild, M., and Stiglitz, J. "Increasing risk: I. A definition," *Journal of Economic Theory, 2* (1970), 225–243.

Rubin, H. "The existence of measurable utility and psychological probability," *Cowles Commission Discussion Paper: Statistics*. no. 332, unpublished.

Ryle, G. "'If', 'so', and 'because'," *Philosophical Analysis*, ed. by M. Black, Ithaca, 1950.

Sahlin, N.-E. "Generalized Bayesian decision models," unpublished paper, Lund, 1980.

Sahlin, N.-E. "On second order probabilities and the notion of epistemic risk," in *Foundations of Utility and Risk Theory with Applications*, ed. by B. P. Stigum and F. Wenstøp, Reidel, Dordrecht, 1983, 95–104.

Sahlin, N.-E. "Levels of aspiration and risk," *Philosophical Studies*, no. 24, Department of Philosophy, Lund University, Lund, 1984.

Sahlin, N.-E. "Three decision rules for generalized probability representation," *The Behavioral and Brain Sciences, 8* (1985), 751–753.

Sahlin. N.-E. "'How to be 100% certain 99.5% of the time'", *Journal of Philosophy, 83* (1986), 91–111.

438   REFERENCES

Sahlin, N.-E. "The significance of empirical evidence for developments in the foundations of decision theory," in *Theory and Experiment*, ed. by D. Batens and J. P. van Bendegem, Reidel, Dordrecht, forthcoming, 1987.

Salmon, W. "The predictive inference," *Philosophy of Science*, 24 (1957), 180–190.

Samuelson, P. "Probability and the attempts to measure utility," *The Economic Review*, July (1950), 169–170.

Samuelson, P. "Probability, utility and the independence axiom," *Econometrica*, 20 (1952), 670–678.

Samuelson, P. "Utility, preference, and probability," (abstract) in *The Collected Scientific Papers of Paul A. Samuelson*, vol. 1, item no. 13, ed. by J. Stiglitz, MIT Press, 1966.

Savage, L. J. "The theory of statistical decision," *American Statistical Association Journal*, 46 (1951), 57–67.

Savage, L. J. *The Foundations of Statistics*, John Wiley, New York, 1954; 2nd revised edition, Dover, New York, 1972. (Reprinted in this volume, Chapters 4 and 9.)

Savage, L. J. "Difficulties in the theory of personal probability," *Philosophy of Science*, 34 (1967), 305–310.

Savage L. J., and others. *The Foundations of Statistical Inference*, London, 1962.

Schick, F. *Explication and Inductive Logic*, Ph.D. dissertation, Columbia University, 1958.

Schick, F. "Consistency," *Philosophical Review*, 75 (1966), 467–495.

Schick, F. "Self-knowledge, uncertainty, and choice," *British Journal for the Philosophy of Science*, 30 (1979), 235–252. (Reprinted in this volume as Chap. 14.)

Schick, F. *Having Reasons*, Princeton University Press, Princeton, N. J., 1984.

Schlaifer, R. *Analysis of Decision Under Uncertainty*, McGraw-Hill, New York, 1969.

Schoemaker, P. J. H. *Experiments on Decision Under Risk*, Martinus Nijhoff, 1980.

Seidenfeld, T. "Levi on the dogma of randomization in experiments," in *Henry E. Kyburg and Isaac Levi*, ed. by R. J. Bogdan, Reidel, Dordrecht, 1981, 263–291.

Seidenfeld, T. "Decision theory without 'independence' or without 'ordering', what is the difference?", forthcoming, 1987.

Sen, A. K. "Rational fools: A critique of the behavioural foundations of economic theory," *Philosophy and Public Affairs*, 6 (1977), 317–344.

Sen, A. K. "Rationality and uncertainty," *Theory and Decision*, 18 (1985), 109–127.

Shackle, G. L. S. *Uncertainty in Economics*, Cambridge University Press, Cambridge, 1955.

Shackle, G. L. S. *The Nature of Economic Thought*, Cambridge University Press, Cambridge, 1966.

Shackle, G. L. S. *Decision, Order and Time in Human Affairs*, Cambridge University Press, Cambridge, 1969.

Shimony, A. "Coherence and the axioms of confirmation," *Journal of Symbolic Logic*, 20 (1955), 1–28.

Skyrms, B. "The role of causal factors in rational decision," in his *Causal Necessity*, Yale University Press, New Haven, 1980*a*, 128–139.

Skyrms, B. *Causal Necessity*, Yale University Press, New Haven, 1980*b*, 128–139.

Skyrms, B. *Pragmatics and Empiricism*, Yale University Press, New Haven, 1984.

Slovic, P., Fischhoff, B., Lichtenstein, S., Corrigan, B., and Coombs, B. "Preference for insuring against probable small losses: Insurance implications," *Journal of Risk and Insurance*, 44 (1977), 237–258.

Slovic, P., Lichtenstein, S., and Fischhoff, B. "Decision making," in *Stevens' Handbook of Experimental Psychology*, 2d ed., ed. by R. C. Atkinson, R. J. Herrnstein, G. Lindzey, and R. D. Luce, Wiley, New York, (1983), to appear.

Slovic, P., and Tversky, A. "Who accepts Savage's axiom?," *Behavioral Science*, 19 (1974), 368–373.

Smith, C. A. B. "Consistency in statistical inference and decision," (with discussion) *Journal of the Royal Statistical Society Ser. B*, 23 (1961), 1–25.

Smith, C. A. B. "Personal probabilities and statistical analysis," *Journal of the Royal Statistical Society Ser. A*, 128 (1965), 469–499.

Sobel, J. H. "Utilitarianisms: Simple and general," *Inquiry*, 13 (1970), 394–449.

Sobel, J. H. *Probability, Chance and Choice: A Theory of Rational Agency*, unpublished paper presented at a workshop on Pragmatism and Conditionals at the University of Western Ontario, May 1978.

Sobel, J. H. "Expected utilities and rational actions and choice," *Theoria*, 49 (1983), 159–183.

Sobel, J. H. "Maximization, stability of decisions and actions in accordance with reason," manuscript, 1986*a*.

Sobel, J. H. "Defenses against and conservative reactions to Newcomb-like problems: Meta-tickles and ratificationism," *PSA 1986*, vol. 1, East Lansing, Michigan, 1986*b*, 342–351.

Sobel, J. H. "Notes on decision theory: Old wine in new bottles," *Australasian Journal of Philosophy*, 64 (1986*c*), 407–437.

Spetzler, C. S. "The development of corporate risk policy for capital investment decisions," *IEEE Transactions on Systems Science and Cybernetics*, SSC–4 (1968), 279–300.

Spielman, S. "Levi on personalism and revisionism," *Journal of Philosophy*, 62 (1975), 785–793.

Stalnaker, R. "A theory of conditionals," in *Studies in Logical Theory*, American Philosophical Quarterly Monograph Series, no. 2, 1968.

Stalnaker, R. "Letter to David Lewis" in *IFS*, ed. by W. L. Harper, R. Stalnaker and G. Pearce, Reidel, Dordrecht, 1978, 153–190.

Stalnaker, R., and Thomason, R. "A semantical analysis of conditional logic," *Theoria*, 36 (1970), 23–42.

Stigum, B. P., and Wenstøp, F. (eds.) *Foundations of Utility and Risk Theory with Applications*, Reidel, Dordrecht, 1983.

Suppes, P. "Concept formation and Bayesian decision," in *Aspects of Inductive Logic*, ed. by J. Hintikka and P. Suppes, Humanities Press, New York, and North-Holland, Amsterdam, 1966*a*.

Suppes, P. "Probabilistic inference and the concept of total evidence," in

440   REFERENCES

Aspects of Inductive Logic, ed. by J. Hintikka and P. Suppes, Humanities Press, New York, and North-Holland, Amsterdam, 1966*b*.

Suppes, P., Davidson, D., and Siegel, S. *Decision-Making*, Stanford University Press, Stanford, 1957.

Swalm, R. O. "Utility theory – Insights into risk taking," *Review of Economic Studies*, 44 (1966), 123–136.

Teller, P. "Conditionalization and observation," *Synthese*, 26 (1973), 218–258.

Thrall, R. M., Coombs, C. H., and Davis, R. L. (eds.). *Decision Processes*, Wiley, New York, 1954.

Tobin, J. "Liquidity preferences as behavior towards risk," *Review of Economic Studies*, 26 (1958), 65–86.

Todhunter, I. *A History of the Mathematical Theory of Probability*, Chelsea Publishing Company, New York, 1965 (first published 1865).

Tversky, A. "Additivity, utility, and subjective probability," *Journal of Mathematical Psychology*, 4 (1967), 175–201.

Tversky, A. "Intransitivity of preferences," *Psychological Review*, 76 (1969), 31–48.

Tversky, A. "Elimination by aspects: A theory of choice," *Psychological Review*, 79 (1972), 281–299.

Tversky, A. "A critique of expected utility theory: Descriptive and normative considerations," *Erkenntnis*, 9 (1975), 163–173.

Tversky, A., and Kahneman, D. "Judgment under uncertainty: Heuristics and Biases," *Science*, 185 (1974), 1124–1131.

van Dam, C. "Another look at inconsistency in financial decision-making," presented at the Seminar on Recent Research in Finance and Monetary Economics, Cergy-Pontoise, March, 1975.

van Frassen, B. "Singular terms, truth-value gaps and free logic," *Journal of Philosophy*, 63 (1966), 481–495.

Vickers, J. "Coherence and the axioms of confirmation," *Philosophy of Science*, 23 (1965), 32–38.

von Neumann, J., and Morgenstern, O. *Theory of Games and Economic Behavior*, Princeton University Press, Princeton, 1944; 2d ed. John Wiley and Sons, 1953.

Wald, A. *Statistical Decision Functions*, John Wiley and Sons, New York, 1950.

Watkins, J. W. N. "Decisions and uncertainty," *British Journal for the Philosophy of Science*, 6 (1955), 66–78.

Watkins, J. W. N. "Towards a unified decision theory: a non-Bayesian approach," in *Foundational Problems in the Special Sciences*, ed. by R. E. Butts and J. Hintikka, Reidel, Dordrecht, 1977, 381–392.

Watson, S. R., Weiss, J., and Donell, M. "Fuzzy decision analysis," *IEEE Transactions on Systems, Man, and Cybernetics*, SMC–9, no. 1, 1–9.

Weatherford, R. *Philosophical Foundations of Probability Theory*, Routledge & Kegan Paul, London, 1982.

Weirich, P. "Decision instability," *Australasian Journal of Philosophy*, 63 (1985), 465–471.

Weirich, P. "Decisions in dynamic settings," *PSA 1986*, vol. 1, East Lansing, Michigan, 1986, 438–449.

Williams, A. C. "Attitudes toward speculative risk as an indicator of attitudes toward pure risk," *Journal of Risk and Insurance, 33* (1966), 577–586.

Williams, J. S. "The role of probability in fiducial inference," *Sankhyā*, A, *28* (1966), 271–296.

Wold, H. "Ordinal preference or cardinal utility?" *Econometrica, 20* (1952), 661–663.

Yates, J. F., and Zukowski, L. G. "Characterization of ambiguity in decision making," *Behavioral Science, 21* (1976), 19–25.

Name index

Achilles, 129
Aczél, P., 215
Allais, M., 8, 11, 163–4, 167, 179, 186, 190, 199, 204, 207, 215–16, 226–30, 232, 335
Allen, 231
Anderson, H. H., 199
Anscombe, G. E. M., 89
Archimedes, 34
Aristotle, 19, 126, 272
Arrow, K. J., 13, 14, 57, 59, 69, 146–7, 166, 172, 176, 181, 183–4, 215, 217, 219, 222–3, 233, 237–8, 245, 255
Austin, J. L., 90, 91

Bar-Hillel, M., 321, 369
Barnes, J. D., 202
Becker, G. M., 170
Becker, S. W., 242, 330, 333
Berger, J., 243
Bernoulli, D., 13, 40, 164, 166
Bernoulli, J., 57
Bernstein, A. R., 319
Blake, W., 19
Boethius, 272
Bolker, E., 10, 11, 87
Braithwaite, R. B., 19
Brams, S. J., 376
Brewer, K. R. W., 170
Brownson, F. O., 242, 330, 333

Campbell, N., 27, 336
Cardano, J., 119
Cargile, J., 118
Carnap, R., 101–2, 106, 109, 111, 116, 127, 133, 288, 291, 293–300, 321
Carroll, L., 129
Cartwright, N., 377, 394–5
Chadwick, J. A., 25

Chellas, B. F., 396
Chernoff, H., 54, 55, 59, 64–5, 68, 71, 172
Chew, S. H., 215, 226
Chipman, J. S., 247, 253, 256, 260, 264
Coombs, C. H., 170, 199
Courant, R., 120
Cox, R. T., 119, 124

Dalkey, N., 256
Davidson, D., 198, 247
Dean, G. W., 202
Debreu, G., 213, 218, 256
de Finetti, B., 4, 73, 77, 97, 101, 104–6, 111, 114–16, 118, 122, 124–5, 132, 247, 281, 288, 298, 300, 316
DeGroot, M. H., 170
DeMorgan, A., 38
Dempster, A. P., 243, 301–2, 318
de Neufville, R., 231–2
Donkin, W. F., 19
Dreyfus, H. L., 167
Dreyfus, S. E., 167
Dreze, J., 216

Edman, M., 318
Edwards, W., 199, 224
Eells, E., ix, 337, 382, 408, 423
Einhorn, H. J., 242–3
Einstein, A., 26
Ellsberg, D., 12, 169–70, 203, 212, 241, 327–30, 333, 335

Farquhar, P., 231
Fellner, W., 170, 199, 203, 212, 268, 327
Fischhoff, B., ix, 6
Fishburn, P. C., 188, 199, 202, 215, 226, 311
Fisher, R. A., 299

444   Name index

Forster, E. M., 383
Freeling, A., 321
Freeman, E., 166
Friedman, M., 176–7, 183, 227
Fuchs, V. R., 190

Galanter, G., 200–1
Gärdenfors, P., 315, 318, 321, 324, 329, 389, 413, 417
Georgescu–Roegen, N., 260, 264
Gibbard, A., 10, 337, 339, 377, 385, 395, 397–400, 402, 407–10, 413–14, 419
Gibbs, J., 41
Ginet, 273
Goldman, A. I., 273
Goldsmith, R. W., 242, 313–14, 325
Good, I. J., 78, 133, 243, 301–2, 304–5, 311, 318
Grayson, C. J., 202
Green, P. E., 202
Grether, D. M., 207, 219, 239
Grofman, B., 376

Hacking, I., 99, 123, 126–8, 130, 241, 298–9
Hadar, J., 221
Hagen, O., 167, 179, 215, 226, 228–9
Halldén, S., ix, 313, 318
Halter, A. N., 202
Hampshire, S., 279
Handa, U., 215
Hansson, B., x, 13, 14, 15, 99, 160, 167, 199, 313, 324
Harper, W. L., 10, 337, 339, 377, 385, 395, 397–400, 402, 407–10, 413–14, 418–19, 423
Harsanyi, J. C., 13, 141–2, 148, 215
Hayes, B., 215
Helson, H., 199
Herstein, I., 218
Hicks, J., 231
Hintikka, J., 127, 276
Hodges, J. L., 77, 79, 248, 265, 327
Hogart, R., 242–3
Höög, V., x
Horgan, T., 397
Hume, D., 46, 271
Hunter, D., 377, 403–4
Hurwicz, L., 55–7, 59, 64, 66, 69, 71–2, 77–8, 248, 257–8, 265, 327

Jackson, F., 377
Jeffrey, R. C., ix, 4, 9, 10, 11, 15, 87–9, 91, 96, 99, 116, 124, 144, 273–5, 278, 317, 337, 338, 341, 342, 347, 348–50, 357, 377, 382–3, 386, 406–12, 414, 417–22
Jeffreys, H., 101, 123, 288, 298–9

Kahneman, D., 6, 11, 14, 15, 159, 167, 173, 212, 215, 219, 224, 228–9, 239, 333–4

Kant, I., 38, 42
Karmarkar, U., 232
Kavka, G., 377
Keeney, R. L., 183
Kemeny, J. G., 7, 101
Keynes, J. M., 20–5, 37, 38, 40–3, 46, 102, 109, 298–9, 321
Knight, F., 245, 248, 254–5, 261
Kochenberger, G. A., 188, 202
Kolmogorov, A. N., 122
Koopman, B. O., 102, 301–2
Krantz, D. H., 212–13, 317
Kuhn, T., 292
Kunreuther, H., 208
Kyburg, H. E., 4, 98, 242–3, 299–300, 302, 332

Lakatos, I., 86
Langford, C. H., 120, 126
Laplace, P. S., 71
Larson, J. R., 242
Larsson, S., 167, 186, 188, 204, 229
Lehman, E. L., 77, 79, 248, 265, 327
Lehman, R. S., 101
Levi, I., ix, 104, 166, 175, 241, 281, 284, 286, 303, 306, 313, 315–16, 318, 326, 330–3, 341, 369, 416
Lewis, C. I., 91, 120, 126, 129
Lewis, D., 339, 341, 343–4, 346, 379, 387, 390, 394, 396, 399–400, 402, 407–9
Lichtenstein, S., 6, 207
Lindley, D. V., 123–4
Little, J., 166
Luce, R. D., 8, 48, 49, 59, 166, 167, 247, 251, 309–10, 317, 332
Lucretius, 271
Lyon, P., 166

MacCrimmon, K. R., 167, 186, 188, 204, 215–16, 226, 229
Machina, M., 161, 166, 173–4, 179–80, 216–17, 221–2, 224–7, 231, 234
Madansky, T., 245
Margalit, A., 369
Markowitz, H., 173, 188, 190, 198, 201, 204, 209, 227
Marschak, J., 170, 179, 256
Maugham, S., 373
McClennen, E., ix, 11, 215
McCord, M., 231–2
McGlothlin, W. H., 210
Meyer, M., 166
Mill, J. S., 47
Milnor, J. W., 8, 59, 70–1, 218, 251
Mohs, 27
Morgenbesser, S., 287
Morgenstern, O., 9, 73, 77, 139, 145, 170, 183, 202, 221–2, 225–6, 231, 233, 238, 251, 262, 306

Morlat, G., 163
Morrison, D., 227
Moskowitz, H., 227
Mosteller, F., 92, 198

Nagel, E., 57, 106, 287
Newcomb, W., 336, 343, 368, 376, 380–1, 386
Neyman, J., 299–300
Nogee, P., 198
Nozick, R., 336, 343, 352, 363, 368–9, 376, 381, 382, 407–8
Nute, D., 389

O'Hara, J., 373
O'Neill, B., 341, 372

Peano, G., 128
Pears, D., 273
Peirce, C. S., 19, 38, 44, 294, 322
Plateau, J., 120
Pliner, P., 200–1
Plott, C. R., 207, 219, 239
Pollock, J. L., 389
Popper, K. R., 94–5, 278, 320
Porebski, C., 341
Pratt, J. W., 14, 146, 184, 217, 219, 222–3, 233, 238
Price, H., 408

Quine, W. V. O., 84, 126

Rabinowicz, W., x, 338, 406, 409, 414, 417, 424
Rader, T., 166
Raiffa, H., 8, 48, 49, 59, 141, 169, 171–4, 180, 183, 206, 216, 227, 248, 251–2, 257, 309–10, 332
Ramsey, F. P., 2, 6–7, 10–11, 13, 15, 86–8, 99, 101–6, 122, 139, 202, 242, 247–9, 253, 281, 284
Rapoport, A., 376
Reichenbach, H., 102, 117, 299–300
Reinmuth, J. E., 202
Richter, R., 377, 403, 404
Ritchie, A. D., 37
Rosenkrantz, R. D., 321
Roskies, R., 213
Rothschild, M., 218
Rubin, H., 8, 59, 63–4, 69, 170, 173, 175, 251
Runyon, D., 246
Russel, B., 22, 27, 28
Russel, W., 221
Ryle, G., 129, 272

Sahlin, N.-E., 5, 6, 14, 15, 242, 314, 318, 324–5, 329, 332
Salmon, W., 109, 117, 300

Samuelson, P., 161, 167, 169, 170, 251, 256, 286
Savage, L. J., 2, 5–11, 13, 15, 49, 53, 57, 59, 71, 73, 75–8, 98–9, 101, 104–5, 107–9, 118–23, 125–6, 128, 131, 133–5, 159, 166, 169, 175–8, 183, 202, 216, 227, 243, 247–53, 255–7, 260–4, 266–7, 269, 281, 298, 300, 311, 316–17, 328–9, 342, 348, 350–1, 367–8, 386, 397, 407–9, 414, 416–23
Schelling, T., 245, 269
Schick, F., 166, 242–3, 277, 282, 287, 302
Schlaifer, R., 144, 175, 256
Schoemaker, P. J. H., 167
Seidenfeld, T., 161, 166, 175, 180–1, 287
Sen, A. K., 6, 13
Shackle, G. L. S., 246, 248–9, 261, 270–1, 273–5, 279, 280–1, 286
Shanteau, J. C., 199
Shapley, L., 245
Shimony, A., 7, 101, 124–6, 132
Siegel, S., 247
Sinclair–Wilson, J., ix
Skyrms, B., 339, 377, 382, 385, 389, 391–4, 402, 407–8, 414, 418
Slovic, P., 6, 186, 207–9, 216, 227
Smith, C. A. B., 243, 301–2, 308–9, 317–18
Sobel, J. H., 215, 339, 341, 349, 355, 377, 385, 389–90, 391, 402–3, 406–9, 412, 423–5
Socrates, 128
Sowden, L., 336
Spetzler, C. S., 210
Spielman, S., 287
Stalnaker, R., 341, 343–5, 352, 377, 389, 397–9
Stein, H., 287, 302
Stiglitz, J., 218
Stigum, B. P., 166, 215
Suppes, P., 124, 128, 167, 247
Swalm, R. O., 202

Tarski, A., 95
Tartaglia, 119
Tchebycheff, 112–13
Thomason, R., 344
Tobin, J., 190
Tversky, A., 6, 11, 14, 15, 159, 167, 173, 186, 192, 197, 199, 212, 215, 216, 219, 224, 227–9, 239, 333–4

van Dam, C., 199
van Fraassen, B., 399, 412
Vickers, J., 118, 126, 129
von Neumann, J., 9, 73, 77, 139, 145, 170, 183, 202, 221–2, 225–6, 231, 233, 238, 251, 262, 306

Wald, A., 71, 85, 306, 326–7
Watkins, J. W. N., 141–2, 148–50, 281

Watson, S. R., 321
Wattenberg, F., 319
Weirich. P., 423
Wenstøp, F., 166, 215
White, H., 215
Williams, A. C., 188
Williams, J. S., 122
Winter, S., 245

Wittgenstein, L., 25, 33, 38, 39, 41, 42
Wold, H., 216

Yates, J. F., 242

Zeckhauser, 206
Zukowski, L. G., 242

Subject index

admissible alternatives *(see also* E-, P-, and S-admissibility), 60, 289, 304
Allais's paradox, 11, 13, 159, 163–5, 186, 204, 207, 227, 229–30, 335
ambiguity, 12, 203, 212, 242, 258–69

Bayes's theorem, 123
Bergen paradox, 216, 228
Bernoulli's theorem, 40

causality, 10, 47, 271, 336–7, 394
certainty effect, 186–8, 216
certainty equivalent, 236
choice rules, *see* decision rules
coherence, 7, 88, 132, 161, 297–8, 317
common consequence effect, 216, 227–8, 231, 234
common ratio effect, 216, 228–9, 231, 234
complete ignorance, *see* ignorance
complete information, 316
conditional excluded middle, 344, 398–400, 402, 409
confirmation function, 102, 293
confirmational commitment, 293–302, 311
confirmational conditionalization, 293, 297
confirmational tenacity, 294–5, 299
conjoint measurement, 139, 151–5, 212
contraction of beliefs (credal states), 291
contraction of decision situation, 68–71, 332
cost of thinking, 133–4
counterfactuals, 337, 341–56, 360, 390, 395–403, 409, 413, 419
  back-tracking, 396–7
  causal, 10, 396–7, 401

decision making:
  under risk, 5, 12, 48, 50, 73, 183, 195, 326
  under uncertainty, 5, 12, 48–79, 269, 309, 326
decision rules (other than maximal expected utility)
  Hurwicz's α-criterion, 55–66, 69–73, 78–9, 258
  minimax (regret) strategy, 52–5, 62–3, 68, 71–2, 249, 258, 265, 284, 327
  maximal conditional expected utility (MCEU), 9–10, 336, 338
  maximax strategy, 258, 264, 275
  maximin criterion for expected utilities (MMEU), 324–33
  maximin strategy (rule), 51–3, 58–9, 63–4, 67–72, 79, 263–4, 275, 306, 326
de Finetti's representation theorem, 4, 105, 111, 114, 116, 125
deleting column of decision matrix, *see* contraction of decision situation
dependence *(see also* independence):
  counterfactual, 395–403
  hypotheses, 383–405
determinism, 271, 278, 281
direct inference, 298–9
dominance, 60–1, 70–2, 168, 176, 181–2, 196, 199, 206, 336–7, 348, 353, 359, 363–7, 376, 418, 420–3
Dutch book, 7, 104, 106, 122, 124–6, 131–3, 160–1, 234–5, 241, 309, 316–17

E-admissibility, 303–10, 330–2
editing (of decision situations), 195–7, 205, 211

Ellsberg's paradox, 11–13, 241–2, 255–7, 335
epistemic reliability, 243, 314, 318–33
epistemic risk, see risk, epistemic
ethically neutral proposition, 33, 103
evidentiary value, 408, 410
exchangeable events, 105–6, 108, 111–2, 114, 116
expansion of beliefs (credal states), 281
expected value (utility):
  estimation of, 264
  indeterminate, 285
  minimal, 324–5, 328, 331–2
  range of, 285

foreknowledge, 271–9, 391, 402

Hurwicz's α-criterion, 55–66, 69–73, 78–9, 258
Hypothesis II, 232–9

ignorance, 51, 57–9, 66–8, 72, 79, 258–60, 348
  partial, 72–9
imaging function, 388–91
independence:
  assumption, 407–9
  axiom (of utility theory), 161, 167–9, 174–6, 179–80, 187–8, 207, 215–39
  causal, 337, 359–69, 376, 396–7, 407, 409, 411, 419
  probabilistic (stochastic), 5, 9, 336, 346–7, 352–3, 359–68
independence of irrelevant alternatives, 54, 61–3, 309–10, 332
independence of probabilities and pay-offs (Savage's postulate 4), 251
indeterminism, 399–400
inductive logic, 37–9, 41–7, 115, 291, 294–300, 310
  rule of acceptance for, 115–16
intransitive preferences (choices), 197, 199 217
intuitionism, 119
isolation effect, 192–5, 211, 333
  level of aspiration (see also reference point), 323, 325, 333–4

maximal conditional expected utility (MCEU), 9–10, 336, 338
maximal expected utility (MEU):
  definition of, 5
maximax strategy, 258, 264, 275
maximin criterion for expected utilities (MMEU), 324–33
maximin strategy (rule), 51–3, 58–9, 63–4, 67–72, 79, 263–4, 275, 306, 326

MCEU, see maximal conditional expected utility
MEU, see maximal expected utility
minimax (regret) strategy, 52–5, 62–3, 58, 71–2, 249, 258, 265, 284, 327
mixed alternatives, see randomized alternatives
MMEU, see maximin criterion for expected utilities
Monte Carlo simulation, 119, 134

Newcomb's problem, 10, 335–8, 343, 368–72, 376, 377–8, 380–4, 391, 393, 397, 408, 416, 418

observational propositions, 89–90
observation by candlelight, 90–1
option preservation, 304–5, 331
outlying probabilities, 229–30, 235
oversensitivity to changes in probability, 229–31

P-admissibility, 304–5, 308, 331
paradox of ideal evidence, 95, 314, 320
pessimism-optimism index criterion, see Hurwicz's α-criterion
potential surprise, 281
principle of insufficient reason (indifference), 40–1, 57–9, 63, 66–72, 255
prisoner's dilemma, 375–6, 408, 418
probabilistic independence, see independence, probabilistic
probabilistic insurance, 190–2, 208
probability:
  after-choice, 419, 424
  axioms of, see probability, laws of
  indeterminate, 169, 283–4, 295, 302
  intervals, 117, 242–3, 301–2, 311, 318
  laws of, 35–6, 103–6, 119, 125
  objective, 4, 9, 77, 97
  operational definition of, 25–35, 251
  overweighting of, 160, 203–205
  personal, 4, 8, 73–7, 97–9, 101–2, 107–16, 118–35, 288
  qualitative, 250, 256
  second order, 319
  subjective, see probability, personal
  unreliable, 109, 313–35
probability distribution:
  confidence in (see also reliability of), 264–9, 329
  convex set of, 297, 301–3, 307, 310, 318, 330
  epistemically possible, 317–19, 322, 325–30
  estimated, 262, 269
  permissible, 296–7, 300–11, 318, 330

reasonable, 78, 259, 261–5
reliability of, 98, 318–33
prospect theory, 160, 183–214, 224

Ramsey's rule, 7, 33
randomized alternatives (acts), 64–5, 304
ratifiability of decisions, 338, 406–24
reference level (point), 14–15, 200–1, 207,
    209–11, 334
reflection effect, 188–90, 195, 333
relevance, 322
    causal, 392
    confirmational, 303
revisions of beliefs (credal states), 288–9,
    292, 303
risk:
    aversion, 13–14, 136–9, 143–58, 184,
        190–2, 195, 202, 208, 217, 219, 222–4,
        227–8, 232, 239, 332
    epistemic, 323–4, 328, 332
    function, 326
    seeking, 188–90, 195, 202, 208–11
risky prospect, 184, 193, 211, 215–16
Rubin's axiom, 63–5, 69

S-admissibility, 305–10, 331–2
St. Petersburg paradox, 13, 144
Savage's postulate 2, see sure-thing
    principle
screening off, 338, 410–1, 413

security level, 51–3, 262, 264, 305
self–knowledge, 270–81, 311, 379, 383
stability of decisions, 372–5, 406, 409, 414,
    422–5
states of nature, 2, 48–50
    act-independent, 348–52, 367–8, 386,
        397
stochastic dominance, 224–225, 228, 233
subadditivity of probabilities, 203–5, 208
substitution axiom (of utility theory), see
    independence axiom
sure–thing principle (Savage's postulate
    2), 8, 11–12, 80–5, 139–41, 159–61, 163–
    4, 167–9, 176–81, 250–7, 263, 266, 328–9,
    357–63

tickle defence, 382–5
total knowledge (evidence), 108, 297, 377

undecidability, 119
utility:
    evaluation effect, 216, 231–2, 235
    indeterminate, 283
    operational definition of, 7, 100
    saturation of, 156–8
    uncertainty of, 284

value function, 199–202, 211

weight of evidence, 107, 109–10, 321–2